OPTIMIZATION OF
UNIT OPERATIONS

BÉLA G. LIPTÁK

OPTIMIZATION OF UNIT OPERATIONS

Boilers
Chemical Reactors
Chillers
Clean Rooms
Compressors
Condensers
Cooling Towers
Fans
Fired Heaters
Heat Exchangers
HVAC Systems
Pumping Stations
Reboilers
Vaporizers

CRC Press
Taylor & Francis Group
Boca Raton London New York

CRC Press is an imprint of the
Taylor & Francis Group, an **informa** business

First published 1987 by Chilton Book Company

Published 2019 by CRC Press
Taylor & Francis Group
6000 Broken Sound Parkway NW, Suite 300
Boca Raton, FL 33487-2742

ISBN 13: 978-0-367-45143-1 (pbk)
ISBN 13: 978-0-8019-7706-0 (hbk)

Visit the Taylor & Francis Web site at
http://www.taylorandfrancis.com

and the CRC Press Web site at
http://www.crcpress.com

Designed by William E. Lickfield
Illustrations by Adrian J. Ornik

Library of Congress Cataloging in Publication Data

Lipták, Béla G.
  Optimization of unit operations.
  Includes index.
  1. Manufacturing processes. 2. Industrial equipment.
3. Efficiency, Industrial. I. Title.
TS183.L57   1987   660.2'842   86-47960
ISBN 0-8019-7706-1

SCIENCE *and its offspring, technology, must not be void of quality, purpose or direction, must not be a "value-free" force, out of control, amoral, dangerous. It must be subordinated to subjective human values in order to serve as the tool of a better tomorrow. This book is dedicated to the engineers who will convert this vision into reality.*

Contents

Preface

A DICTIONARY DEFINITION of optimization might read: "A strategy giving the best result obtainable under a set of specific conditions." To the mathematician, optimization might mean the task of finding a mountain peak in a multidimensional space. To the practicing engineer, optimization usually suggests a highly theoretical exercise, which is not very relevant in the real world, where pipes leak, sensors plug, and pumps cavitate. Optimization, as used in this book, is the integration of process control know-how to maximize industrial productivity.

After some decades of process control experience, instrument engineers realize that it is desirable to control that which the plant produces. Plants do not produce levels, flows, or temperatures; therefore, these variables should be only constraints, whereas the controlled variables should relate to the productivity or efficiency of the plant. Accepting this conclusion requires an acceptance of the end of the era of the single control loops and the beginning of multivariable envelope control. The envelope is a polygon, with its sides representing the constraints of pressure, temperature, etc.; inside this envelope, the process is continuously moved to maximize efficiency. While studying this book, the reader will realize that multivariable optimization is the approach of common sense. It is the control technique applied by nature, and frequently it is also the simplest and most elegant method of control.

The control systems described in this book have one common characteristic: they work. Each chapter is a snapshot of the state of the art of controlling a particular unit operation. It is only a snapshot, because our understanding of process control is improving every day. The control systems described (with few exceptions) were not invented by the writer, but evolved through the contributions of the Eckmans, Shinskeys and others less well known in the process control field. There could be more chapters in this book, as there are more unit operations of interest. One such unit operation, distillation, was left out because it is well covered in other books. Other unit operations may be discussed in later books or in future editions of this one.

In spite of the many shortcomings and limitations of this book, the bottom line is that the know-how in these nine chapters is enough to eliminate our trade deficit.

Chapter 1

OPTIMIZATION
OF
AIRHANDLERS

THE AIRHANDLER is the basic unit operation of space conditioning. It is used to keep occupied spaces comfortable (figure 1.1) or unoccupied spaces at desired levels of temperature and humidity. In addition to supplying or removing heat and/or humidity from the conditioned space, the airhandler also provides ventilation and fresh air makeup. Depending on the type of space involved, from 75,000–300,000 Btu/yr. (19,000–76,000 Cal/yr.) are required to condition one square foot (0.092 m²) of office space. Depending on the energy sources used, this corresponds to a yearly operating cost of a few dollars per square foot of floor space.

Whereas other unit operations have benefited substantially from the advances in process control, airhandlers have not. Airhandlers today are frequently controlled the same way as they were twenty or thirty years ago. For this reason, airhandler optimization can result in much greater percentages of savings than can the optimization of any other unit operation. Optimization can cut the cost of airhandler operation in half—a savings that can seldom be achieved in any other type of unit operation.

Some of the optimization goals and strategies include the following:

- Let the building heat itself
- Use free cooling and/or free drying
- Benefit from gap control or zero energy band (ZEB)
- Eliminate chimney effect
- Optimize start-up timing
- Optimize air make-up (CO_2)
- Optimize supply air temperature
- Minimize fan energy use
- Automate the selection of operating modes
- Minimize reheat
- Automate balancing of air distribution

Components of an Airhandler

The purpose of heating, ventilation, and air conditioning (HVAC) controls is to provide comfort in laboratories, clean rooms, warehouses, offices, and manufacturing spaces. Supply air is the means of providing comfort in the conditioned zone. The air supplied to each zone must provide heating or cooling, raise or lower humidity, and provide air refreshment. To satisfy these requirements, it is necessary to control the temperature, humidity, and fresh air ratio in the supply air.

3

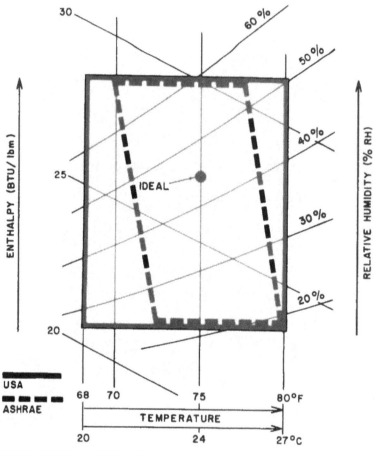

Fig. 1.1 "Comfort zones" are defined in terms of temperature and humidity. (From Lipták [5])

Figure 1.2 illustrates the main components of an airhandler. The term *airhandler* refers to the total system, including fans, heat-exchanger coils, dampers, ducts, and instruments. The system operates as follows: outside air is admitted by the outside air damper (OAD-05) and is then mixed with the return air from the return air damper (RAD-04). The resulting mixed air is filtered (F), heated (HC) or cooled (CC), and humidified (H) or dehumidified (CC) as required. The resulting supply air is then transported to the conditioned zones (groups of offices) by the variable-volume supply fan station. Variable volume means that the air flow rate generated by the fan(s) is variable.

In each zone, the variable air volume damper (VAV-23) determines the amount of air required, and the reheat coil (RHC) adjusts the air temperature as needed. The

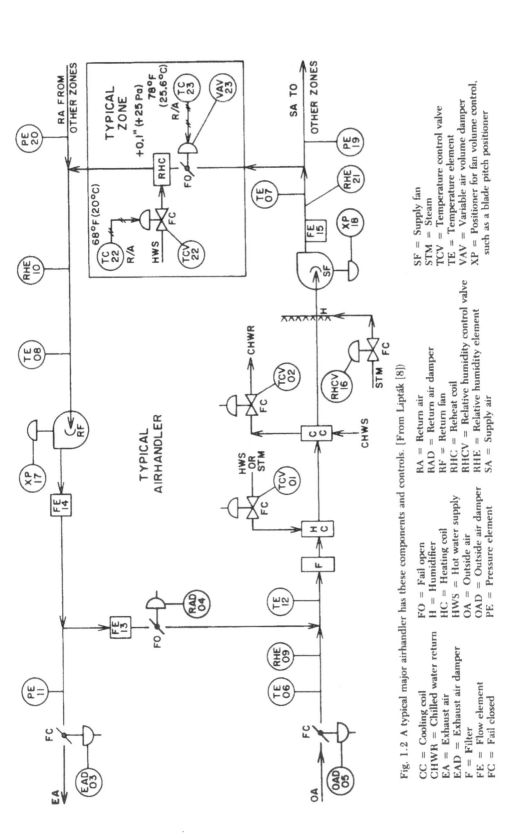

Fig. 1.2 A typical major airhandler has these components and controls. [From Lipták [8]]

CC = Cooling coil	FO = Fail open	RA = Return air
CHWR = Chilled water return	H = Humidifier	RAD = Return air damper
EA = Exhaust air	HC = Heating coil	RF = Return fan
EAD = Exhaust air damper	HWS = Hot water supply	RHC = Reheat coil
F = Filter	OA = Outside air	RHCV = Relative humidity control valve
FE = Flow element	OAD = Outside air damper	RHE = Relative humidity element
FC = Fail closed	PE = Pressure element	SA = Supply air

SF = Supply fan
STM = Steam
TCV = Temperature control valve
TE = Temperature element
VAV = Variable air volume damper
XP = Positioner for fan volume control, such as a blade pitch positioner

return air from the zones is transported by the variable-volume return-air fan station. If the amount of available return air exceeds the demand for it, the excess air is exhausted by the exhaust air damper (EAD-03). The conditioned spaces are typically pressurized to about $0.1"H_2O$ (25 Pa), relative to the barometric pressure on the outside. This pressurization results in some air leakage through the walls and windows, which varies with the quality of construction. Therefore, the air balance around the system is:

$$OA = EA + Pressurization\ Loss$$

Under "normal" operation, the airhandler operates with about 10 percent outside air. In the "purge" or "free cooling" modes, RAD is closed, OAD is fully open, and the airhandler operates with 100 percent outside air.

As can be seen, the HVAC process is rather simple. Its process material is clean air, its utility is water or steam, and its overall system behavior is slow, stable, and forgiving. For precisely these reasons, it is possible to obtain acceptable HVAC performance using inferior-quality instruments, which are configured into poorly designed loops. Yet there is an advantage in applying the state of the art of process control to the HVAC process, because it can provide a drastic reduction in operating costs, attributable to increased efficiency of operation. Some of the more efficient control concepts are described in the sections below.

Operating Mode Selection

The correct identification and timing of the various operating modes can contribute to the optimization of the building. The *normal* operating modes include start-up, occupied, night, and purge.

Optimizing the time of *start-up* will guarantee that the minimum required cost is invested in getting the building ready for occupancy. This is done by automatically calculating the amount of heat that needs to be transferred and dividing it according to the capacity of the start-up equipment. A computer-optimized control system will serve to initiate the unoccupied (night) mode of operation; it will also recognize weekends and holidays and, in general, provide a flexible means of time-of-day controls.

The *purge* mode is another convenient tool of optimization. Whenever the outside air is preferred to the return air, the building is automatically purged. In this way, "free cooling" can be obtained on dry summer mornings or "free heating" can be provided on warm winter afternoons. Purging is the equivalent of opening the windows in a home. In computer-optimized buildings, an added potential is to use the building structure as a means of heat (or coolant) storage. In this case, the purge mode can be automatically initiated during cold nights prior to hot summer days, thereby bringing the building temperature down and storing some free cooling in the building structure.

Summer/Winter Mode Reevaluation

Another important mode selection involves switching from summer to winter mode and vice versa. Conventional systems are switched according to the calendar, whereas optimized ones recognize that there are summerlike days in the winter and winterlike hours during summer days. Seasonal mode switching is therefore totally inadequate. Optimized building operation can be provided only by making the summer/winter selection on an enthalpy basis: if heat needs to be added, it is "winter"; if heat needs to be removed, it is "summer," regardless of the calendar. In those airhandlers which serve a variety of zones, it is essential to first determine if the unit is in a "net" cooling (summer) or "net" heating (winter) mode before the control system can decide if free cooling (or free heating) by outside air can be used to advantage. Figure 1.3 illustrates the heat balance evaluation that is required to determine the prevailing overall mode of operation. This type of heat balance calculation, which must be reevaluated every 15 to 30 minutes, can be implemented only through the use of computers.

Emergency Mode

In addition to the above operating modes, the airhandler can also be placed in an *emergency* mode, if fire, smoke, freezing temperature, or pressure conditions require it. Figure 1.4 shows the status of each fan, damper and valve in each of the operating

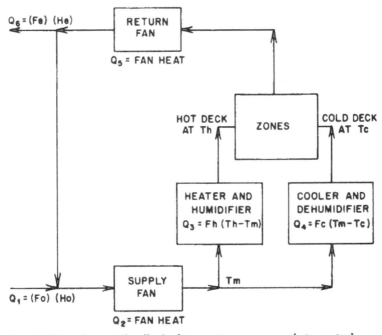

Fig. 1.3 When the net airhandler load is negative, summer mode is required; when it is positive, winter mode is required.

Fc = Cold deck flow
Fe = Exhaust air flow
Fh = Hot deck flow
He = Exhaust air enthalpy

Ho = Outside air enthalpy
Tc = Cold deck temperature
Th = Hot deck temperature
Tm = Mixed air temperature

Net airhandler load = $Q_0 = Q_1 + Q_2 + Q_3 - Q_4 + Q_5 - Q_6$

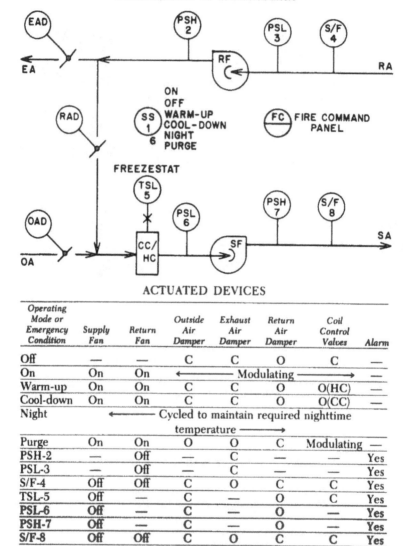

ACTUATED DEVICES

Operating Mode or Emergency Condition	Supply Fan	Return Fan	Outside Air Damper	Exhaust Air Damper	Return Air Damper	Coil Control Valves	Alarm
Off	—	—	C	C	O	C	—
On	On	On	←————— Modulating ————→				—
Warm-up	On	On	C	C	O	O(HC)	—
Cool-down	On	On	C	C	O	O(CC)	—
Night	←————— Cycled to maintain required nighttime temperature ————→						
Purge	On	On	O	O	C	Modulating	—
PSH-2	—	Off	—	C	—	—	Yes
PSL-3	—	Off	—	C	—	—	Yes
S/F-4	Off	Off	C	O	C	C	Yes
TSL-5	Off	—	C	—	O	C	Yes
PSL-6	Off	—	C	—	O	—	Yes
PSH-7	Off	—	C	—	O	—	Yes
S/F-8	Off	Off	C	O	C	C	Yes

Fig. 1.4 Safety and operating mode selection instruments. S/F = Smoke and fire detector(s). FC = Fire command panel, which provides the fire chief with access to all fans and dampers in the building. This panel is used during fire fighting and building evacuation. (From Lipták [8])

modes. In a computer-optimized control system, both the mode selection and the setting of the actuated devices is done automatically.

When a smoke or fire condition is detected (S/F-4 and 8), the fans stop, the OADs and RADs close, the EAD opens, and an alarm is actuated. The operator can switch the airhandler into its purge mode, so that the fans are started, OAD and EAD are opened, and RAD is closed. If the smoke/fire emergency requires the presence of fire fighters, the fire command panel is used. From this panel, the fire chief can operate all fans and dampers as needed for safe and orderly evacuation and protection of the building.

In another emergency condition, a freezestat switch on one of the water coils is actuated. These switches are usually set at approximately 35°F (1.5°C) and serve to protect from coil damage resulting from freeze-ups. Multistage freezestat units might operate as follows:

- At 38°F (3°C) - close OAD
- At 36°F (2°C) - fully open water valve
- At 35°F (1.5°C) - stop fan

If single-stage freezestats are used, they will stop the fan, close the OAD, and activate an alarm.

Yet another type of emergency is signaled by excessive pressures in the ductwork on the suction or discharge sides of the fans, resulting from operation against closed dampers or from other equipment failures. When this happens, the associated fan is stopped and an alarm is actuated.

Fan Controls

The standard fan controls are shown in figure 1.5. Each zone shown in figure 1.2 is supplied with air through a thermostat modulated damper, also called a variable air volume box (VAV-23).

The VAV box openings in the various zones determine the total demand for supply air. The pressure in the supply air (SA) distribution header is controlled by PIC-19, which modulates the supply-air fan station to match demand (figure 1.5). When the PIC-19 output has increased the fan capacity to its maximum, PSH-19 actuates and starts an additional fan. Inversely, as the demand for supply air drops, FSL-15 will stop one fan unit whenever the load can be met by fewer fans than the number in operation. The important point to remember is that in cycling fan stations, fan units are started on pressure and are stopped on flow control. The operating cost of such a fan station is 20 to 40 percent lower than if constant-volume fans with conventional controls were used (figure 1.6).

Because the conditioned zones are pressurized slightly, some of the conditioned air will leak into the atmosphere, creating pressurization loss. Being able to control the pressurization loss is one of the advantages of the control system described in figure 1.5. The flow ratio controller FFIC-14 is set at 90 percent, meaning that the return-air fan station is modulated to return 90 percent of the air supplied to the zones. Therefore, pressurization loss is controlled at 10 percent, which corresponds to the minimum fresh-air make-up requirement, resulting in a minimum-cost operation.

Because the conditioned zones represent a fairly large capacity, a change in supply air flow will not immediately result in a need for a corresponding change in the return air flow. Thus, PIC-20 (figure 1.5) is included in the system to prevent the flow-ratio

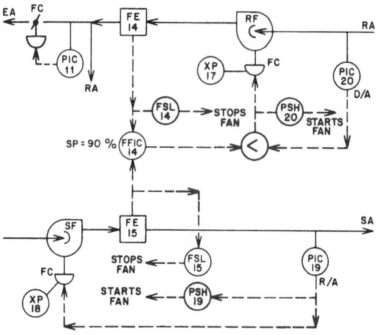

Fig. 1.5 Variable-volume fan controls operate as shown here. (From Lipták [8])

controller from increasing the return air flow rate faster than required. This dynamic balancing eliminates cycling and protects against collapsing the ductwork under excessive vacuum. Closure of the exhaust-air damper by PIC-11 indicates that the control system is properly tuned and balanced and is operating at maximum efficiency. Under such conditions, the outside air admitted into the airhandler exactly matches the pressurization loss, and no return air is exhausted.

Dampers as Control Elements

To maximize the benefits of such an efficient configuration, the dampers must be of tight shutoff design. With a pressure difference of 4 in. H_2O (996 Pa), a closed conventional damper will leak at a rate of approximately 50 cfm/ft^2 (15.2(m^3/min)m^2). In the HVAC industry, a 5 cfm/ft^2 (1.52 (m^3/min)m^2) leakage rate is considered to represent a tight shutoff design. Actually, it is cost-effective to install tight shutoff dampers with leakage rates of less than 0.5 cfm/ft^2 (0.15(m^3/min)m^2), because the resulting savings over the life of the buildings will be much greater than the increase in initial investment for better dampers (figure 1.7).

In order for dampers to give good control, a fair amount of pressure drop should be assigned to them. They should be sized for a ΔP of about 10 percent of the total system drop. On the other hand, excessive damper drops should also be avoided,

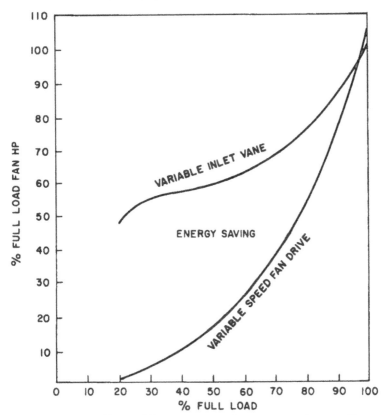

Fig. 1.6 Using variable speed fans can save significant amounts of energy. (Courtesy of Dana Corp.)

because they will increase the operating costs of the fans. A good sizing basis for outside and return air fans is to size them for 1500 fpm (457 m/min) velocity at maximum flow.

In locations where two air streams are mixed, such as when outside and return airs are ratioed (RAD-04 & OAD-05 in figure 1.2), it is important that the damper ΔP be relatively constant as the ratio is varied. Figure 1.8 shows that parallel blade dampers give a superior performance in this service.

Figure 1.9 illustrates the typical pressure profiles in airhandlers. It can be seen that the kind of pressure drops that would be required by opposed blade dampers (figure 1.8) are simply not available. Therefore, if such dampers were installed, the airhandler would be starved for air, (the dampers could not pass the design flow) whenever the ratio was near 50–50 percent.

Figure 1.9 also shows that in traditional airhandlers more fan energy is used than necessary. This is because the return air fan is sized to generate the pressure needed to exhaust the air from the building. A consequence of this is that the pressure drop

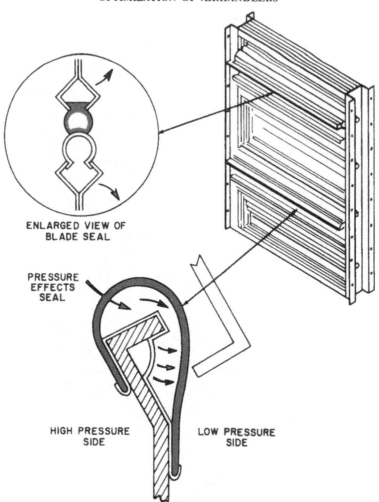

ENLARGED VIEW OF
BLADE SEAL

PRESSURE
EFFECTS
SEAL

HIGH PRESSURE   LOW PRESSURE
SIDE   SIDE

Fig. 1.7 Low-leakage damper designs increase the efficiency of HVAC systems.
(From Lipták [7])

of $1\frac{1}{2}''$ H$_2$O (375 Pa) across the return air damper is three times greater than what is necessary ($\frac{1}{2}''$ H$_2$O or 125 Pa).

The alternate system also shown in figure 1.9 eliminates this waste of fan energy. Here, only the supply fan (SF) operates continuously, which reduces the pressure drop across the return air damper to $\frac{1}{2}''$ H$_2$O (125 Pa). The return fan (RF) is started only when air needs to be relieved, and its speed is varied to adjust the amount of air to be exhausted. Relocating RF also removes its heat input, which, in the traditional system, represents an added load on the cooling coil.

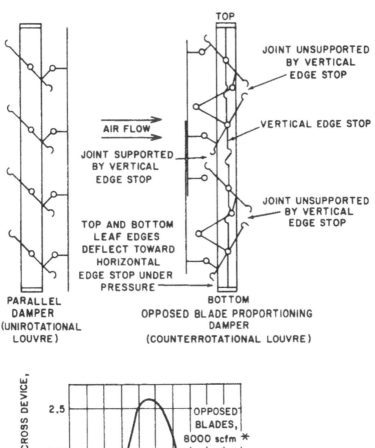

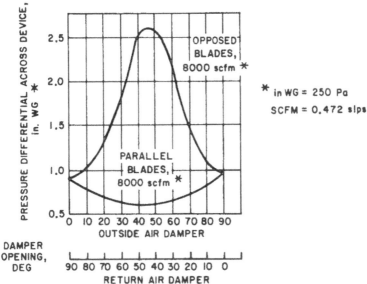

Fig. 1.8 When outside and return air dampers are throttled to vary their ratio at constant total flow, the required pressure drop varies with damper design (see upper portion of figure). The lower portion of this figure shows the results of American Warming and Ventilating Co. tests (per AMCA Standard 500) of pressure drops across parallel-blade and opposed-blade outside and return air damper sets. (From Avery [1])

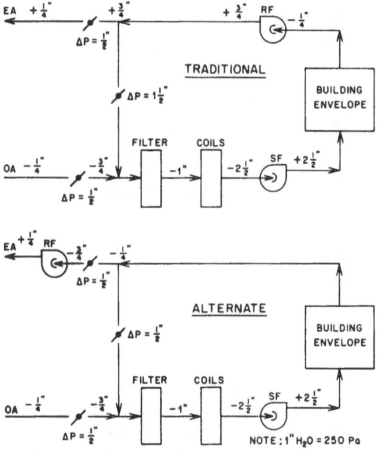

Fig. 1.9 Pressure profiles and damper drops in various airhandlers. (Data from Avery [1])

Temperature Controls

Space temperatures are controlled by thermostats. The traditional thermostat is a proportional-only controller. The pressure of the output signal from a pneumatic "stat" is a near straight-line function of the measurement, described by the following relationship:

$$O = K_c (M - M_o) + O_o$$

where

O = output signal

K_c = proportional sensitivity (K_c can be fixed or adjustable depending on the design)

M   = measurement (temperature)

M_o = "normal" value of measurement, corresponding to the center of the throttling range

O_o = "normal" value of the output signal, corresponding to the center of the throttling range of the control valve (or damper)

Another term used to describe the sensitivity of thermostats is *throttling range*. As shown in figure 1.10, this term refers to the amount of temperature change that is required to change the thermostat output from its minimum to its maximum value, such as from 3 to 13 PSIG (21 to 90 kPa). The throttling range is usually adjustable from 2 to 10°F (1 to 5°C).

One important point to remember is that thermostats *do not* have set points, in the sense of having a predetermined temperature to which they would seek to return the controlled space. (Integral action must be added in order for a controller to be able to return the measured variable to a set point after a load change.) M_o does not represent a set point; it only identifies the space temperature that will cause the cooling damper in figure 1.11 to be 50-percent open. This can be called a "normal" condition, because relative to this point the thermostat can both increase and decrease the cooling air flow rate as space temperature changes. If the cooling load doubles, the damper will need to be fully open, which cannot take place until the controlled space temperature has risen to 73°F (23°C). As long as the cooling load remains that high, the space temperature must also stay up at the 73°F (23°C) value. Similarly, the only way this thermostat can

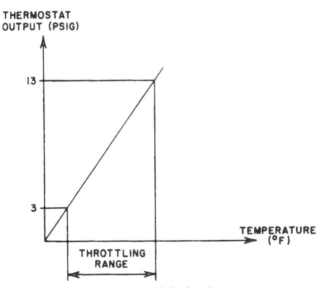

Fig. 1.10 Throttling range can be defined as the temperature change required to change the thermostat output from its minimum to its maximum value. (From Lipták [10])

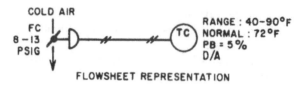

FLOWSHEET REPRESENTATION

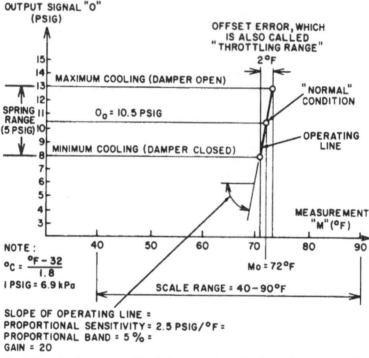

Fig. 1.11 A fixed proportional band thermostat has a fixed throttling range, and no set point. (From Lipták [10])

reduce the opening of the cooling damper below 50 percent is to first allow the space temperature to drop below 72°F (22°C). Thus, thermostats have throttling ranges, not set points. If a throttling range is narrow enough, this gives the appearance that the controller is keeping the variable near set point, when in fact the narrow range allows the variable to drift within limits.

Special-Purpose Thermostats

Night-day, set-back, or *dual room thermostats* will operate at different "normal" temperature values for day and night. They are provided with both a "day" and a "night" setting dial, and the change from day to night operation can be made automatic for a group of thermostats. The pneumatic day-night thermostat uses a two-pressure air supply system, the two pressures often being 13 and 17 PSIG (89.6 and 117 kPa) or 15 and 20 PSIG (103.35 and 137.8 kPa). Changing the pressure at a central point from one value to the other actuates switching devices in the thermostat and indexes them from day to night or vice versa. Supply air mains are often divided into two or more circuits so that switching can be accomplished in various areas of the building at different times. For example, a school building may have separate circuits for classrooms, offices and administrative areas, the auditorium, and the gymnasium and locker rooms. In some of the electric designs, dedicated clocks and switches are built into each thermostat.

The *heating-cooling* or *summer-winter thermostat* can have its action reversed and, if desired, can have its set point changed by means of indexing. This thermostat is used to actuate controlled devices, such as valves or dampers, that regulate a heating source at one time and a cooling source at another. It is often manually indexed in groups by a switch, or automatically by a thermostat that senses the temperature of the water supply, the outdoor temperature, or another suitable variable.

In the heating-cooling design, there are frequently two bimetallic elements, one being direct acting for the heating mode, the other being reverse acting for the cooling mode. The mode is switched automatically in response to a change in the air supply pressure, much as the day-night thermostats operate.

The *limited control range thermostat* usually limits the room temperature in the heating season to a maximum of 75°F (24°C), even if the occupant of the room has set the thermostat beyond these limits. This is done internally, without placing a physical stop on the setting knob.

A *slave or submaster thermostat* has its set point raised or lowered over a predetermined range, in accordance with variations in the output from a master controller. The master controller can be a thermostat, manual switch, pressure controller, or similar device. For example, a master thermostat measuring outdoor air temperature can be used to adjust a submaster thermostat controlling the water temperature in a heating system. Master-submaster combinations are sometimes designated as single-cascade action. When such action is accomplished by a single thermostat having more than one measuring element, it is known as *compensated control*.

Multistage thermostats are designed to operate two or more final control elements in sequence.

A *wet-bulb thermostat* is often used for humidity control, as the difference between wet- and dry-bulb temperature is an indication of moisture content. A wick or other means for keeping the bulb wet and rapid air motion to assure a true wet-bulb measurement are essential.

A *dew-point thermostat* is a device designed to control humidity on the basis of dew point temperatures.

A *smart thermostat* is usually a microprocessor-based unit with an RTD type or a transistorized solid state sensor. It is usually provided with its own dedicated memory and intelligence, and it can also be equipped with a communication link (over a shared data bus) to a central computer. Such units can minimize building operating costs by combining time-of-day controls with intelligent comfort gap selection and maximized self-heating.

Zero Energy Band Control

A recent addition to the available thermostat choices is the zero energy band (ZEB) design. The idea behind ZEB control is to conserve energy by not using any when the room is comfortable. As illustrated by figure 1.12, the conventional thermostat wastes energy by continuing to use it when the area's temperature is already comfortable. The "comfort gap" or "ZEB" on the thermostat is adjustable and can be varied to match the nature of the particular space.

ZEB control can be accomplished in one of two ways. The single set point and single output approach is illustrated on the left side of figure 1.13. Here the cooling valve fails closed and is shown to have an 8–11 PSIG (55–76 kPa) spring range, while the heating valve is selected to fail open and has a 2–5 PSIG (14–34 kPa) range. Therefore, between 5 and 8 PSIG (34 and 55 kPa), both valves are closed; no pay energy is expended while the thermostat output is within this range. The throttling range is usually adjustable from 5 to 25°F (3 to 13°C). Thus, if the ZEB is 30 percent of the throttling range, it can be varied from a gap size of 1.5°F (.85°C) to 7.5°F (4.2°C) by changing the throttling range (or gain).

Although the split-range approach is a little less expensive than the dual set point scheme (shown on the right of figure 1.13), it is also less flexible and more restrictive. The two basic limitations of the split-range approach are the following:

- The gap width can be adjusted only by also changing the thermostat gain; maximum gap width is limited by the minimum gain setting of the unit.
- The heating valve must fail open, which is undesirable in terms of energy conservation.

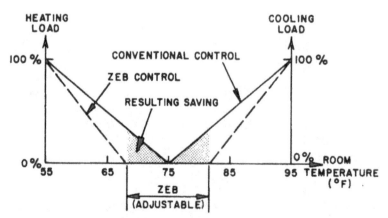

Fig. 1.12 Zero energy band (ZEB) control is designed to save energy by not using any when the room is comfortable. (From Lipták [10])

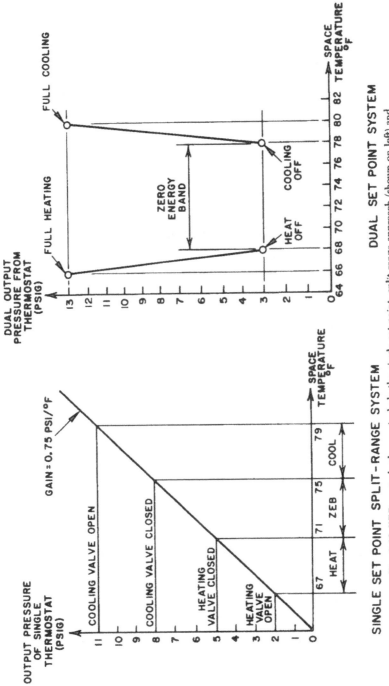

SINGLE SET POINT SPLIT-RANGE SYSTEM   DUAL SET POINT SYSTEM

Fig. 1.13 ZEB control schemes include the single set point–split range approach (shown on left) and the dual set point approach (shown on right). (From Lipták [10])

19

These limitations are removed when a dual set point, dual output thermostat is used. Here both valves can fail closed, and the band width and the thermostat gain are independently adjustable. The gains of the heating and cooling thermostats are also independently adjustable. In figure 1.13, the heating thermostat is reverse-acting and the cooling thermostat is direct-acting.

The most recent advances in thermostat technology are the microprocessor-based units. These are programmable devices with memory capability. They can be monitored and reset by central computers, using pairs of telephone wires as the communication link. Microprocessor-based units can be provided with continuously recharged back-up batteries and accurate room temperature sensors. They can also operate without a host computer (in the "stand-alone" mode). In this case, the user manually programs the thermostat to maintain various room temperatures as a function of the time of day and other considerations.

Gap Control and the Self-Heating Building

The winter and summer enthalpy settings of a building are illustrated in figure 1.14. The concept of gap control is simple: when comfort level in a zone is somewhere between acceptable limits, the use of "pay energy" is no longer justified. Allowing the

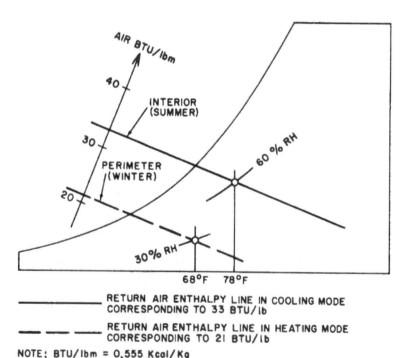

RETURN AIR ENTHALPY LINE IN COOLING MODE
CORRESPONDING TO 33 BTU/lb

RETURN AIR ENTHALPY LINE IN HEATING MODE
CORRESPONDING TO 21 BTU/lb

NOTE: BTU/lbm ≈ 0.555 Kcal/Kg

Fig. 1.14 The air returning from the interior can transport about 10 Btu/1bm to the perimeter, making the building self-heating. (From Lipták [9])

zones to float between limits instead of maintaining them at arbitrarily fixed conditions can substantially reduce the operating cost of the building. The savings come from two sources. First, there is a direct trade-off between the selected acceptable limits of discomfort and the yearly total of required degree days of heating and cooling. Second, there is the added side benefit that the building becomes self-heating during winter conditions. This occurs because during winter conditions, gap control automatically transfers the heat generated in the inside of the building to the perimeter areas, where heating is needed. This can result in long periods of building operation without the use of any pay energy.

Gap control can be looked upon as an override mode of control that is superimposed on the operation of the individual zone thermostats. As such, it can be implemented by all levels of automation, but the flexibility and ease of adjustment of the computerized systems make them superior in those applications in which the gap limits of the various zones are likely to change frequently.

Figure 1.1 illustrates the concept of comfort envelopes. Any combination of temperature and humidity conditions within such envelopes is considered to be comfortable. Therefore, as long as the space conditions fall within this envelope, there is no need to spend money or energy to change those conditions. This comfort gap is also referred to as zero energy band (ZEB), meaning that if the space is within this band, no pay energy of any type will be used. This concept is very cost-effective. When a zone of the airhandler in figure 1.2 is within the comfort gap, its reheat coil is turned off and its VAV box is closed to the minimum flow required for air refreshment. When all the zones are inside the ZEB, the HW, CHW, and STM supplies to the airhandler are all closed and the fan is operated at minimum flow. When all other airhandlers are also within the ZEB, the pumping stations, chillers, cooling towers, and HW generators are also turned off.

With larger buildings that have interior spaces that are heat-generating even in the winter, ZEB control can make the building self-heating. Optimized control systems in operation today are transferring the interior heat to the perimeter without requiring any pay heat until the outside temperature drops below 10°F to 20°F ($-12.3°C$ to $-6.8°C$). In regions in which winter temperature does not drop below 10°F ($-12.3°C$), ZEB control can eliminate the need for pay heat altogether. In regions farther north, ZEB control can lower the yearly heating fuel bill by 30 to 50 percent.

Supply Air Temperature Control

A substantial source of inefficiency in conventional HVAC control systems is the uncoordinated arrangement of temperature controllers. Two or three separate temperature control loops in series are not uncommon. For example, one of these uncoordinated controllers may be used to control the mixed air temperature, another to maintain supply (SA) temperature, and a third to control the zone-reheat coil. Such practice can result in simultaneous heating and cooling and therefore in unnecessary waste. Using a fully coordinated split-range temperature control system, such as that shown in figure 1.15, will reduce yearly operating costs by more than 10 percent.

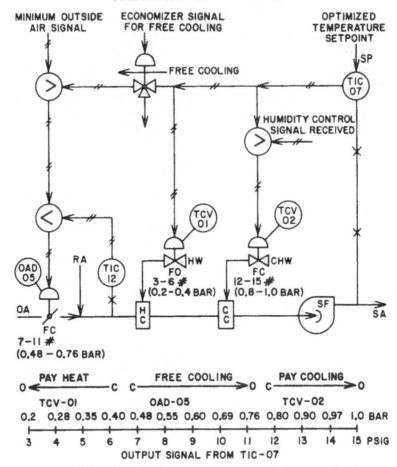

Fig. 1.15 A fully coordinated split-range temperature control system will reduce yearly operating costs by more than 10 percent. (From Lipták [8])

In this control system, the SA temperature set point (set by the temperature controller, TIC-07) is continuously modulated to follow the load. The methods of finding the correct set point will be discussed under "Optimized Temperature Control." The loop automatically controls all heating or cooling modes. When the TIC-07 output signal is low—3 to 6 PSIG (20.7 to 41.3 kPa)—heating is done by TCV-01. As the output signal reaches 6 PSIG (41.3 kPa), heating is terminated; if free cooling is available, it is initiated at 7 PSIG (48.2 kPa). When the output signal reaches 11 PSIG (75.8 kPa)—the point at which OAD-05 is fully open—the cooling potential represented by free cooling is exhausted, and at 12 PSIG (82.7 kPa), "pay cooling" is started by opening TCV-02. In such split-range systems, the possibility of simultaneous heating and cooling is eliminated. Also eliminated are interactions and cycling.

Figure 1.15 also shows some important overrides. TIC-12, for example, limits the allowable opening of OAD-05, so that the mixed-air temperature will never be allowed to drop to the freezing point and permit freeze-up of the water coils.

The minimum outdoor air requirement signal guarantees that the outside air flow will not be allowed to drop below this limit.

The economizer signal allows the output signal of TIC-07 to open OAD-05 only when "free cooling" is available. (A potential for free cooling exists when the enthalpy of the outdoor air is below that of the return air.)

Finally, the humidity controls will override the TIC-07 signal to TCV-02 when the need for dehumidification requires that the supply-air temperature be lowered below the set point of TIC-07.

Humidity Controls

Humidity in the zones is controlled according to the moisture content of the combined return air (see figure 1.16). The process controlled by RHIC-10 is slow and contains large dead time and transport-lag elements. In other words, a change in the SA humidity will not be detected by RHIC-10 until some minutes later. During the winter, it is possible for RHIC-10 to demand more and more humidification. To prevent possible saturation of the supply air, the RHIC-10 output signal is limited by RHIC-21. In this way, the moisture content of the supply air is never allowed to exceed 90 percent RH.

For best operating efficiency, a nonlinear controller with a neutral band is used at RHIC-10. This neutral band can be set to a range of humidity levels—say, between 30 percent RH and 50 percent RH. If the RA is within these limits, the output of RHIC-10 is at 9 PSIG (62 kPa), and neither humidification nor dehumidification is demanded. This arrangement can lower the cost of humidity control during the spring and fall by approximately 20 percent.

The same controller (RHIC-10) controls both humidification (through the relative humidity control valve, RHCV-16) and dehumidification (through the temperature control valve, TCV-02) on a split-range basis. As the output signal increases, the humidifier valve closes, between 3 and 9 PSIG (20.7 and 62 kPa). At 9 PSIG (62 kPa), RHCV-16 closes and remains so, as the output signal increases to 12 PSIG (82.7 kPa). At this condition, TCV-02 starts to open. Dehumidification is accomplished by cooling through TCV-02. This chilled-water valve is controlled by humidity (RHIC-10) or temperature (TIC-07). The controller that requires more cooling will be the one allowed to throttle TCV-02.

Subcooling the air to remove moisture can substantially increase operating costs if this energy is not recovered. The dual penalty incurred for overcooling for dehumidification purposes is the high chilled-water cost and the possible need for reheat at the zone level. The savings from a pump-around economizer can eliminate 80 percent of this waste. In this loop, whenever TDIC-07 detects that the chilled water valve

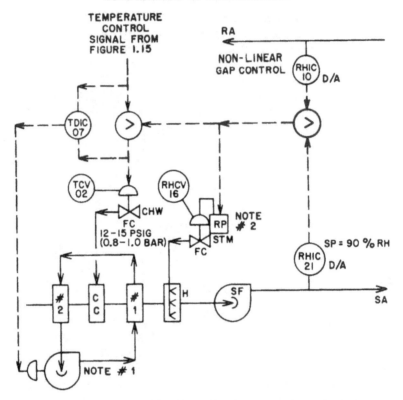

Fig. 1.16 Humidity is controlled in the combined return air. (From Lipták [8])

Note 1: When the need for dehumidification (in the summer) overcools the supply
air and therefore increases the need for reheat at the zones, this pump-around
economizer loop is started. TDIC-07 will control the pump to "pump around"
only as much heat as is needed.

Note 2: This reversing positioner functions as follows:

Input from RHIC-10	Output to RHCV-16
3	15
9	3

(TCV-02) is open more than would be necessary to satisfy TIC-07, the pump-around
economizer is started. This loop in coil #1 reheats the dehumidified supply air, using
the heat that the pump-around loop removed from the outside air in coil #2 before it
entered the main cooling coil.

In this way, the chilled-water demand is reduced in the cooling coil (TCV-02), and
the need for reheating at the zones is eliminated. Although figure 1.16 shows a modulating controller setting the speed of a circulating pump, it is also possible to use a
constant-speed pump operated by a gap switch.

Outdoor Air Controls

Outdoor air is admitted to satisfy requirements for fresh air or to provide free cooling. Both control loops are shown in figure 1.17.

The minimum requirement for fresh outdoor air while the building is occupied is usually 10 percent of the airhandler's capacity. In more advanced control systems, this value is not controlled as a fixed percentage but as a function of the number of people in the building or of the air's carbon dioxide content. In the most conventional systems, the minimum outdoor air is provided by keeping 10 percent of the area of the outdoor air damper always open when the building is occupied. This method is inaccurate, because a constant damper opening does not result in a constant air flow. This flow varies with fan load, because changes in load will change the fan's suction pressure and will therefore affect ΔP across the damper. This conventional design results in waste of air-conditioning energy at high loads and insufficient air refreshment at low loads.

The control system depicted in figure 1.17 reduces operating costs while maintaining a constant minimum rate of air refreshment, which is unaffected by fan loading. Direct measurement of outdoor air flow is usually not possible because of space limitations. For this reason, figure 1.17 shows the outdoor air flow as being determined as

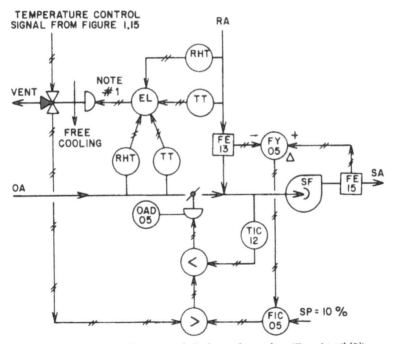

Fig. 1.17 Outside air control loops provide fresh air or free cooling. (From Lipták[8])

Note: The enthalpy logic unit (EL) compares the enthalpies of the outside and return airs and vents its output signal if free cooling is avaiable. Therefore, the economizer cycle is initiated whenever Hoa < Hra.

the difference between FE-15 and FE-13. FIC-05 controls the required minimum outdoor air flow by throttling OAD-05.

CO_2 Based Ventilation

In conventional installations the amount of outdoor air admitted is usually based on one of the following criteria:

- 0.1 to 0.25 cfm/ft^2 (30 to 76 lpm/m^2) of floor area
- 10 to 25 percent of total air supply rates
- About 5 cfm (25 lps) volumetric rate per person

These criteria all originated at a time when energy conservation was no serious consideration; therefore, their aim was to provide simple, easily enforceable, rules that will guarantee that the outdoor air intake always exceeds the required minimum. Today the goal of such systems is just the opposite: it is to make sure that air quality is guaranteed at minimum cost. As the floor does not need oxygen—only people do— some of the above rules make little sense.

There is a direct relationship between savings in building operating costs and reduction in outdoor air admitted into the building. According to one study in the U.S. [3] infiltration of outdoor air accounted for 55 percent of the total heating load and 42 percent of the total cooling load. Another survey [11] showed that 75 percent of fuel oil consumed in New York City schools was devoted to heating ventilated air. Because building conditioning accounts for nearly 20 percent of all the energy consumed in the U.S. [13], optimized admission of outdoor air can make a major contribution to reducing our national energy budget. This goal can be well served by CO_2 based ventilation controls.

The purpose of ventilation is not to meet some arbitrary criteria, but to maintain a certain air quality in the conditioned space. Smoke, odors, and other air contaminant parameters can all be correlated to the CO_2 content of the return air [12]. This then becomes a powerful tool of optimization, because the amount of outdoor air required for ventilation purposes can be determined on the basis of CO_2 measurement, and the time of admitting this air can be selected so that the air addition will also be energy efficient. With this technique, health and energy considerations will no longer be in conflict, but will complement each other.

CO_2 based ventilation controls can easily be integrated with the economizer cycle and can be implemented by use of conventional or computerized control systems. Because the rate of CO_2 generation by a sedentary adult is 0.75 cfh (21 lpm), control by CO_2 concentration will automatically reflect the level of building occupancy [2]. Energy savings of 40 percent have been reported [4] by converting conventional ventilation systems to intermittent CO_2 based operation.

Economizer Cycles

The full use of free cooling can reduce the yearly air-conditioning load by more than 10 percent. The enthalpy logic unit (EL) in figure 1.17 will allow the temperature-controller signal (TIC-07 in figure 1.15) to operate the outdoor air damper whenever free cooling is available. This economizer cycle is therefore activated whenever the enthalpy of the outdoor air is below that of the return air.

Free cooling can also be used to advantage while the building is unoccupied. Purging the building with cool outdoor air during the early morning results in cooling capacity being stored in the building structure, reducing the daytime cooling load.

The conventional economizers—such as the one shown in figure 1.17—are rather limited devices for two reasons. First, they determine the enthalpy of the outdoor and return air streams, using somewhat inaccurate sensors. Secondly, although they consider the free cooling potential of the outside air, they disregard all the other possibilities of using the outdoor air to advantage.

Advanced, microprocessor-based economizers overcome both of these limitations. They use accurate sensors and memorize the psychrometric chart to evaluate all potential uses of outside air, not only free cooling. Figure 1.18 illustrates the various zones of operation, based on the relative conditions of the outside and return airs.

If the enthalpy of the outside air falls in zone #1 (that is, if its Btu content exceeds 33), no free cooling is available; therefore, the use of outside air should be minimized in the summer. In the winter or fall, it is possible that the enthalpy of the outside air on sunny afternoons will exceed the return air enthalpy, which in the winter is about

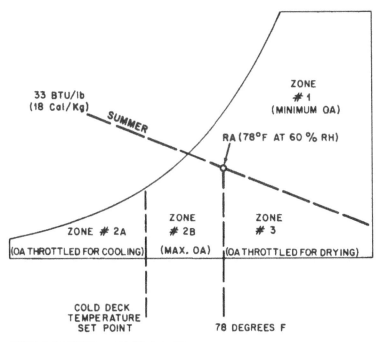

Fig. 1.18 Free cooling and drying can often be obtained in the summer, depending on the zone where the outside air falls relative to the return air.

21 Btu/lbm. (11.6 Cal/Kg). Under such conditions, "free heating" can be obtained by admitting the outside air in zone #1.

If the condition of OA corresponds to zone #2 (Btu<33 and temp.<78°F), free cooling is available.

If the condition of OA corresponds to zone #3 (Btu<33 and temp.>78°F), free dehumidifying (latent cooling) is available.

When there are both cold and hot air ducts in the building (dual duct system), the control system in zone #2 will function differently depending on whether the outside air temperature is above or below the cold deck temperature. If it is above that temperature (zone 2B), maximum (100 percent) outside air can be used; if it is below that temperature (zone 2A) the use of outside air needs to be modulated or time-proportioned. Therefore, in zone 2A, where the outside air is cooler than the cold deck temperature, free cooling is available, but only some of the total potential can be used. The OA damper therefore must be either positioned or cycled to admit enough cold outside air to lower the mixed air temperature to equal the cold deck temperature set point. This way the need for "pay cooling" is eliminated.

When the outside air is in zone #3 (hot but dry), the control system should admit some of it to reduce the overall need for latent cooling. The amount of outside air admitted under these conditions should be controlled so that the resulting mixed air dew point temperature will equal the desired cold deck dew point. This method takes full advantage of the free drying potential of the outside air.

As can be seen from the above, the use of the economizer cycles is similar to opening the windows in a room (partially or fully) to achieve maximum comfort with minimum use of "pay energy."

Optimizing Strategies

The main goal of optimization in airhandlers is to match the demand for conditioned air with a continuous, flexible, and efficient supply. This requires the floating of both air pressures and temperatures to minimize operating costs while following the changing load.

In traditional HVAC control systems, the set point of TIC-07 in figure 1.15 is set manually and is held as a constant. This practice is undesirable, because for each particular load distribution the optimum SA temperature is different. If the manual setting is less than that temperature, some zones will become uncontrollable. If it is more than optimum, operating energy will be wasted.

Temperature Optimization in the Winter

In the winter, the goal of optimization is to distribute the heat load between the main heating coil (HC in figure 1.15) and the zone reheat coils (RHC in figure 1.19) most efficiently. The highest efficiency is obtained if the SA temperature is high enough to meet the load of the zone with the *minimum* load. In this way, the zone reheat coils are used only to provide the difference between the loads of the various zones, whereas the base load is continuously followed by TIC-07. In figure 1.19, the set point of TIC-

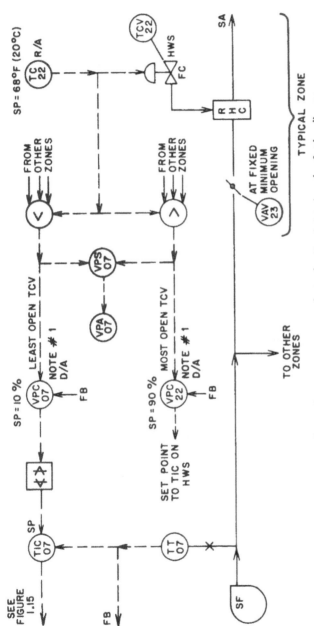

Fig. 1.19 Optimization of air and water temperatures during heating (winter) mode of airhandler operation is designed to distribute the heat load efficiently between the main heating coil and the zone reheat coils. (From Lipták [8])

Note: Valve position controllers (VPCs) are provided with integral action only for stable floating control. The integral time is set to be ten times the integral time of the associated TIC. External feedback is provided to eliminate reset wind-up when VPC output is overruled by the set point limit on TIC.

29

07 is adjusted to keep the least-open TCV-22 only about 10 percent open. First, the least open TCV-22 is identified and its opening (in the valve position controller, VCP-07) is compared with the desired goal of 10 percent. If the valve opening is more than 10 percent, the TIC-07 set point is increased; if less, the set point is lowered. Stable operation is obtained by making VPC-07 an integral-only controller with an external feedback to prevent reset wind-up.

In figure 1.19 it is assumed that the hot-water supply (HWS) temperature is independently adjustable and will be modulated by VPC-22 so that the most-open TCV-22 will always be 90 percent open. The advantages of optimizing the HWS temperature include:

- Minimizing heat-pump operating costs by minimizing HWS temperature
- Reducing pumping costs by opening all TCV-22 valves in the system
- Eliminating unstable (cycling) valve operation, which occurs in the nearly closed position, by opening all TCV-22 valves.

In figure 1.19, a valve position alarm (VPA-07) is also provided to alert the operator if this "heating" control system is incapable of keeping the openings of all TCV-22 valves between the limits of 10 percent and 90 percent. Such alarm will occur if the VPCs can no longer change the TIC set point(s), because their maximum (or minimum) limits have been reached. This condition will occur only if the load distribution was not correctly estimated during the design phase of the project or if the mechanical equipment was not correctly sized.

If the HWS temperature cannot be modulated to keep the most-open TCV-22 from opening to more than 90 percent, then the control loop depicted in figure 1.20 should be used. In this loop, as long as the most-open valve is less than 90 percent open, the SA temperature is set to keep the least-open TCV-22 at 10 percent open (zone A). When the most-open valve reaches 90 percent opening, control of the least-open valve is abandoned and the loop is dedicated to keeping the most-open TCV-22 from becoming fully open (zone B). This, therefore, is a classic case of herding control, in which a single constraint envelope "herds" all TCV openings to within an acceptable band and thereby accomplishes efficient load following.

Temperature Optimization in the Summer

In the cooling mode during the summer, the SA temperature is modulated to keep the most-open variable volume box (VAV-23) from fully opening. Once a control element is fully open, it can no longer control; therefore, the occurrence of such a state must be prevented. On the other hand, it is generally desirable to open throttling devices such as VAV boxes to accomplish the following goals:

- Reduce the total friction drop in the system
- Eliminate cycling and unstable operation (which is more likely to occur when the VAV box is nearly closed)
- Allow the airhandler to meet the load at the highest possible supply-air temperature

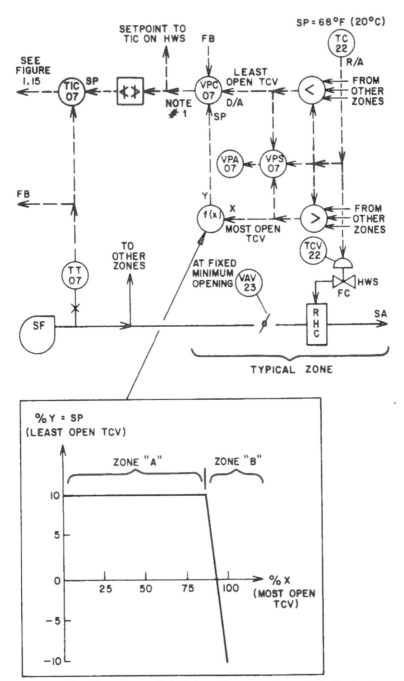

Fig. 1.20 This alternative method of air supply temperature optimization in winter should be used when the HWS temperature cannot be modulated. (From Lipták [8])

Note: Valve position controllers (VPCs) are provided with integral action only for stable floating control. The integral time is set to be ten times the integral time of the associated TIC. External feedback is provided to eliminate reset wind-up when VPC output is overruled by the set point limit on TIC.

This statement does not apply if air transportation costs exceed cooling costs (for example, undersized ducts, inefficient fans). In this case, the goal of optimization is to transport the minimum quantity of air. The amount of air required to meet a cooling load will be minimized if the cooling capacity of each unit of air is maximized. Therefore, if fan operating cost is the optimization criterion, the SA temperature is to be kept at its achievable minimum, instead of being controlled as in figure 1.21.

If the added feature of automatic switchover between winter and summer modes is desired, the control system depicted in figure 1.22 should be used. When all zones require heating, this control loop will behave exactly as does the one shown in figure 1.20; when all zones require cooling, it will operate as does the system shown in figure 1.21. In addition, this control system will operate automatically with maximum energy efficiency during the transitional periods of fall and spring. This high efficiency is a result of the exploitation of the self-heating effect. If some zones require heating (perimeter offices) and others require cooling (interior spaces), the airhandler will auto-

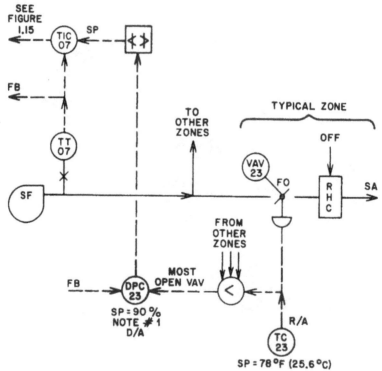

Fig. 1.21 This method of air supply temperature optimization in summer (cooling mode) should be used if fan operating costs are less than cooling costs. (From Lipták [8])

Note: The damper position controller (DPC) has integral action only, with its setting being ten times the integral time of TIC-07. External feedback is provided to eliminate reset wind-up.

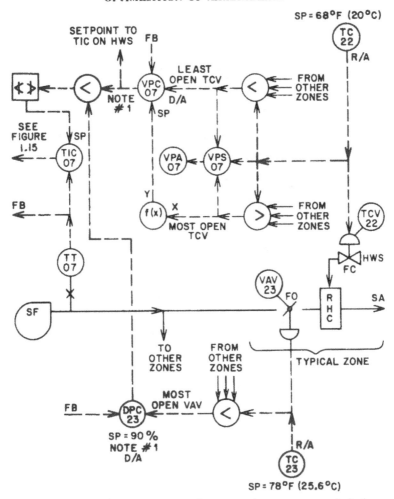

Fig. 1.22 This control system optimizes the water and air temperatures in both summer and winter. See also figures 1.20 and 1.21. (From Lipták [8])

matically transfer this free heat from the interior to the perimeter zones by intermixing the return air from the various zones and moving it through the 10-degree zero energy band (ZEB) between the settings of TC-22 and TC-23. When the zone temperatures are within this comfort gap of 68°F (20°C) to 78°F (26°C), no pay energy is used and the airhandler is in its self-heating, or free-heating, mode. This is an effective means of reducing operating costs in buildings. The savings can amount to more than 30 percent during the transitional seasons.

When the temperature in one of the zones reaches 78°F (26°C), the air supply temperature set point will be lowered by DPC-23 and the air-side controls will be automatically switched to cooling (as depicted in figure 1.21). If, at the same time, some other zone temperatures drop below 68°F (20°C), requiring heating, their heat demand will have to be met by the heat input of the zone-reheat coils only. This mode of operation is highly inefficient because of the simultaneous cooling and reheating of the air. Fortunately, this combination of conditions is highly unlikely, because under proper design practices, the zones served by the same airhandler should display similar load characteristics. The advantage of the control loop in figure 1.22 is that it can automatically handle any load or load combination, including this unlikely, extreme case.

Auto-Balancing of Buildings

In computerized building control systems, the optimization potentials are greater than those that have been discussed up to this point. When all zone conditions are detected and controlled by the computer, it can optimize not only the normal operation but also the start-up of the building.

The optimization of airhandler fans is directed at two goals simultaneously. The first goal is to find the optimum value for the set point of the supply air pressure controller (PIC-19 in figure 1.5). Generally, the supply air pressure is at an optimum value when it is at the lowest possible value, while all loads are satisfied. As the supply pressure is lowered, the fan operating cost is reduced, but with lowered supply pressures the VAV boxes serving the individual zones (VAV-23 in figure 1.2) will have to open up so that the air flow to the zones will not be reduced. Therefore, the optimum setting for PIC-19 is that pressure at which the most-open VAV box is nearly 100 percent open, while all other VAV boxes are less than 100 percent open.

The second goal of optimization is to rebalance the air distribution in the building automatically as the load changes. If the VAV boxes (VAV-23 in figure 1.2) are not pressure independent (are not able to maintain constant air flow when the supply pressure changes), manual rebalancing is required every time the load distribution changes. Naturally this is a very labor-demanding and inefficient operation. The optimization strategy described below serves the multiple purposes of automatic rebalancing and finding the optimum set points for the supply air pressure and temperature.

Figure 1.23 illustrates an airhandler serving several zones. The abbreviations used in that figure and in the algorithm tables that follow are listed below:

     AI-1 to AI-N: Analog inputs (zone temperatures)
     AI-AT: Analog input (air supply temperatures)
     AI-RT: Analog input (return air temperature)
     AO-1 to AO-N: Analog output (zone VAV opening)
     AT: Supply air temperature
     EA: Exhaust air
     OA: Outside air
     PIC: Pressure controller (supply air)
     PT: Pressure transmitter
     RA: Return air
     RT: Return air temperature

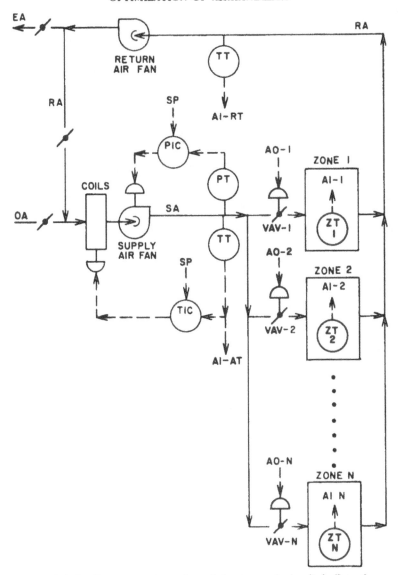

Fig. 1.23 Airhandler optimization and auto-balancing can be handled efficiently by computer.

SA: Supply air
SP: Set point
TH: Upper limit of comfort zone (figure 1.1) (max. allowable zone temperature)
TIC: Temperature controller (air supply)
TL: Lower limit of comfort zone (figure 1.1) (min. allowable zone temperature)

VAV-1 to VAV-N: Variable air volume boxes
ZT-1 to ZT-N: Zone temperatures
ZT5-1 to ZT5-N: ZT five minutes after start-up
ZT10-1 to ZT10-N: ZT ten minutes after start-up
XMIN: Min. VAV opening required for ventilation
XSET-1 to XSET-N: Initial VAV opening after "start-up"
XMAX-1 to XMAX-N: Max. limit on VAV opening during normal operation

Start-Up Algorithm

All VAV boxes are set to their minimum openings required for ventilation purposes (XMIN), such as 25 percent. Therefore, at the time of start-up, AO-1 through AO-N

Table 1.1
ALGORITHM TO DETERMINE START-UP OPENINGS
OF INDIVIDUAL VAV BOXES (XSET)

Input Conditions			Output
Operating Mode	Approach between zone and supply temperatures	Amount of temperature change during last 5 minutes of start-up	Initial value of XSET to be used for AO-1 to AO-N (%)
Heating (AT>RT)	(TL − ZT10)>5°F	(ZT10-ZT5) <0.5°F	100
		(ZT10-ZT5) 0.5–1°F	90
		(ZT10-ZT5) 1–1.5°F	80
		(ZT10-ZT5) 1.5–2°F	70
		(ZT10-ZT5) 2–3°F	60
		(ZT10-ZT5) 3–4°F	50
		(ZT10-ZT5) 4–5°F	40
		(ZT10-ZT5) >5°F	30
	(TL − ZT10)<5°F	Disregard	25
Cooling (AT<RT)	(ZT10 − TH)>5°F	(ZT5-ZT10) <0.5°F	100
		(ZT5-ZT10) 0.5–1°F	90
		(ZT5-ZT10) 1–1.5°F	80
		(ZT5-ZT10) 1.5–2°F	70
		(ZT5-ZT10) 2–3°F	60
		(ZT5-ZT10) 3–4°F	50
		(ZT5-ZT10) 4–5°F	40
		(ZT5-ZT10) >5°F	30
	(ZT10 − TH)<5°F	Disregard	25

Note: C° = (°F − 32)/1.8

are all set for 25 percent. PIC is set to midscale; therefore, the start-up value of its SP = 50 percent. TIC is set for $(TL+25)°F$ in the heating mode and for $(TH-25)°F$ in the cooling mode.

After five minutes of operation, the zone temperatures are detected (ZT5-1, ZT5-2, etc.), and after ten minutes of operation they are detected again (ZT10-1, ZT10-2, etc.). At the end of the first ten minutes of operation, the supply air temperature is also measured as AT10 and the return air temperature as RT10.

Once the above readings are obtained, they are entered into a table, such as table 1.1, which serves as the basis for determining the required start-up openings of each of the VAV boxes (XSET-1, XSET-2, etc.). The purpose of this table is to select the initial opening for each VAV box in a logical manner. Therefore, if the zone temperature after ten minutes of operation is already within 5°F of reaching the comfort zone, the VAV box can be left at its minimum opening. If comfort is not yet within 5°F, a higher opening is needed. The initial VAV opening is increased on the basis of the zone performance during the previous five minutes. The larger the temperature change experienced by the zone during the previous five minutes, the sooner it will reach the comfort zone and therefore the smaller the opening that is required. By this logic, the VAV boxes on those zones which are furthest from comfort and which are moving most slowly toward comfort will be given the highest openings.

Normal Algorithm for VAV Throttling

The initial VAV opening for each zone (XSET), which is determined by the methods above, is then used as the maximum limit on the VAV opening (XMAX) during the first five minutes of normal operation. The value of XMAX is reevaluated every five minutes, as shown in table 1.2. The logic here is to increase the maximum limit on VAV opening (XMAX) to any zone in which the VAV has been open to its maximum limit for five minutes. Similarly, this logic will lower the XMAX limit if the VAV damper was at its XMIN during the previous five minutes. If a VAV damper has been throttled somewhere

Table 1.2
REEVALUATION OF VALUE OF XMAX

Input Conditions		Output
Has VAV been continuously open to its XMAX during last 5 minutes?	Has VAV been continuously throttled to its XMIN during the last 5 minutes?	Incremental change in value of XMAX at the end of 5-minute period
Yes	Yes	Leave XMAX = XMIN
	No	Increase by 10%
No	Yes	Decrease by 10%
	No	Leave as is

in between these two limits (XMAX and XMIN), its limit will not be altered. The change increment of 10 percent shown in table 1.2 is adjustable for maximum flexibility.

The algorithm described above and illustrated in table 1.2 guarantees that changes in load distribution will not result in starving some zones; the building will be automatically rebalanced in an orderly manner. The value of XMAX from table 1.2 and the permanent values of XMIN, determined by ventilation requirements are used to re-evaluate the individual VAV openings every two minutes, as described by the algorithm in table 1.3.

The main optimizing and auto-balancing feature of this algorithm is that whenever a zone is inside the comfort gap, its VAV opening is reduced to XMIN. This reduces the load on the fans and also provides more air to the zones experiencing the highest loads.

Optimization of Air Supply Pressure and Temperature

Optimization means that the load is met at minimum cost. The cost of operating an airhandler is the sum of the cost of air transportation and conditioning. These two cost factors tend to change in opposite directions; minimizing the cost of one will increase the cost of the other. Therefore, it is important to monitor both the transportation and the conditioning costs continuously and to minimize the larger one when optimizing the system. Computerized control systems allow these costs to be calculated readily on the basis of utility costs and quantities.

For example, if the transportation cost exceeds the conditioning cost, the optimization goal is to minimize fan operation. This is achieved by conditioning the space with as little air as possible. The quantity of air transported can be minimized if each pound of air is made to transport more conditioning energy, that is, if each pound of air carries more cooling or heating Btu's. Therefore, when the goal is to minimize fan costs, the air supply pressure is held as low as possible, and the supply temperature is maximized in the winter and minimized in the summer. Fan costs tend to exceed

Table 1.3
ALGORITHM TO DETERMINE ANALOG OUTPUTS,
SETTING THE OPENINGS OF VAV BOXES

Input Conditions		Output
Operating Mode	Control Criteria	Required VAV opening: AO-1 to AO-N is to equal
Heating (AT>ZT)	$ZT<(TL-1)$	XMAX
	$(TL-1)<ZT<(TL+1)$	No change
	$ZT>(TL+1)$	XMIN
Cooling (AT<ZT)	$ZT>(TH+1)$	XMAX
	$(TH-1)<ZT<(TH+1)$	No change
	$ZT<(TH-1)$	XMIN

Table 1.4

OPTIMIZATION OF SUPPLY AIR PRESSURE AND
TEMPERATURE, WHEN FAN COSTS EXCEED
CONDITIONING COSTS (Frequency = 5 minutes)

VAV Status	Airhandler Mode	Is TIC SP at its Limit?	Incremental Ramp Adjustment in the Set Points of	
			TIC	PIC
None at 100% for 15 minutes continuously	Heating (AT>RT)	Yes, @ Max.	−2°F	N.C. (@ Min.)
		No	−1°F	N.C. (@ Min.)
	Cooling (AT<RT)	Yes, @ Min.	+2°F	N.C. (@ Min.)
		No	+1°F	N.C. (@ Min.)
Not more than one at 100% for more than 30 minutes continuously	Heating (AT>RT)	Yes, @ Max.	N.C. (@ Max.)	N.C.
		No	N.C.	N.C.
	Cooling (AT<RT)	Yes, @ Min.	N.C. (@ Max.)	N.C.
		No	N.C.	N.C.
More than one at 100% for more than 30 minutes continuously	Heating (AT>RT)	Yes, @ Max.	(@ Max.)	+¼"H2O
		No	+1°F	N.C.
	Cooling (AT<RT)	Yes, @ Min.	(@ Min.)	+¼"H2O
		No	−1°F	N.C.
More than one at 100% for more than 60 minutes continuously	Heating (AT>RT)	Yes, @ Max.	(@ Max.)	+¼"H2O
		No	+2°F	N.C.
	Cooling (AT<RT)	Yes, @ Min.	(@ Min.)	+¼"H2O
		No	−2°F	N.C.

N.C. = No change is made at the end of that 5 minute period.

conditioning costs when the loads are low, such as in the spring or fall, or when the economizer cycle is used to provide free cooling.

Table 1.4 describes the algorithm used to achieve this goal. When none of the VAV boxes (figure 1.23) are fully open, indicating that all loads are well satisfied, the air pressure (PIC set point) is kept at a minimum, and the air temperature (TIC set point) is lowered in the winter and raised in the summer. When more than one VAV box is fully open, the air supply temperature is increased in the winter (lowered in the summer). When its limit is reached, the algorithm will start raising the PIC set point.

Table 1.5 describes the algorithm used when the conditioning costs are higher than the fan operating costs. This is likely to be the case when the loads are high, such as in the summer or the winter. Under such conditions, the supply pressure is maximized before the supply air temperature is increased in the winter or lowered in the summer. When none of the VAV boxes in figure 1.23 are fully open, the PIC set point is lowered, while the TIC set point is at or near minimum in the winter (maximum in summer). When more than one VAV box is fully open, the PIC set point is increased to its

Table 1.5
OPTIMIZATION OF SUPPLY AIR PRESSURE AND
TEMPERATURE, WHEN CONDITIONING COSTS EXCEED
FAN COSTS (Frequency = 5 minutes)

VAV Status	Airhandler Mode	Is PIC Set Point at its Maximum?	Incremental Ramp Adjustment in the Set Points of	
			TIC	PIC
None at 100% for 15 minutes continuously	Heating (AT>RT)	Yes	$-1°F$	$-\frac{1}{2}"$ H2O
		No	N.C. @ Min.	$-\frac{1}{4}"$ H2O
	Cooling (AT<RT)	Yes	$+1°F$	$-\frac{1}{2}"$ H2O
		No	N.C. @ Max.	$-\frac{1}{4}"$ H2O
Not more than one at 100% for more than 30 minutes continuously	Heating (AT>RT)	Yes	N.C.	N.C. (@ Max.)
		No	N.C.	N.C.
	Cooling (AT<RT)	Yes	N.C.	N.C. (@ Max.)
		No	N.C.	N.C.
More than one at 100% for more than 30 minutes continuously	Heating (AT>RT)	Yes	$+1°F$	(@ Max.)
		No	N.C.	$+\frac{1}{4}"$ H2O
	Cooling (AT<RT)	Yes	$-1°F$	(@ Max.)
		No	N.C.	$+\frac{1}{4}"$ H2O
More than one at 100% for more than 60 minutes continuously	Heating (AT>RT)	Yes	$+2°F$	(@ Max.)
		No	N.C.	$+\frac{1}{2}"$ H2O
	Cooling (AT<RT)	Yes	$-2°F$	(@ Max.)
		No	N.C.	$+\frac{1}{2}"$ H2O

N.C. = No Change.

maximum setting. When that is reached, the supply temperature is started to be increased in the winter (decreased in the summer).

The algorithms described above provide the dual advantages of automatic balancing and minimum operating cost. They eliminate the need for manual labor or for the use of pressure-independent VAV boxes, while reducing operating cost by about 30 percent. They also provide the flexibility of assigning different comfort envelopes (different TL and TH values) to each zone. Thereby as occupancy or use changes, the comfort zone assigned to the particular space can be changed automatically.

Elimination of Chimney Effects

In high-rise buildings, the natural draft resulting from the chimney effect tends to pull in ambient air at near ground elevation and to discharge it at the top of the building. Although eliminating the chimney effect can lower the operating cost by approximately 10 percent, few such systems are yet in operation.

Figure 1.24 shows the required pressure controls. The key element of this control system is the reference riser, which allows all pressure controllers in the building to be referenced to the barometric pressure of the outside atmosphere. Using this pressure

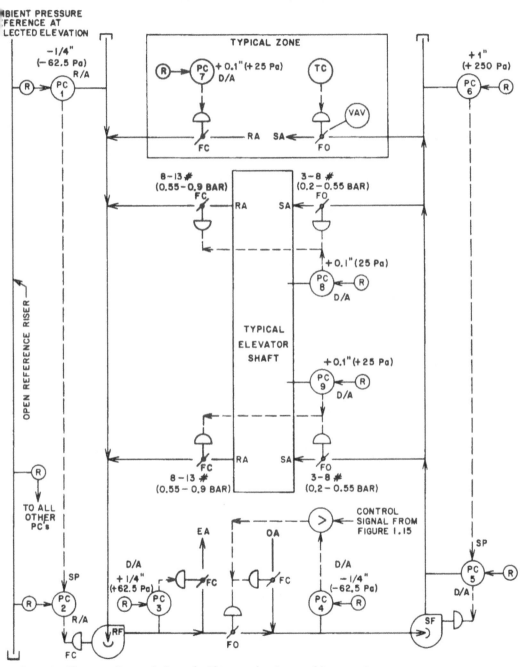

Fig. 1.24 Chimney effects in high-rise buildings can be eliminated by using the proper pressure controls. (From Lipták [8])

reference allows all zones to be operated a 0.1 in H_2O (25 Pa) pressure (PC-7) and permits this constant pressure to be maintained at both ends of all elevator shafts (PC-8 and 9). If the space pressure is the same on the various floors of a high-rise building, there will be no pressure gradient to motivate the vertical movement of the air, and as a consequence, the chimney effect will have been eliminated. A side benefit of this control strategy is the elimination of all drafts or air movements between zones, which also minimizes the dust content of the air. Another benefit is the capability of adjusting the "pressurization loss" of the building by varying the settings of PC-7, 8, and 9.

Besides reducing operating costs, the use of pressure-controlled elevator shafts increases comfort because drafts and the associated noise are eliminated.

Figure 1.24 also shows the use of cascaded fan controls. The set points of the cascade slaves (PC-2 and PC-5) are programmed so that the air pressure at the fan is adjusted as the square of flow and the pressure at the end of the distribution headers (cascade masters PC-1 and PC-6) remains constant. This control approach results in the most efficient operation of variable-air volume fans.

Conclusions

The airhandler is just one of the industrial unit operations. The process of air conditioning is similar to all other industrial processes. Fully exploiting state-of-the-art instrumentation and control results in dramatic improvements. There are few other processes in which the use of optimization and of instrumentation know-how alone can halve the operating cost of a process.

The control and optimization strategies described in this chapter can be implemented by pneumatic or electronic instruments and controlled by analog or digital systems. The type of hardware used in optimization is less important than the understanding of the process and of the control concepts that are to be implemented. The main advantage of digital and computerized systems is their flexibility and convenience in making changes, without the need to modify equipment or wiring.

REFERENCES

1. Avery, G., "VAV Economizer Cycle," *Heating, Piping, Air Conditioning*, August, 1984.

2. Department of Defense, "Environmental Engineering for Shelters," TR-20-Vol. 3, Department of Defense, Office of Civil Defense, May 1969.

3. Kovach, E. G., "Technology of Efficient Energy Utilization: The Report of a NATO Science Committee Conference Held at Les Arcs, France, 8–12 October, 1973," Scientific Affairs Division, NATO, Brussels, Belgium.

4. Kusuda, T., "Intermittent Ventilation for Energy Conservation," NBS report of ASHRAE Symposium in Dallas, Texas, February 1976.

5. Lipták, B. G., "Applying the Techniques of Pro-cess Control to the HVAC Process," *ASHRAE Transactions*, paper no. 2778, vol. 89, part 2A, 1983.

6. Lipták, B. G., "CO_2 Based Ventilation," *ASH-RAE Journal*, July 1979.

7. Lipták, B. G., "Dampers," in *Instrument Engineer's Handbook, Process Measurement* (Radnor, PA: Chilton Book Co., 1982), section 5.1.

8. Lipták, B. G., "HVAC Controls," in *Instrument Engineer's Handbook, Process Control* (Radnor, PA: Chilton Book Co., 1985), section 8.14.

9. Lipták, B. G., "Reducing the Operating Costs of Buildings by the Use of Computers," *ASHRAE Transactions* 83:1, 1977.

10. Lipták, B. G., "Thermostats," in *Instrument En-*

gineer's Handbook, Process Measurement (Radnor, PA: Chilton Book Co., 1982), section 4.12.

11. Liu, S. T., et al., "Research, Design, Construction and Evaluating an Energy Conservation School Building in New York City," NBSIR.

12. Nomura, G., and Y. Yamada, "CO_2 respiration rate for the ventilation calculation," *Transaction of Japanese Architectural Society Meetings*, October 1969.

13. Stanford Research Institute, "Patterns of Energy Consumption in the United States," prepared for the Office of Science and Technology, Washington, D.C.

Chapter 2

OPTIMIZATION
OF
BOILERS

Boiler Efficiency

The efficiency of a steam generator is defined as the ratio of the heat transferred to the water (steam) to the higher heating value of the fuel. The purpose of optimization is to continuously maximize the boiler efficiency, as variations occur in the load, fuel, ambient, and boiler conditions.

The boiler efficiency is influenced by many factors. A fully loaded large boiler, which is clean and properly tuned (with blowdown losses and pump and fan operating costs disregarded) is expected to have the following efficiencies [10]:

88 percent on coal with 4 percent excess O_2
87 percent on oil with 3 percent excess O_2
82 percent on gas with 1.5 percent excess O_2

Boiler efficiencies seldom exceed 90 percent or drop below 65 percent. Efficiencies will tend to vary with individual design and with loading, as shown in figure 2.1. Efficiencies will also vary as a function of excess air, flue gas temperature, and boiler maintenance. A 1-percent loss in efficiency on a 100,000 lbs./hr. (45,454 Kg./hr.) boiler will increase its yearly operating cost by about $20,000. A 1-percent efficiency loss can result from a 2-percent increase in excess oxygen [4] or from about a 50°F increase in flue gas temperature [21].

In this chapter, the various strategies for the automatic and continuous optimization of boilers will be discussed. In order to save space, the text assumes that the reader is already familiar with conventional boiler controls. If that is not the case, supplemental texts should be studied first (see [10] and [19], for example).

The Role of Sensors

A unit operation cannot be optimized unless the control system inputs are accurate and the control system outputs are achievable by the final control elements used. Therefore, the first step toward optimization must be to select the most accurate sensors and to install the best dampers for control. Figure 2.2 shows the in-line instruments used on a boiler, together with some advice on their selection.

The most important sensors are the flow detectors, which provide the basis for both material and heat balance controls. For successful control of the air-fuel ratio, combustion air flow measurement is important. In the past it was impossible to obtain ideal flow detection conditions. Therefore, the practice was to provide some device in

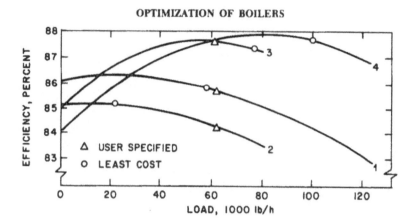

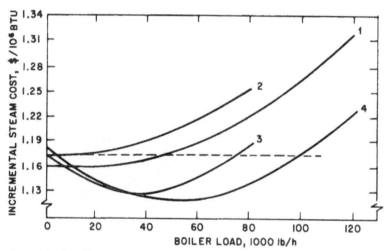

Fig. 2.1 Boiler efficiencies and steam costs vary with both the design of the boiler and its loading. (From Cho [2])

the flow path of combustion air or combustion gases and to field-calibrate it by running combustion tests on the boiler.

These field tests, carried out at various boiler loads, used fuel flow measurements (direct or inferred from steam flow) and measurements of percent of excess air by gas analysis; they also used the combustion equations to determine air flow. Because what is desired is a relative measurement with respect to fuel flow, the air flow measurement is normally calibrated and presented on a relative basis. Flow versus differential pressure characteristics, compensations for normal variations in temperature, and variations in desired excess air as a function of load—all are included in the calibration. The desired result is to have the air flow signal match the steam or fuel flow signals when combustion conditions are as desired.

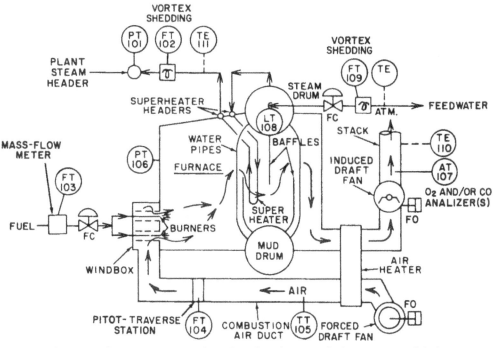

Fig. 2.2 The main in-line instruments are shown here for a drum-type boiler. (From Lipták [10])

The following sources of pressure differential are normally considered:

Burner differential (windbox pressure minus furnace pressure)
Boiler differential (differential across baffle in combustion gas stream)
Air heater differential (gas side differential)
Air heater differential (air side differential)
Venturi section or flow tube (installed in stack)
Piezometer ring (at forced draft fan inlet)
Venturi section (section of forced draft duct)
Orifice segment (section of forced draft duct)
Air foil segments (section of forced draft duct)

None of these sensors meets the dual requirement of high accuracy and rangeability. Actually, they are of little value at 30-percent flow or less. Table 2.1 lists some better flow sensors, such as the multipoint thermal flow probe or the area-averaging pitot stations provided with "hexcel"-type straightening vanes and with membrane-type pressure balancing d/p cells. These represent a major advance in combustion air flow detection. This table shows the measurement errors that can be anticipated. At full load (100-percent flow), these errors are quite acceptable. Unfortunately, as the flow is reduced, the error—in percentage of actual measurement—increases in all cases except

Table 2.1
FLOW SENSOR ERRORS ON BOILERS

Flow Streams Measured	Type of Flowmeter	Error at 100% Flow in % of Measurement	Error at 10% Flow in % of Measurement
Fuel Oil	Coriolis mass flow	0.5	0.5
Steam and Water	Vortex shedding*	1	1
	Orifice	1	Useless (10% with two d/p cells)
	Area averaging pitot traverse station	0.5	Useless if one d/p cell is used; 5% with two
Air	Multipoint thermal	2	20 (better with dual range)
	Piezometer ring, orifice segment Venturi section airfoil section	3	Useless

From Lipták [12].
*Limited to 750°F (400°C)

the first two. With linear flow meters, the error increases linearly with turndown. Therefore, the thermal flow meter error increases tenfold at a turndown of 10:1. In case of nonlinear flow meters, the error increases exponentially with turndown. Therefore, at a turndown of 10:1, the orifice or pitot error increases a hundredfold and causes these devices to become useless. This situation can be alleviated somewhat by the use of two d/p cells on the same element.

Based on the data in table 2.1, if the boiler efficiency is to be monitored on the basis of time-averaged fuel and steam flows, the lowest error that can be hoped for is around 1 percent. If the efficiency is determined on the basis of flue gas composition, steam temperature, and fuel type, the error will be lower. Boiler efficiency equations can consider as many as 7 sources of heat loss and require up to 44 pieces of input data [5]. If the lesser factors are omitted, the efficiency can be calculated using the following equation [19]:

$$E = 100 \left[1 - 10^{-3} \left(0.22 + \frac{K''y}{1 - y/0.21} \right) (T_s - T_a) - \frac{\Delta H_c}{H_c} \right]$$

Where y is the mole fraction of oxygen in the flue gas, and K'' is a coefficient assigned to each fuel: 1.01 for coal, 1.03 for oil, and 1.07 for natural gas. The term $\Delta H_c/H_c$ is about 0.02 for coal, 0.05 for oil, and 0.09 for gas. The terms T_s and T_a are the stack and ambient temperatures.

Similarly, the air-fuel ratio cannot be measured to a greater accuracy than the air flow. At high turndown ratios, this error can be very high. Considering that a 2-percent reduction in excess oxygen will increase the boiler efficiency by 1 percent, both the accurate measurement and the precise control of air flows is essential in boiler optimization.

The demand for sensor accuracy elsewhere is not so stringent. Standard instrumentation allows for the control of steam pressure within $\pm 1\%$, furnace pressure within $\pm 0.1''$ H_2O (25 Pa), water level within $\pm 1''$ of desired level and steam temperature to within $\pm 10°F$ (5.6°C).

Boiler Dynamics

Boiler response to load changes is usually limited by both equipment design and dead-time considerations. Usually the maximum rate of load change that can be handled is from 20 percent to 100 percent per minute. This limitation is due to the maximum rates of change in burner flame propagation and to the "shrink"—"swell" effects on the water level. The period of oscillation of a typical boiler is between 2 and 5 minutes. This is the result of a dead time of 30 to 60 seconds. The transportation lag in the boiler is partially due to the displacement volume of the furnace. For example, if the air-fuel ratio is changed, the furnace volume will have to be displaced before the flue gas composition can reflect that change. The lower the air flow (the lower the load), the longer it will take to displace this fixed volume. Therefore, dead time increases as load is lowered on a boiler.

The transportation lag described above is only one component of the total dead time. The oxygen analyzer also contributes to the total delay, because of its location and its flyash filter. In addition to the dead time contribution of instruments, another important dead time component is caused by cross limited parallel metering (figure 2.3). This is a safety feature guaranteeing that no change in the firing rate (up or down) can result in a "fuel-rich" mixture. The consequence of this feature is that when a signal is received to increase the firing rate, it will first increase the air flow; fuel flow will be allowed to rise only afterward. The total dead time of a boiler can be reduced if the above cross limited safety system is disabled when the amount of change in the firing rate is small.

Another way to reduce dead time and thereby increase boiler response is by using feedforward loops. The firing rate demand signal can be made more responsive by feed forward off steam flow, which responds to load changes faster than does steam pressure. Similarly, the induced draft loop can be made more responsive by adding feedforward off the forced draft position (figure 2.3). In this system, as soon as the air flow into the furnace changes, the outflow is also modified in the same direction, so that furnace draft pressure is relatively unaffected.

Traditional Air Controls

Air controls on traditional boilers were just as inaccurate and unreliable as were the air flow sensors when air flow rates were less than 25 percent of maximum rates. This was unfortunate, because it is precisely at low loads that the boiler tends to be

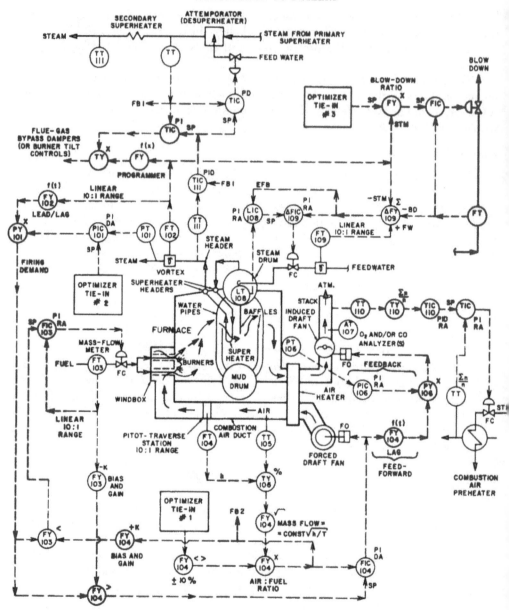

Fig. 2.3 This diagram shows good boiler controls without optimization.

the least efficient to start with. Yet, for reasons of equipment inadequacy, some manufacturers will turn off the oxygen trim of the air-fuel ratio at low loads. The reasons for these problems were inaccurate sensors at low flows; leaking, non-linear dampers with hysteresis and dead band; and the use of constant speed fans.

For many years the industry accepted these problems as if they were unalterable facts of life and built their controls around them. In many cases, the firing rate signal itself was characterized to set the excess air (figure 2.4); it was also used as the oxygen set point on most air-fuel ratio trim controls. Frequently, the excess air requirement curve was directly calibrated into the combustion system, and feedback trim based on excess oxygen was not used at all. The advocates of this open loop control strategy argue that the predetermined excess oxygen curve is rather permanent and if an unmeasured effect necessitates a change in it, that change will be the same throughout the full firing range. This view of the process neglects nonlinearities, hysteresis, dead time, the play in linkages, sticking dampers, and other effects that are now understood.

In many traditional systems, the nonlinearity of dampers (figure 2.5) was taken into account by characterizing the fuel valve, the linkage, and/or the signal to the fan damper actuator. This is no easy task, because louver-type dampers are not only nonlinear, they also lack repeatability. In these systems, the air-fuel ratio trim was disabled below 25-percent load, because the conventional dampers could not be controlled. It was argued that closed dampers leak as much as 10 percent of their full capacity and thus need to be opened only to 2 percent to deliver 25 percent flow. Therefore, according

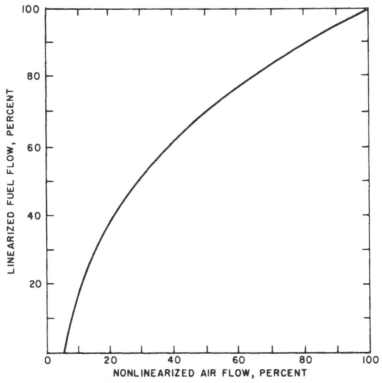

Fig. 2.4 To apply air flow characterization, the boiler's fuel/air relationship must be determined empirically for various boiler load levels. (From Hurlbert [8])

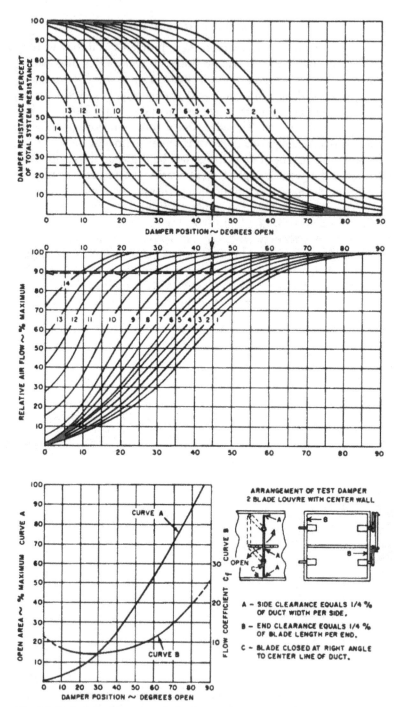

Fig. 2.5 Multiple-leaf louvers with dividing partitions are used in many traditional systems. (From Dickey [6])

to this argument, if the oxygen trim signal was not disabled and it did request a 1-percent change in air-fuel ratio, this would mean a 1-percent change in the 2-percent opening of the damper, or a 0.02-percent change in its stroke or rotation; this the damper positioner could not handle. Naturally, the argument is correct in all respects except in its assumption that such dampers are a necessary limitation.

Other observations [15] include that PID-type controls cannot handle frequent load changes, because the speed of response of the fuel flow control loop is much faster than that of the air flow control loop. Suppliers attempted to correct this by making the air actuator twice as fast as the fuel actuator.

In general, the nonlinear nature of the loops has created difficulties for the traditional boiler control designs. The need to change the firing rate in exact proportion to load changes was difficult to meet, because it required characterization in the field of the fuel valve and air damper actuators.

As will be discussed on the following pages, distributed microprocessor-based intelligence can linearize the nonlinear systems. It can also memorize their dead-band and speed of response. Still, this capability of state-of-the-art control equipment should not be used as a justification for installing inferior quality dampers.

Dampers

The recommendations for the use of accurate and high-rangeability sensors have already been made in connection with table 2.1. Now the desirable damper features will be discussed from a control quality viewpoint.

The reasons why dampers are undesirable as final control elements include their hysteresis, non-linearity, and leakage. If the flow is accurately measured, the consequences of these undesirable features are easier to handle. For example, assuming that even after proper maintenance the damper displays some hysteresis or dead band, with a reliable flow sensor, the opening and closing characteristics of the damper can be determined. If the controls are provided with digital memory, this characteristic curve can be remembered and used later. The damper characteristic curve can also be automatically updated and corrected, as it might change because of wear, dirt build-up, or other causes.

Figure 2.5 shows the nonlinear characteristics of a multiple-blade damper [6]. Curve A on the bottom gives the relationship between damper rotation and open area, while curve B relates damper rotation and flow coefficient. The upper portion of figure 2.5 describes the *installed* performance of the same damper, considering its pressure drop relative to the total system pressure drop. For example, if the damper resistance is 25 percent of the total, when the damper rotation is 45 percent, curve #7 will give the damper characteristics and actual air flow will be about 88 percent of maximum. From such curves as these, both the damper leakages and the nonlinearities can be estimated before actual installation.

It can be seen from figure 2.5 that an increase of 1° in damper rotation that occurs at a 10° opening causes a much greater increase in the actual air flow than a 1° increase that occurs at a 60° or 70° opening. This is undesirable. Stable controls would require

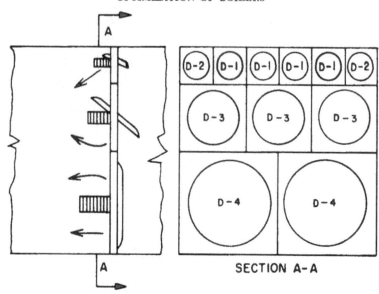

Fig. 2.6 Improved dampers can be designed to provide low leakage (0.001 percent of maximum flow). (Courtesy of Mitco Corp.)

constant gain in the loop. Ideally, the change in air flow per increment change in damper opening should be uniform from the tightly closed to the wide open position. Such inherently linear dampers are hard to manufacture, although the design illustrated in figure 2.6 does hold such potentials. A more often used solution is the use of compensators and characterizer positioners, so that as the damper gain drops off—as it opens—these compensators introduce more gain into the loop, keeping its total gain nearly constant.

While it is possible to compensate for nonlinearity and hysteresis, leakage must be eliminated by selecting the correct design. Figures 1.7 and 2.6 illustrate some of the tight shut-off designs. If these are used, it is no longer necessary to turn off the excess oxygen based air-fuel ratio trim when the load is below 25 percent. With such compensated, low-leakage dampers, trim need not be turned off, and the resulting increase in efficiency need not be abandoned, until load drops to 10 percent.

Fans

Even if the best damper is used, damper control will always have a disadvantage, namely that it burns up fan energy in order to control. Therefore, the best method of controlling air flow is to eliminate the damper completely and throttle the fan itself. Instead of burning up the unnecessary air transportation energy in the form of damper pressure drop, this unnecessary energy is just not introduced in the first place. This goal can be met by the use of variable speed fans (figure 1.6) or by varying the pitch of the fan blades (figure 2.7).

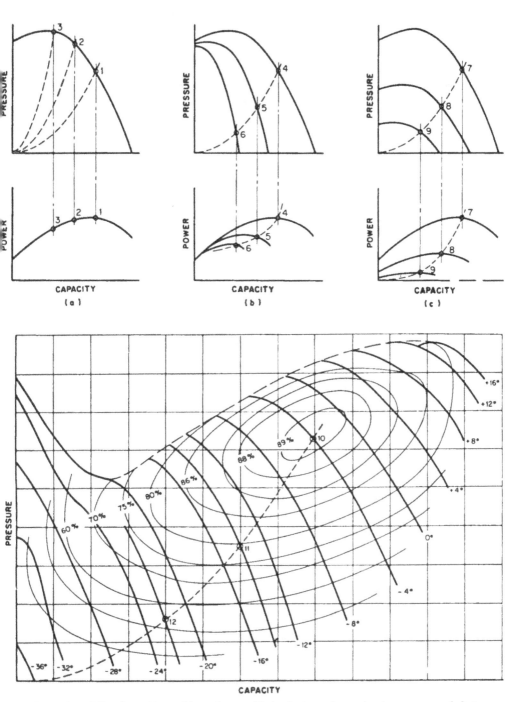

Fig. 2.7 Axial-flow fans with variable pitch control help eliminate the need to burn up unneeded air transportation energy as damper pressure drops. Fan characteristics are shown with (a) damper control, (b) variable-inlet vane control, and (c) variable speed control. (From Jorgensen [9])

The use of variable speed or variable pitch fans will do more than conserve air transportation energy, thereby reducing the overall cost of boiler operation. Their other advantages are the better linearity, hysteresis, dead band, and leakage characteristics that they can provide. For these reasons, the overall control diagram in figure 2.3 above shows fan controls instead of damper controls.

Basic Boiler Controls

As the emphasis is on optimization, the basic boiler control loops can be mentioned only briefly here. Readers who require an in-depth discussion of these control loops might be advised to consult other sources ([10] and [19], for example).

Figure 2.3 (above) shows a possible configuration of the basic boiler control loops and the tie-in points for optimization. Needless to say, although this is a well-designed control system, it is just one of many possible configurations. Boiler size, steam pressure, and number and type of fuel(s) all can vary, necessitating variations in this scheme. As the major loops shown in figure 2.3 are numbered, they will be described here in their numerical order.

Loops 101 and 102 measure the pressure and flow of the generated steam respectively. The flow transmitter (FT-102) is a high rangeability and linear device, capable of accurate measurements even at low loads. The firing rate demand signal is generated in a feedback manner by the pressure controller PIC-101. In order to speed up the response of this loop to load changes, a feedforward trim is added. This trim is based on steam flow, because this flow is the first to respond to load changes. Therefore, as soon as the demand for steam changes, FY-102 will trim the firing demand signal, without waiting for the steam pressure to change first. The dynamics of FY-102 are adjusted to reflect the time constants of the boiler, recognizing the time displacement between a change in firing rate and the resulting change in the rate of steam generation some time later.

The fuel flow is detected by a high rangeability, accurate mass flow meter (FT-103). The set point for the fuel flow controller (FIC-103) is the smaller of either the biased airflow or the firing rate (FY-103). The set point of the air flow controller (FIC-104) is selected as the higher of either the biased fuel flow or the firing rate (FY-104).

These high and low selectors provide the so-called cross limited parallel metering system [4]. The coefficients in the system are adjusted so that the pairs of signals provided to the high and low selectors are equal under steady state conditions. When the firing rate demand increases, the high selector provides it as an air flow set point while the low selector transmits the air flow process variable as the fuel flow set point. Air flow therefore starts to increase immediately; fuel flow rises only after the air flow responds. When the firing rate demand signal decreases, the low selector provides it as the fuel flow set point while the high selector transmits the fuel flow process variable as the air flow set point. Fuel flow starts decreasing immediately; air flow drops only after fuel flow responds. Likewise, if a disturbance causes air flow to drop, the low selector transmits the air flow signal to the fuel flow controller. Fuel flow then decreases regardless of the steam demand, preventing a fuel-rich condition.

The disadvantage of the system is that overall response is constrained by the slower of air flow or fuel flow response to changing demand signals. For example, if the firing rate demand rises slightly, the system first positions the damper to increase air flow; then, as air flow rises, the system opens the fuel valve.

The bias and gain modules FY-103 and 104 were added in figure 2.3 to improve the system response to *small* load changes. These reduce the effective fuel flow signal presented to the high selector while raising the air flow variable provided to the low selector.

Under steady state conditions, the firing rate demand is presented as the set point to both controllers. Likewise, if the firing rate demand changes only slightly, it will still be transmitted by both signal selectors and will cause the fuel and air flows to increase or decrease accordingly. If the changes in firing rate demand exceed the offset introduced by the bias and gain modules, the system will operate like a "cross limited" system. As a result, fuel flow can respond to small increases in firing rate demand without raising air flow. Similarly, air flow may be adjusted slightly downward if firing rate demand falls, without having to decrease fuel flow.

Feedforward Control of Air Flow

The air flow is also detected by a high rangeability flow meter (FT-104), which is compensated for temperature variations to approximate mass flow [19]. The air-fuel ratio is adjusted by applying a gain to the air flow (FY-104). The continuous optimization of this ratio is one of the major tools of boiler optimization.

The air flow controller throttles the speed or blade pitch of the forced draft fan and the signal to the induced draft fan in a feedforward manner. Dynamic compensation is provided by the lag module (FY-104). Thus, as soon as the air inflow to the furnace is changed, the outflow will also start changing. This improves the control of the furnace draft, as the feedback pressure controller (PIC-106) will need only to trim the feedforward signal at PY-106 to account for measurement and other errors.

Feedwater and Blowdown Controls

The control system shown in figure 2.3 is an improved version of the three-element feedwater system [4 and 19]. It is a cascade loop with level being the master (LIC-108) and flow balance being the slave (Δ FIC-109). In order to protect against reset windup, LIC-108 is provided with an external feedback signal, as shown in figure 2.3.

The overall functioning of the process is as follows [19]. Feedwater is always colder than the saturated water in the drum. Some steam is thus necessarily condensed when it is contacted by the feedwater. As a consequence, a sudden increase in feedwater flow tends to collapse some bubbles in the drum and temporarily reduce their formation in the water pipes. Then, although the mass of liquid in the system has increased, the apparent liquid level in the drum falls. Equilibrium is restored within seconds and the level will begin to rise. Nonetheless, the initial reaction to a change in feedwater flow tends to be in the wrong direction. This property, called *inverse response*, causes an effective delay in control action, making control more difficult. Liquid level in a vessel lacking these thermal characteristics can typically be controlled with a proportional band of 10 percent or less. By contrast, the drum-level controller needs a proportional band more in the neighborhood of 100 percent to maintain stability [19]. Integral action is then necessary, whereas it can usually be avoided when very narrow proportional band settings can be used.

The cascade loop in figure 2.3 provides feedforward action to maintain the steam-water balance and thereby to minimize the impact of the inverse-response and of the shrink-swell phenomena. The summer Δ FY-109 subtracts steam and blowdown flows from the feedwater flow and, when their scaled sum is zero, sends a 50-percent signal to Δ FIC-109. Thus, an increase in steam or blowdown flow will increase the feedwater flow immediately, without depending on the level controller. This means that the feedback portion of the loop (LIC-108) will need only to trim the Δ FIC-109 set point to correct for flow meter errors. Because its role is reduced from manipulating feedwater flow across its entire range to adjusting only for flow meter errors, deviations in level from the set point will be minimized. Controller mode settings are not as critical in this situation, and incorrect actions caused by shrink, swell, and inverse response are reduced [19].

Temperature Controls

Steam temperature is usually controlled to meet the needs of the steam users. Flue gas temperature is important for two reasons: first, as an indicator of boiler efficiency and, second, because if it drops below the dew point, the condensate formed will dissolve the oxides of sulfur in the flue gas and cause corrosion.

Figure 2.3 shows both of these temperature control loops (TIC-110 and TIC-111). The purpose of TIC-110 is to keep the flue gas dry and above its dew point. The flue gas temperature at the cold end of the stack is usually arrived at as the average of several sensor outputs, measuring the temperature at various points in the same plane. When this averaged temperature drops near the dew point, TIC-110 will start increasing the set point of the air preheaters. Through the use of steam or glycol coils in the combustion air, the inlet air temperature to the boiler is increased, which in turn raises the flue gas temperature.

The control of steam temperature (TIC-111) is also important, because superheat increases the available work in the steam above that which is available at saturation. Therefore, work efficiencies are maximized by operating at the highest steam temperature at which the metals are capable of operating. In central stations, this limit is 1050°F (566°C); in industrial applications it is lower [19]. If improved control can elevate the TIC-111 set point from 1000°F to 1040°F, this will increase the available work by 17 Btu/lbm (9.4 CAl/kg) in the steam [19].

Flame temperature does not vary much with load, but the hottest gases do tend to propagate further at higher loads. To increase the steam temperatures at low loads, recirculation blowers or tilting burners are used, which will direct the heat at the superheater sections. At high loads, the rise in the steam temperature is prevented by opening up some of the flue gas bypass dampers or by desuperheating the steam through attemporation with water. For each boiler, the relationship between load and the required damper, recirculation fan, or burner tilt position is well established and therefore can be preprogrammed.

Figure 2.3 shows the input signal to the programmer as coming from FT-102, the steam flow transmitter. (Vortex shedding meters are limited to about 750°F or 400°C.) In other designs the input to the programmer is taken from the combustion air flow signal (FY-104). In either case, the preprogrammed relationship does need to be adjusted to overcome inaccuracies and changes in boiler characteristics. This feedback trim is applied through a multiplier. As the feedback loop gain varies directly with load, whereas the process gain varies inversely with load, they tend to cancel each other out [19].

TIC-111 throttles the set points of both slave controllers in cascade. Reset wind-up in TIC-111 is prevented by the external feedback (FB). Without it, wind-up would occur at low loads, when high temperatures cannot be maintained [20]. The task of temperature control is split between two slaves. The slower of these slaves—the PI controller—modulates burner or damper positions for long-term control. The desuperheater controller is faster; actually, it is faster than its cascade master. In order to use that speed, to minimize water usage (an irreversible waste of available work), and to avoid conflict with the PI controller, only P & D control modes are used [19]. This controller is usually biased, so that at zero deviation it will return to the same position, delivering some nominal amount of feedwater.

The relationship between boiler efficiency and the temperatures of flue gas and steam will be discussed later.

Excess Air Optimization

If a boiler is operating on a particular fuel at a specific load, it is possible to plot the various boiler losses as a function of air excess or efficiency, as shown in figure 2.8. The sum total of all the losses is a curve with a minimum point. Any process that has an operating curve of this type is an ideal candidate for instrumental optimization. Such process control systems operate by continuously determining the minimum loss point of the system at that particular load and then shifting the operating conditions until that point is reached.

As shown in figure 2.8, the radiation and wall losses are relatively constant. Most heat losses in a boiler occur through the stack. Under air-deficient operations unburned fuel leaves, and when there is an air excess, heat is lost as the unused oxygen and its accompanying nitrogen are heated up and then discharged into the atmosphere. The goal of optimization is to keep the total losses at a minimum. This is accomplished by minimizing excess air and by minimizing the stack temperatures.

The minimum loss point in figure 2.8 is not where excess oxygen is zero. This is because no burner is capable of providing perfect mixing. Therefore, if only as much oxygen would be admitted into the furnace as is required to convert each carbon molecule into CO_2, some of the fuel would leave unburned, as not all O_2 molecules would find their corresponding carbon molecules. This is why the theoretical minimum

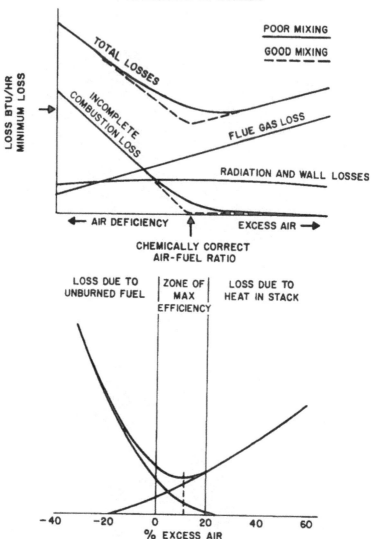

Fig. 2.8 Boiler losses can be plotted as a function of excess air. (*Top*) the minimum point of the total loss curve of a boiler is where optimized operation is maintained. (*Bottom*) Most efficient operation of a boiler occurs when the amount of excess air in the stack balances the losses in unburned fuel. (From Lipták [10] and Walsh [21])

loss point shown by the dotted line in figure 2.8 is to the left of the actual one. This actual minimum loss or maximum efficiency point is found by lowering the excess oxygen as far as possible, until opacity or CO readings indicate that the minimum has been reached. At this minimum loss point the flue gas losses balance the unburned fuel losses.

Flue Gas Composition

Figure 2.9 shows the composition of the flue gas as a function of the amount of air present. The combustion process is usually operated so that enough air is provided to convert all the fuel into CO_2, but not much more. This percentage of excess oxygen is *not* a constant. It varies with boiler design, burner characteristics, fuel type, air infiltration rates, ambient conditions, and load.

The top portion of figure 2.10 shows that the percentage of excess air must be increased as the load drops off. This is because at low loads the burner velocities drop off and the air flow is reduced, while the furnace volume remains constant. This reduces turbulence and lowers the efficiency of mixing between the fuel and the air. This loss of mixing efficiency is compensated for by the higher percentage of excess oxygen admitted at low loads.

The upper curve shown in figure 2.10 theoretically illustrates the relationship between the excess O_2 requirement and load, and the lower plot provides actual test data for a specific boiler. Because each boiler has its own unique personality, this relationship must be experimentally determined. Once established, it can be used with a fair degree of confidence, although small shifts are still likely to occur as the equipment ages.

The Effect of the Fuel Used

In a boiler furnace (where no mechanical work is done), the heat energy evolved from the union of combustible elements with oxygen depends on the ultimate products of combustion.

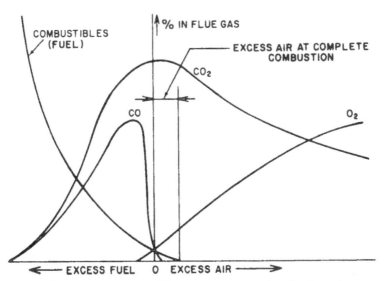

Fig. 2.9 The major components of flue gas are oxygen, carbon dioxide, carbon monoxide, and unburned hydrocarbons. (Adapted from Walsh [21])

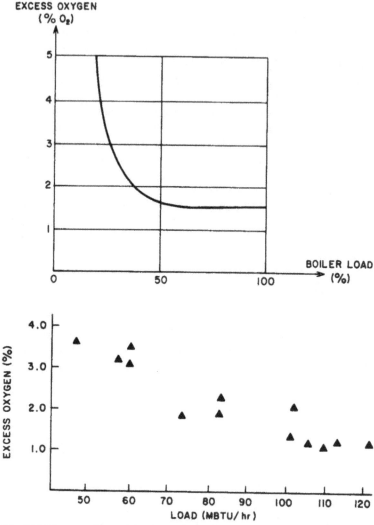

Fig. 2.10 The top portion of this figure shows theoretical relationships between load and O_2; the bottom portion shows the test-based relationships between load and O_2. (From Lipták [10] and Schwartz [18])

A simple demonstration of this law is the union of 1 pound of carbon with oxygen to produce a specific amount of heat (about 14,100 Btu or 3.553 Cal.) The gaseous product of combustion, CO_2, can be formed in one or two steps. If CO is formed first, it produces a lesser amount of heat (about 4,000 BTU or 1,008 Cal), which, when it is converted to form CO_2, releases an additional 10,100 Btus. The sum of the heats released in each of these two steps equals the 14,100 Btu evolved when carbon is burned in one step to form CO_2 as the final product.

Because of the weight ratios of oxygen and nitrogen in air, (.2315 and .7685 respectively), to supply 1 lb. (0.45 kg.) of oxygen for combustion, it is necessary to supply 1/.2315 = 4.32 lb. (1.96 kg.) of air. In this amount of air, there will be 4.32 × .7685 = 3.32 lb. (1.5 kg.) of nitrogen, which does not enter directly into the combustion process but which nevertheless remains present.

When burning carbon to carbon dioxide, 12 parts by weight of carbon (the approximate molecular weight of C) combine with 32 parts by weight of oxygen (the molecular weight of O_2) to form 44 parts by weight of carbon dioxide (the approximate molecular weight of CO_2). By simple division, 1 lb. of carbon plus 2.66 lbs. of oxygen will yield 3.66 lbs. of carbon dioxide.

The theoretical amount of air required for the combustion of a unit weight of fuel can be calculated as follows [2]:

$$11.3\ C + 34.34\ (H_2 - O_2/8) + 4.29\ S$$

Where C, H_2 and S represent the unit weights of the combustible elements—carbon, hydrogen, and sulfur—in the fuel. The factor $O_2/8$ is the correction for the hydrogen, which is already combined in the form of water. The amount of moisture per unit weight of air is around 0.013, on the average.

If theoretical or total air is defined as the amount required on the basis of the above equation, then excess air is the percentage over that quantity. Table 2.2 lists the excess air ranges required to burn various fuels. It can be seen that excess air require-

Table 2.2
USUAL AMOUNT EXCESS AIR SUPPLIED TO
FUEL-BURNING EQUIPMENT

Fuel	Type of Furnace or Burners	Excess Air %
Pulverized coal	Completely water-cooled furnace for slag-tap or dry-ash-removal	15–20
	Partially water-cooled furnace for dry-ash-removal	15–40
Crushed coal	Cyclone furnace—pressure or suction	10–15
Coal	Stoker-fired, forced-draft, B&W chain-grate	15–50
	Stoker-fired, forced-draft, underfeed	20–50
	Stoker-fired, natural draft	50–65
Fuel oil	Oil burners, register type	5–10
	Multifuel burners and flat flame	10–20
Acid sludge	Cone and flat flame type burners, steam-atomized	10–15
Natural, coke oven and refinery gas	Register type burners	5–10
	Multifuel burners	7–12
Blast-furnace gas	Intertube nozzle type burners	15–18
Wood	Dutch oven (10–23% through grates) and Hofft type	20–25
Bagasse	All furnaces	25–35
Black liquor	Recovery furnaces for kraft and soda pulping processes	5–7

From Babcock & Wilcox Co. [2]

ment increases with the difficulty to atomize the fuel for maximum mixing. Figure 2.11 also illustrates that gases require the lowest and solid fuels the highest percentage of excess oxygen for complete combustion. The ranges in table 2.2 and the curves in figure 2.11 also illustrate that as the load drops off, the percentage of excess oxygen needs to be increased. Figure 2.12 illustrates the relationship between excess air and excess oxygen for a particular fuel. The optimum excess oxygen percentages for gas, oil, and coal are around 1 percent, 2 percent, and 3 percent respectively.

Detectors of Flue Gas Composition

As shown in figure 2.9, excess air can be correlated to O_2, CO, CO_2, or combustibles present in the flue gas. Combustibles are usually detected either as unburned hydrocarbons or in the form of opacity [12]. These measurements are not well suited as the basis for optimization, because the goal is not to maintain some optimum concentration, but to eliminate combustibles from the flue gas. Therefore, such measurements are usually applied as limit overrides.

The measurement of CO_2 [12] is not a good basis for optimization either, because, as shown in figure 2.13, its relationship to excess O_2 is very much a function of the type of fuel burned. The CO_2 concentration of the flue gas also varies slightly with the CO_2 content of the ambient air. It can also be noted from figure 2.9 that CO_2 is not a very sensitive measurement. Its rate of change is rather small at the point of optimum excess air. In fact, the CO_2 curve is at its maximum point when the combustion process is optimized.

Excess O_2 as the basis of boiler optimization is also a relatively insensitive measurement, but it is popular. It uses zirconium oxide probes [12]. In order to minimize duct leakage effects, the probe should be installed (figure 2.14) close to the combustion zone, but still at a point where the gas temperature is below that of the electrically

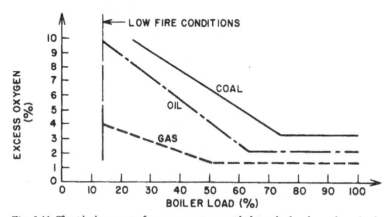

Fig. 2.11 The ideal amount of excess oxygen provided to a boiler depends on load as well as fuel properties. (From Lipták [10])

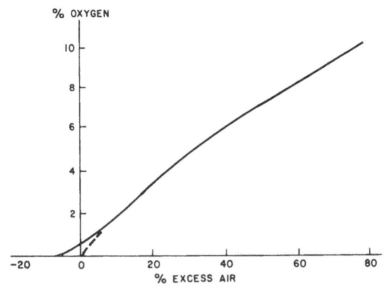

Fig. 2.12 The amounts of oxygen and excess air in flue gas can be correlated as shown. (From Walsh [21])

heated zirconium oxide detector. The flow should be turbulent at the sensor location, if possible, to ensure that the sample will be well mixed and representative of flue gas composition. The output signal of these zirconium oxide probes is logarithmic. According to Shinskey [19], this is desirable. The correct location of the probe will reduce but will not eliminate the bias error caused by air infiltration. Ambient tramp air enters the exhaust ductwork (which is under vacuum) not only through leakage but also to cool unused burners and registers. The O_2 probe cannot distinguish the oxygen that entered through leakage from excess oxygen left over after combustion.

Another limitation of the zirconium-oxide fuel cell sensor is that it measures *net* oxygen. In other words, if there are combustibles in the flue gas, they will be oxidized on the hot surface of the probe and the instrument will register only that oxygen, which remains *after* this reaction. This error is not substantial when the total excess oxygen is around 5 percent, but in optimized boilers, in which excess oxygen is only 1 percent, this difference between total and net O_2 can cause a significant error. As infiltration tends to cause an error towards the high side while the fuel-cell effect results in a low reading, the amount of uncertainty is too high to rely on O_2 sensors alone when maximum efficiency is desired.

Other limitations of excess oxygen based optimization include the problem that local problems at the burners can result in incomplete combustion, even when the excess oxygen in the flue gas is normal. Another limitation is the precision and accuracy of such excess oxygen curves, as shown in figure 2.10. This precision is a function not

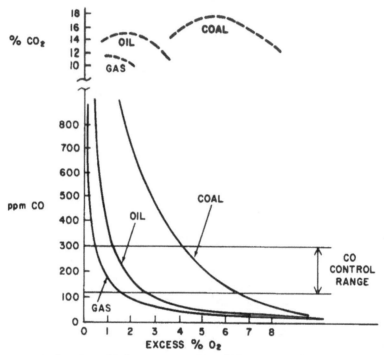

Fig. 2.13 The relationship between excess O_2 and CO or CO_2 in the flue gas of a boiler operated at a constant load is a function of the type of fuel burned. (From McFadden [14])

only of the resolution at which the curve was prepared but also of changes in fuel composition and boiler conditions.

CO Measurement

As shown in figure 2.9, the most sensitive measurement of flue gas composition is the detection of carbon monoxide [12]. As can be seen from figures 2.13 and 2.15, optimum boiler efficiency can be obtained when the losses due to incomplete combustion *equal* the effects of excess air heat loss. These conditions prevail at the "knee" of each curve. While the excess O_2 corresponding to these "knee" points varies with the fuel, the corresponding CO concentration is relatively constant.

Theoretically, CO should be zero whenever there is oxygen in the flue gas. In actual practice, maximum boiler efficiency can usually be maintained when the CO is between 100 and 400 ppm. CO is a very sensitive indicator of improperly adjusted burners; if its concentration rises to 1000 ppm, that is a reliable indication of unsafe conditions. Because CO is a direct measure of the completeness of combustion and nothing else, it is also unaffected by air infiltration, other than the dilution effect.

For these reasons, control systems utilizing the measurement of both excess O_2

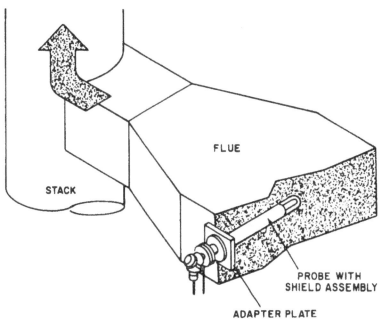

STACK

FLUE

PROBE WITH
SHIELD ASSEMBLY

ADAPTER PLATE

Fig. 2.14 The probe-type oxygen analyzer should be installed close to the combustion zone, but at a point where the temperature is below the limit for the zirconium oxide detector. (From Lipták [10])

and carbon monoxide can optimize boiler efficiency, even if load, ambient conditions, or fuel characteristics vary. Also, when these systems detect a shift in the characteristic curve of the boiler, that shift can be used to signal a need for maintenance of the burners, heat transfer units, or air and fuel handling equipment.

Non-dispersive infrared (IR) analyzers [12] can be used for simultaneous in-situ measurement of CO and other gases or vapors such as that of water. This might signal incipient tube leakage. Most IR sensors use a wavelength of 4.7 micron for CO detection, because the absorption of CO peaks at this wavelength, whereas that of CO_2 and H_2O does not. CO_2 is also measured and is used to determine the dilution compensation factor for CO.

The CO analyzers cannot operate at high temperatures and therefore are usually located downstream of the last heat exchanger or economizer. At these points, the flue gas dilution due to infiltration is frequently high enough to require compensation. The measurement of CO_2 is used to calculate this compensation factor.

Setting the Air-Fuel Ratio

In figure 2.3, the set point of the air-fuel ratio relay is designated as the #1 tie-in point for the optimizer controls. This was done to emphasize the importance of this

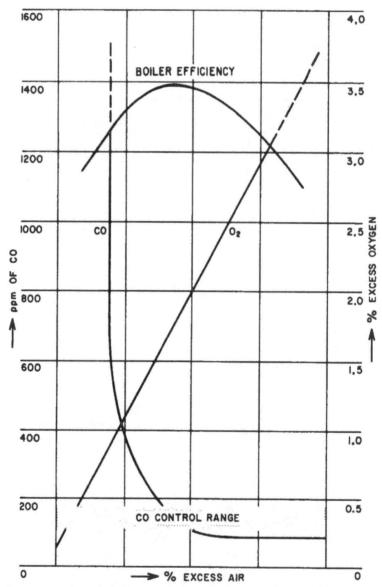

Fig. 2.15 Gas burning boiler efficiency is maximum when CO is within control range. (Courtesy of Econics Corp.)

setting and to show that the method used to determine the correct air-fuel ratio will determine if the boiler is optimized or not.

In the early designs of small boilers, the air-fuel ratio was set by mechanical linkages between valve, damper, and the common jackshaft, as illustrated in figure 2.16. Even at fixed loads, these controls were only as good as the setting of the linkages, which had to be readjusted manually as conditions changed.

When the importance of feedback control based on flue gas analysis was better understood, such mechanically linked boiler controls were retrofitted as shown in figure 2.17. In this system, the excess oxygen content of the flue gas is used to provide a feedback trim on the preset relationship between firing rate (ZT) and damper opening (AZ). The influence of this trimming signal is bounded by the high/low limiter (AY-3) as a safety precaution to prevent the formation of fuel rich mixtures as a result of analyzer or controller failure.

A later development was to incorporate the characteristic load curve of the boiler (figure 2.10) into the control system. This was done because manual readjustment placed an unreasonable burden on the operator. Such a control system is shown in figure 2.18, in which FY-102 represents the relationship between load and excess oxygen. The input to FY-102 is steam flow (in other systems firing rate is used as the input), and the output is the set point of the excess oxygen controller, AIC-107. The summer (HY) provides a bias for the operator, so he can shift the characterizer curve up or down to compensate for changes in air infiltration rates or in boiler equipment performance. The oxygen controller compares the measured flue gas oxygen concentration to the load-programmed set point and applies PI action to correct the offset. Anti-reset wind-up and adjustable output limiting are usually also provided. The oxygen controller is direct-acting; the air/fuel ratio adjustment factor therefore increases if oxygen concentration

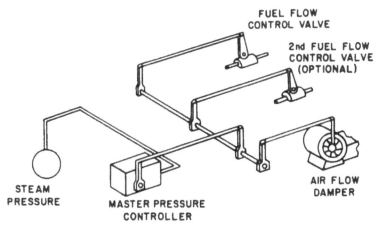

Fig. 2.16 Early designs of small boilers often used systems such as this: a direct positioning jackshaft combustion control system with air/fuel ratio established through fixed mechanical linkages. (From Congdon [4])

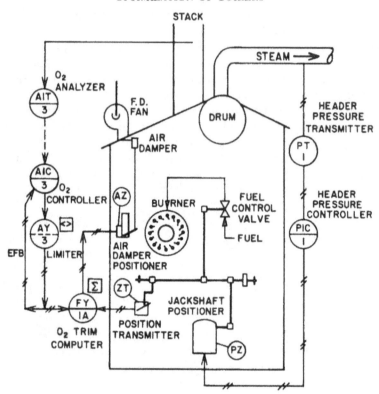

Fig. 2.17 Parallel positioning combustion control systems can be retrofitted with excess oxygen trim. (From O'Meara [16])

in the stack rises because of effects such as reduced fuel heating value at constant flow. Increasing the air/fuel ratio adjustment factor raises the compensated process variable transmitted to the air flow controller, FIC-104 in figure 2.3.

The control system in figure 2.18 can be further improved by increasing its speed of response. The transportation lag in a boiler can be as much as a minute. This lag is the time interval that has to pass after a change in firing rate before its effect can be detected in the composition of the flue gas. This dead time varies with flue gas velocity and usually increases as the load drops. In boilers in which loads are frequently shed or added at irregular intervals, this dead time can cause control problems.

The feedforward strategy [14] shown in figure 2.19 can substantially lower this dead time. Here, the air flow damper position is controlled in a closed loop based on a set point developed through a characterization curve from the firing rate command. This loop acts to adjust the excess oxygen in a feedforward manner. Feedback trim is provided by an oxygen measurement, which modifies the set point to the damper position controller. This system anticipates the need for excess oxygen changes by

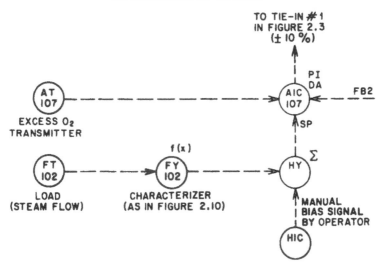

Fig. 2.18 In this system, excess oxygen trim has been characterized for load variations.

responding to load swings, then correcting oxygen concentration to correspond with the excess air curve. A further refinement can be implemented using the corrections provided by the oxygen controller to adapt the damper characterization curve for a particular fuel to the current position in a learning mode.

Multivariable Envelope Control

Multiple measurements of flue gas composition can be used to eliminate the manual bias in figure 2.18 and to obtain more accurate and faster control than what is possible with excess O_2 control alone.

Figure 2.20 shows a control system in which the manual bias is replaced by the output signal of a CO controller, trimming the characterized set point of AIC-107. This trimming corrects the characterized excess O_2 controller set point for changes in the characteristic curve. These changes can be caused by local problems at the burners, resulting in incomplete combustion, or by changes in fuel characteristics, equipment, or ambient conditions. The control systems shown in figure 2.20 could be further improved if the CO measurement signal was corrected for dilution effects due to air infiltration, or if the CO controller set point was also characterized as a function of load. As was shown in figure 2.13, such characteristics can be determined for each fuel and firing rate. The control range for CO tends to remain relatively constant. As CO gives an indication of the completeness of combustion, it is *not* a feasible basis for control if the boiler is in poor mechanical condition or if the fuel does not combust cleanly.

Figure 2.21 provides the added feature of an opacity override to meet environ-

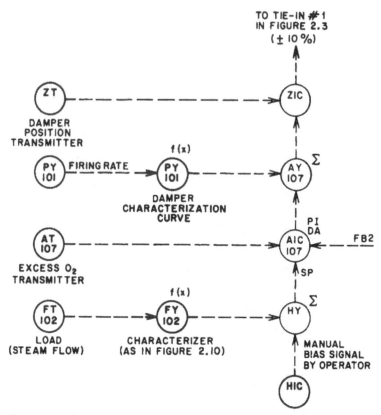

Fig. 2.19 Feedforward based on the characterized damper position increases sensitivity. (Adapted from McFadden [14])

mental regulations. In this system, under normal conditions, the cascade master is CO, just as it was in figure 2.20. Similarly, the cascade slave is excess O_2, but when the set point of the opacity controller is reached, it will start biasing the O_2 set point upward until opacity returns to normal.

With microprocessor-based systems, it is possible to configure a control envelope, such as that shown in figure 2.22. With these control envelopes, several control variables are simultaneously monitored, and control is switched from one to the other depending on which limit of the envelope is reached. For example, assuming that the boiler is on CO control, the microprocessor will drive the CO set point toward the maximum efficiency ("knee" point in figure 2.13), but if in so doing the opacity limit is reached, it will override the CO control and will prevent the opacity limit from being violated. Similarly, if the microprocessor-based envelope is configured for excess oxygen control, it will keep increasing boiler efficiency by lowering excess O_2 until one of the envelope limits is reached. When that happens, control is transferred to that constraint parameter

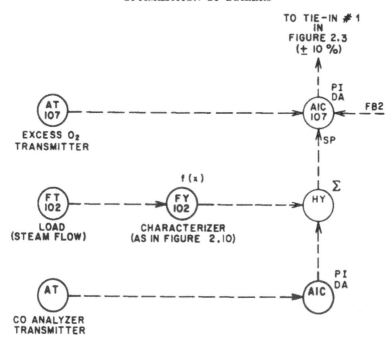

Fig. 2.20 The need for manual biasing is eliminated by the addition of fine trimming based on CO. (Adapted from McFadden [14])

(CO, HC, opacity, etc.); through this transfer, the boiler is "herded" to stay within the envelope defined by these constraints. These limits are usually set to keep CO under 400 ppm, opacity below #2 Ringlemann, and HC or NO_x below regulations.

Microprocessor-based envelope control systems usually also include subroutines for correcting the CO readings for dilution effects or for responding to ambient humidity and temperature variations. As a result, these control systems tend to be both more accurate and faster in response than if control was based on a single variable. The performance levels of a gas burning boiler under both excess O_2 and envelope control are shown on the lower part of figure 2.22.

Envelope control can also be implemented by analog controllers that are configured in a selective manner. This is illustrated in figure 2.23. Each controller measures a different variable and is set to keep that variable under (or over) some limit. The lowest of all the output signals is selected for controlling the air-fuel ratio in figure 2.3, which ensures that the controller that is most in need of help is selected for control. Through this herding technique, the boiler process is kept within its control envelope. As shown in figure 2.23, reset wind-up in the idle controllers is prevented by the use of external reset, which also provides bumpless transfer from one controller to the next.

Operator access is shown to be provided by a single auto/manual (A/M) station. A

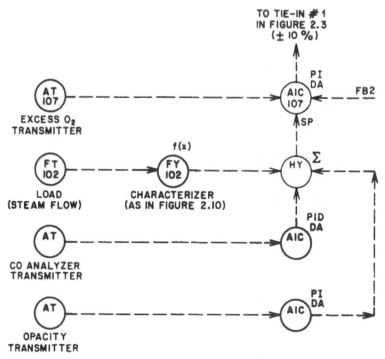

Fig. 2.21 An opacity override has been added to a characterized CO to O_2 cascade system to meet environmental regulations. (Adapted from Am. Tech. Services [1])

better solution is to provide each controller with an A/M station. Then, if a measurement is lost, only the defective loop needs to be switched to manual, not the whole system.

Flue Gas Temperature

The amount of energy wasted through the stack is a function of both the amount of excess air and the temperature at which the flue gases leave. The flue gas temperature is a consequence of load, air infiltration, and the condition of the heat transfer surfaces. Like any other heat exchanger, the boiler will also give its most efficient performance when clean and well maintained.

In optimized boiler controls, a plot of load versus stack temperature is made when the boiler is in its prime condition, and this plot is used as a reference baseline in evaluating boiler performance. If the stack temperature rises above this reference baseline, this indicates a loss of efficiency. Each 50°F (23°C) increase will lower the boiler efficiency by about 1 percent. On a 100,000 lb./hr. (45,450 kg./hr.) boiler, this is a yearly loss of about $20,000.

The reason for rising flue gas temperatures can be fouling of heat transfer surfaces in the air preheater, scale build-up on the inside of the boiler tubes, soot build-up on

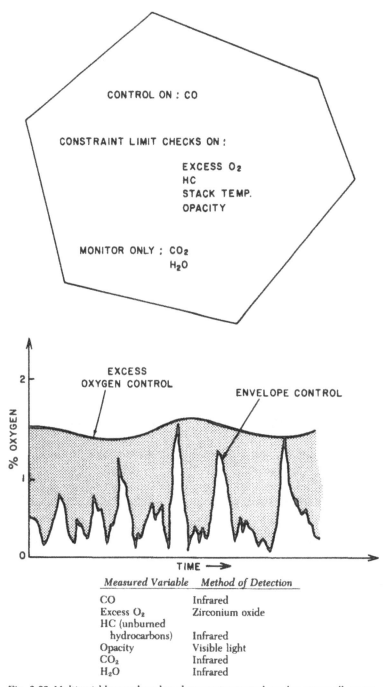

CONTROL ON : CO

CONSTRAINT LIMIT CHECKS ON :

EXCESS O_2
HC
STACK TEMP.
OPACITY

MONITOR ONLY : CO_2
H_2O

Measured Variable	Method of Detection
CO	Infrared
Excess O_2	Zirconium oxide
HC (unburned hydrocarbons)	Infrared
Opacity	Visible light
CO_2	Infrared
H_2O	Infrared

Fig. 2.22 Multivariable envelope-based constraint control can lower overall excess oxygen. This can be achieved by monitoring carbon dioxide and water and by performing constraint limit checks on excess oxygen, hydrocarbons, stack temperature, and opacity. (From Lipták [13])

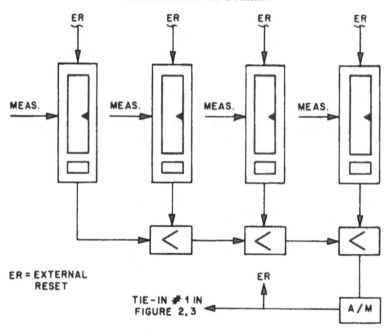

Fig. 2.23 Controllers configured into a selective loop can also implement envelope control.

the outside of the boiler tubes, or deteriorated baffles that allow the hot gases to bypass the tubes. Some microprocessor-based systems will respond to a rise in flue temperature by taking corrective action, such as automatically blowing the soot away, or by giving specific maintenance instructions.

If the stack temperature drops below the reference baseline, this does *not* necessarily signal an increase in boiler efficiency. More likely, it can signal the loss of heat due to leakage. Cold air or cold water can leak into the stack gases if the economizer or the regenerator are damaged. The consequence of this is loss of efficiency and the danger of corrosive condensation, if the temperature drops down to the dew point. A drop in flue gas temperature will also lower the stack effect, thus increasing the load on the induced draft fan.

For the above reasons, the advanced envelope control systems (figure 2.22) include both high and low limit constraints on stack temperature, using the above-described baseline as a reference.

Fuel Savings through Optimization

The overall boiler efficiency is the combined result of its *heat transfer efficiency* and its *combustion efficiency*. Heat transfer efficiency is reflected by stack temperature, which at the hot end should not exceed steam temperature by more than 150°F (65°C) when excess air is near optimum [15]. Combustion efficiency is tied to excess oxygen, which is brought as low as possible without exceeding the limits on CO (usually 400 ppm), opacity (#2 Ringlemann), and unburned carbon and NO_x.

When optimization reduces the flue gas losses, the resulting savings can be estimated from the amount of reduction in these losses. In case of a 100,000 lb./hr. (45,450 kg./hr.) steam boiler, a 1-percent reduction in fuel consumption (a 1-percent increase in efficiency) will lower the yearly operating costs by about $20,000 if the fuel cost is estimated at $2 per million Btu.

The fuel savings resulting from the lowering of excess O_2 can be estimated from graphs, such as those shown in figures 2.24 or 2.25. On these graphs the temperature is the difference between the stack and ambient temperatures. If, in a boiler operating at a 500°F stack temperature difference, optimization lowers the excess O_2 from 5 percent to 2 percent, this will result in a fuel savings of 1.5 percent (2.25 percent minus 0.75 percent), according to figure 2.24. For the same conditions, figure 2.25 gives a savings of 2 percent (10.5 percent minus 8.5 percent). Such differences are acceptable, as the size and designs of the equipment does influence the results. Although optimization has lowered the excess O_2 in some boilers to around 0.5 percent, the value of these last increments of savings tends to diminish in proportion to the cost of accomplishing them. Lowering excess O_2 from 1 percent to 0.5 percent will increase boiler efficiency by only 0.25 percent, a savings of about $5,000/yr. in a 100,000 lb./hr. boiler, and the controls needed to sustain such operation need to be as sophisticated as the ones described in figure 2.22. For these reasons, an optimization target of 1 percent excess O_2 on gas fuel is reasonable.

The fuel savings resulting from improved thermal efficiency can be estimated from figures 2.25 or 2.26. Lowering the stack temperature difference from 500°F to 400°F at an excess O_2 of 2 percent will result in a savings of 1.7 percent (8.5 percent minus 6.8 percent), according to figure 2.25. Assuming an ambient temperature of 60°F, these same conditions are marked in figure 2.26 at 560°F and 460°F, resulting in a savings of about 2 percent. Therefore, as a crude approximation, a fuel savings of 1 percent can be estimated for each 50°F reduction in stack gas temperature.

As was already discussed, stack gas temperature reductions must be limited to achieving a temperature above dew point that is high enough to provide the required stack effect.

Consequently, the total savings from improved thermal and combustion efficiencies can be estimated on the basis of both stack temperature and excess oxygen being lowered to their limits. The resulting total fuel savings can be around 5 percent.

Steam Pressure Optimization

In figure 2.3, the second tie-in point for the optimizer is the set point of the pressure controller PIC-101. In traditional boiler operation, the steam pressure was maintained at a constant value. It is only in recent years that this practice has begun to change.

In cogenerating plants, in which the boiler steam is used to generate electricity

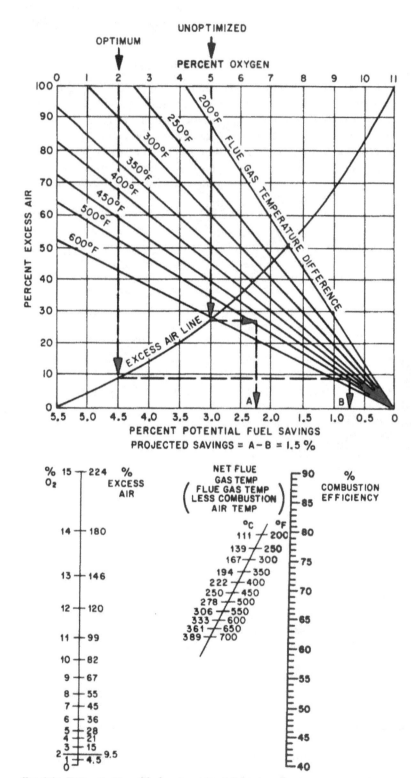

Fig. 2.24 Determination of fuel-saving potential due to reductions in excess oxygen. Gas combustion efficiency is found in the lower chart at the intersection of a straight line drawn through the applicable excess oxygen and flue gas temperature points. (Adapted from Westinghouse [22])

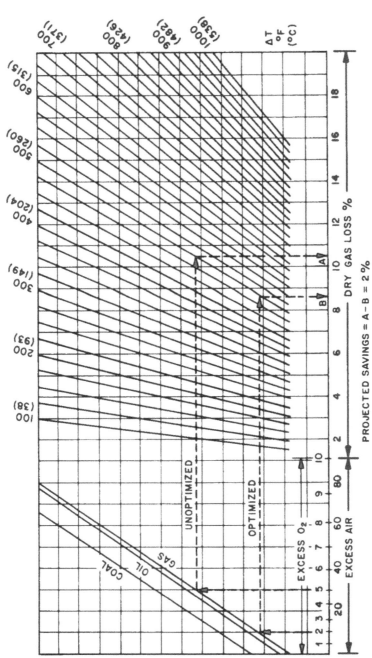

Fig. 2.25 The fuel-saving potential can also be computed as shown here. (Courtesy of Dynatron, Inc.)

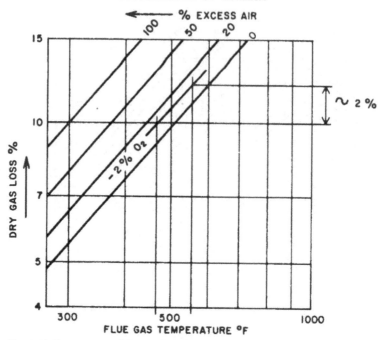

Fig. 2.26 Temperature of flue gas leaving a medium-size boiler has a direct effect
on combustion efficiency. (Adapted from Walsh [21])

and the turbine exhaust steam is used as a heat source, optimization is obtained by
maximizing the boiler pressure and minimizing the turbine exhaust pressure, so as to
maximize the amount of electricity generated (figure 2.27).

In plants that do not generate their own electricity, optimization is achieved by
minimizing boiler operating pressure. This reduces the pressure drops in turbine governors by opening them up further, lowers the cost of operating feedwater pumps
because their discharge pressure is reduced, and generally lowers radiation and wall
losses in the boiler and piping.

Pump power is particularly worth saving in high-pressure boilers, because as much
as 3 percent of the gross work produced by a 2400 PSIG (16.56 MPa) boiler is used to
pump feedwater. Figure 2.28 illustrates the method of finding the optimum minimum
steam pressure, which then becomes the set point for the master controller PIC-101
in figure 2.3.

As long as all steam user valves (including all turbine throttle valves) are less than
fully open, a lowering in the steam pressure will not restrict steam availability, because
the user valves can open further. The high signal selector (TY-1) selects the most-open
valve, and the valve position controller (VPC-2) compares that signal with its set point
of, for example, 80 percent. If even the most-open valve in the plant is less than 80
percent open, the pressure controller set point is slowly lowered. VPC-2 is an "integral-

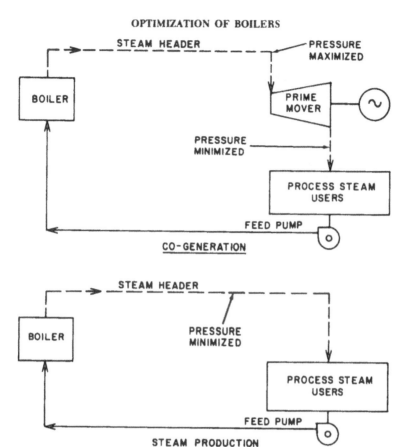

Fig. 2.27 The optimum steam pressure at the boiler is also a function of what the steam is used for.

only" controller; its reset time is at least ten times that of PIC-101. This slow integral action guarantees that only very slow "sliding" of the steam pressure will occur and that noisy valve signals will not upset the system, because VPC-2 responds only to the integrated area under the error curve. The output signal from VPC-2 is limited by PY-4, so that the steam pressure set point cannot be moved outside some preset limits. This necessitates the external feedback to VPC-2, so that when its output is overridden by a limit, its reset will not wind up.

This kind of optimization, in which steam pressure follows the load, not only increases boiler efficiency but also does the following:

- Prevents any steam valve in the plant from fully opening and thereby losing control
- Opens all steam valves in the plant, thereby moving them away from the unstable (near-closed) zone of operation

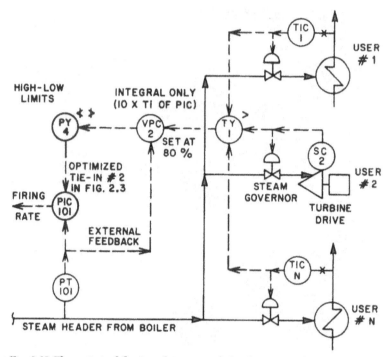

Fig. 2.28 The optimized floating of steam supply header pressure becomes the set point for the master controller PIC-101.

- Reduces valve maintenance and increases valve life by lowering pressure drop
- Increases turbine drive efficiencies by opening up all steam governors

The total savings in yearly operating costs resulting from optimizing the steam pressure to follow the load is a small percentage of the total cost.

Water Side Optimization

The optimization of the water side of a steam generator includes the optimized operation of the feedwater pump at the condensate return system and of the boiler blowdown. As pumping system optimization is the subject of a separate chapter in this book, only the boiler blowdown will be discussed here.

In figure 2.3, the third tie-in point for the optimizer is the set point of the blowdown flow controller. The goal of optimization is to minimize blowdown as much as possible without causing excessive sludge or scale build-up on the inside surfaces of the boiler tubes. The benefits of such optimization include the reduction in the need for make-up water and treatment chemicals and the reduction in heat loss as hot water is discharged. About 90 percent of the blowdown should occur continuously, and 10 percent would result from the periodic blowing down of the mud drum and of the headers.

Blowdown can be optimized by automatically controlling the chloride and conductance of the boiler water. The neutralized conductivity set point is usually around
2,500 micromhos. Automatic control maintains this set point within ± 100 micromhos.
The required rate of blowdown is a function of the hardness, silica, and total solids of
the make-up water and also of the steaming rate and condensate return ratio of the
boiler.

The amount of blowdown can be determined as follows:

$$BD = \frac{S - R}{C - 1}$$

where

BD = Blowdown rate, pounds per hour
R = Rate of return condensate, pounds per hour
S = Steam load, pounds per hour
C = Cycles of concentration based on makeup.

The value for cycles of concentration is generally determined on the basis of the
chloride concentration of the boiler water divided by the chloride content of the make-
up water. The value is also given by dividing the average blowdown rate into the
average rate of make-up water, assuming no mineral contamination in any returned
condensate.

Figure 2.29 illustrates that the rate of blowdown accelerates as the boiler water
conductivity set point is lowered. A reduction of about 20 percent can result from
converting the blowdown controls from manual to automatic [17]. In case of a 100,000
lb./hr. (45,450 kg./hr.) boiler, this can mean a reduction of 1,340 lb./hr. (600 kg./hr.)
in the blowdown rate. If the blowdown heat is not recovered, this can lower the yearly
operating cost by about $10,000 (depending on the unit cost of the fuel).

Overall boiler efficiency can also be increased if the heat content of the hot condensate is returned to the boiler. Pumping water at high temperatures is difficult;
therefore, the best choice is to use pumpless condensate return systems. Figure 2.30
illustrates the operation of such a system, which uses the steam pressure itself to push
back the condensate into the de-aeration tank. This approach eliminates not only the
maintenance and operating cost of the pump but also the flash and heat losses, resulting
in the return of more condensate at a higher temperature.

The performance of the steam and condensate piping system in the plant can also
be improved if steam flows are metered. Such data is helpful not only in accountability
calculations but also in locating problem areas, such as insufficient thermal insulation
or leaking traps.

Load Allocation

The purpose of load allocation between several boilers is to distribute the total
plant demand in the most efficient and optimized manner. Such optimization will reduce
the steam production cost to a minimum. Such computer-based energy management

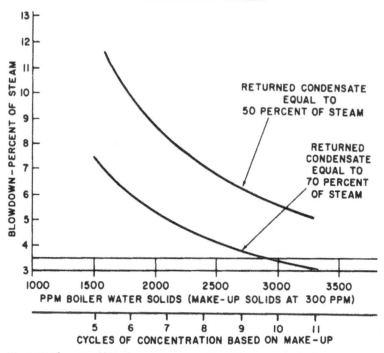

Fig. 2.29 The rate of blowdown increases as the boiler water conductivity set point is lowered. (From Schieber [17])

systems can operate either in an advisory or in a closed-loop mode. The closed-loop systems automatically enforce the load allocation, without the need for operator involvement. The advisory system, on the other hand, provides instructions to the operator but leaves the implementation up to the operator's judgment.

In simple load allocation systems, only the starting and stopping of the boilers is optimized. When the load is increasing, the most efficient idle boiler is started; when the load is dropping, the least efficient one is stopped. In more sophisticated systems, the load distribution between operating boilers is also optimized. In such systems, a computer is used to calculate the real-time efficiency of each boiler. This information is used to calculate the incremental steam cost for the next load change for each boiler.

If the load increases, the incremental increase is sent to the set point of the most cost-effective boiler. If the load decreases, the incremental decrease is sent to the least cost-effective boiler (figure 2.31). The required software packages with proved capabilities for continuous load balancing through the predictions of costs and efficiencies are readily available [3]. With the strategy described in figure 2.31, the most efficient boiler either will reach its maximum loading or will enter a region of decreasing efficiency and will no longer be the most efficient. When the loading limit is reached on one boiler, or when a boiler is put on manual, the computer will select another as the most efficient unit for future load increases.

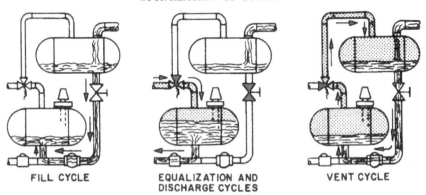

FILL CYCLE   EQUALIZATION AND   VENT CYCLE
DISCHARGE CYCLES

Fig. 2.30 The pumpless condensate return system uses the steam pressure itself to push the condensate back into the de-aeration tank. (Courtesy of Johnson Corp.)

The least efficient boiler will accept all decreasing load signals until its minimum limit is reached. Its load will not be increased unless all other boilers are at their maximum load or in manual. As shown in figure 2.1, some boilers can have high efficiency at normal load while being less efficient than the others at low load. Such units are usually not allowed to be shut down but are given a greater share of the load by a special subroutine.

If all boilers are identical, some will be driven to maximum capacity and others

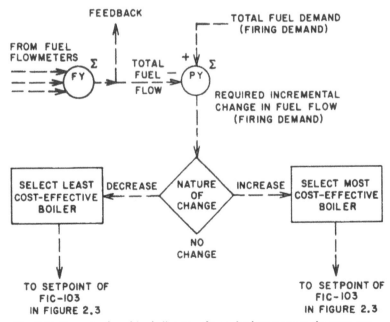

Fig. 2.31 Computer-based load allocation directs load increases to the most cost-effective boiler and sends load decreases to the least cost-effective boiler.

will be shut down by this strategy, and only one boiler will be placed at an intermediate load [19]. Boiler efficiency can be monitored indirectly (by measurement of flue gas composition, temperature, combustion temperature, and burner firing rate) or directly (through time-averaged steam and fuel flow monitoring). For the direct efficiency measurement, it is important to select flow meters with acceptable accuracy and rangeability. In order to arrive at a reliable boiler efficiency reading, the error contribution of the flow meters, based on actual reading, must not exceed $\pm \frac{1}{2}$ percent to $\pm \frac{3}{4}$ percent.

Boiler allocation can be based on actual measured efficiency, on projected efficiency based on past performance, or on some combination of the two. The continuous updating and storing of performance data for each boiler is also a valuable tool in operational diagnostics and maintenance.

Conclusions

The various goals of boiler optimization include the following:

- To minimize excess air and flue gas temperature
- To measure efficiency (use the most efficient boilers; know when to perform maintenance)
- To minimize steam pressure (open up turbine governors; reduce feed pump discharge pressures; and reduce heat loss through pipe walls)
- To minimize blowdown
- To provide accountability (monitor losses; recover condensate heat)
- To minimize transportation costs (use variable speed fans; eliminate condensate pumps; and consider variable speed feedwater pumps)

If the potentials of all of the above optimization strategies are fully exploited, the unit costs of steam generation can usually be lowered by about 10 percent. In larger boiler houses, this can represent a savings that will pay for the optimization system in a year or less [13].

REFERENCES

1. American Technical Services, "Boiler Audits," June 1982
2. Babcock & Wilcox Co., Steam, Its Generation and Use (New York: Babcock & Wilcox Co., 1955).
3. Cho, C. H., "Optimum Boiler Allocation," InTech, October 1978.
4. Congdon, P., "Control Alternatives for Industrial Boilers," InTech, December 1981.
5. DeLorenzi, O., Combustion Engineering, Inc., 1967.
6. Dickey, P. S., "A Study of Damper Characteristics," Bailey Meter Co., Reprint No. A8.
7. Dukelow, S. G., Improving Boiler Efficiency (Kansas State University, 1985).
8. Hurlbert, A. W., "Air Flow Characterization Improves Boiler Efficiency," InTech, March 1978.
9. Jorgensen, R., "Fans," in Marks' Standard Handbook for Mechanical Engineers, 8th ed. (New York: McGraw-Hill Book Co., 1978), p. 14.53.
10. Lipták, B. G., "Boiler Controls," in Instrument Engineer's Handbook, Process Control (Radnor, PA: Chilton Book Co., 1985), section 8.2.
11. Lipták, B. G., "Flow Measurement," in Instrument Engineer's Handbook, Process Measurement (Radnor, PA: Chilton Book Co., 1982).
12. Lipták, B. G., "Analyzers," in Instrument Engineer's Handbook, Process Measurement (Radnor, PA: Chilton Book Co., 1982), chapters 10 and 11.
13. Lipták, B. G., "Save Energy by Optimizing Your Boilers," InTech, March 1981.

14. McFadden, R. W., "Multiparameter Trim in Combustion Control," *InTech*, May 1984.

15. McMahon, J. F., President, Cleveland Controls, verbal communications.

16. O'Meara, J. E., "Oxygen Trim for Combustion Control," *InTech*, March 1979.

17. Schieber, J. R., "The Care for Automated Boiler Blowdown," Universal Interlock, 1969.

18. Schwartz, J. R., "Carbon Monoxide Monitoring," *InTech*, June 1983.

19. Shinskey, F. G., *Energy Conservation through Control* (Academic Press, 1978).

20. Shinskey, F. G., *Process Control Systems* (New York: McGraw-Hill Book Co., 1979).

21. Walsh, T. J., "Controlling Boiler Efficiency," *I&CS*, January 1981.

22. Westinghouse Electric Corp., "Oxygen Trim Control," Westinghouse Electric Corp., AD-106-125, June 1979; January 1985.

Chapter 3

OPTIMIZATION
OF
CHEMICAL REACTORS

THE OPTIMIZATION of batch and continuous chemical reactors has many potential benefits, including increase in productivity and improvement in safety, product quality, and batch-to-batch uniformity. The combined impact of these factors on plant productivity can approach a 25-percent improvement [3]. Such overall results are the consequences of many individual control loops and control strategies. These loops will program temperature and pressure and maintain concentration and safety, while providing sequencing and record-keeping functions. All of these elements of the overall chemical reactor control system are discussed in this chapter.

Reactor Characteristics

In a batch cycle, there is no steady state and therefore no "normal" condition at which controllers could be tuned. The dynamics of the batch process vary with time; thus, the process variables, the process gains, and time constants also vary during the batch cycle. In addition, there are the problems of runaway reactions and batch-to-batch product uniformity. Runaway reactions occur in exothermic reactions, in which an increase in temperature speeds up the reaction, which in turn releases more heat and raises the temperature further. In order to counter this positive feedback cycle, highly self-regulating cooling systems are required. One of the most self-regulating cooling systems is a bath of boiling water, because it needs no rise in temperature to increase its rate of heat transfer. Endothermic reactions are inherently self-regulating.

Batch-to-batch uniformity is a function of many factors, from the purity of reactants, catalysts, and additives to the repeatability of controllers serving to maintain heat and material balance. Before addressing such complex topics, it is necessary to review the basic batch process.

Most batch cycles are started by charging reactants into the reactor and then mixing and heating them until the reaction temperature is reached. The reaction itself is frequently started by the addition of a catalyst. *Exothermic reactions* produce heat, and *endothermic reactions* consume heat. The reactor itself can be *isothermal*, meaning that it is operated at constant temperature, or *adiabatic*, meaning that heat is not added or removed within the reactor; the reaction is controlled by other means, such as the manipulation of pressure, catalyst, reactants, etc. Chemical reactions can follow quite complex paths and sequences, but for engineering purposes such as equipment design and control system analysis, most reactions can be considered as one of four types: irreversible, reversible, consecutive, or simultaneous.

Most reactions are reversible—that is, there is a ratio in product to reactant concentration that brings about equilibrium. Under equilibrium conditions the production rate is zero, because for each molecule of product formed there is one that converts back to its reactant molecules. The equilibrium constant (K) describes this state as the ratio of forward to reverse-rate coefficients. The value of K is also a function of the reaction temperature and the type of catalyst used. K naturally places a limit on the conversion that can be achieved within a particular reactor, but conversion can usually be increased, if at least one of the following conditions can be achieved:

- Reactant concentration can be increased
- Product concentration can be decreased through separation or withdrawal
- Temperature can be lowered by increased heat removal
- A change in operating pressure can be effected (this increases conversion only in certain reactions).

The catalyst does not take part in the reaction, but it does affect the equilibrium constant (K). Some catalysts are solids and are packed in a bed; others are fluidized, dissolved, or suspended. Metal catalysts are frequently formed as flow-through screens. Whatever their shape, the effectiveness of the catalyst is a function of its active surface, because all reactions take place on that surface. When it is fouled, the catalyst must be reactivated or replaced.

The time profiles of heat release, operating temperature, and chemical concentrations are illustrated in figure 3.1 for a consecutive reaction [11], in which first ingredient A is converted into intermediate product B, and then intermediate product B reacts to form final product C. The reaction temperature is controlled so as to maximize the production of C while minimizing the cycle period.

Reaction Rates and Kinetics

The reaction rate coefficient exponentially increases with temperature. The activation energy (E) determines its degree of temperature dependences [15] according to the Arrhenius equation:

$$k = \alpha e^{-(E/RT)}$$

where

k = specific reaction rate (min^{-1})
α = pre-exponential factor (min^{-1})
E = activation energy of reaction (Btu/mole)
R = perfect gas constant (1.99 Btu/mole^{-}°R)
T = absolute temperature (°R)

Figure 3.2 illustrates the strong dependency of K on reaction temperature [15] for the values of $\alpha = e^{29}$ and E/R = 20,000.

Figure 3.3 illustrates the three basic reactor types and defines their fractional conversion of reactant into product (y).

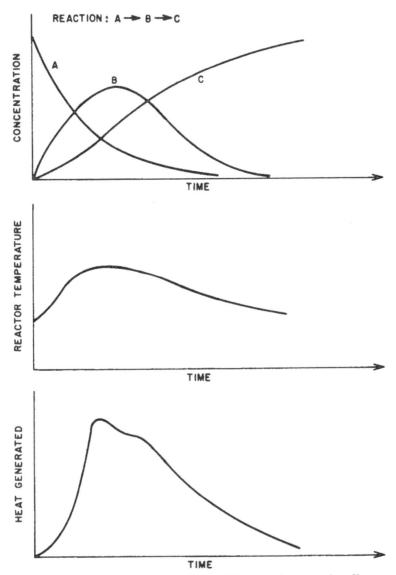

Fig. 3.1 In a consecutive reaction, process variables show these typical profiles. Heat generated by the reaction is greatest during the conversion of A to B, but the temperature of the reactor is controlled at an optimum setting to insure maximum conversion in a minimum amount of time. (From Luyben [11])

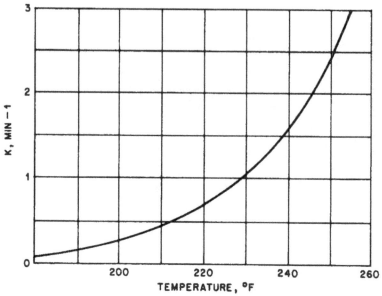

Fig. 3.2 The influence of temperature on reaction rate is substantial. (From Shin-skey [15])

Because the continuous plug flow type reactor is dominated by dead time, its temperature control is difficult. On the other hand (as shown in figure 3.3), the plug flow reactor gives higher conversion than a back-mixed reactor operating under the same conditions. If the reaction rate is low, a long tubular reactor or a large back-mixed reactor is required to achieve reasonable conversions.

In a batch reactor, after the initial charge there is no inflow or outflow. Therefore, an isothermal batch reactor is similar in its conversion characteristics to a plug-flow tubular reactor. If the residence times are similar, both reactors will provide the same conversions. Batch reactors are usually selected when the reaction rates are low, when there are many steps in the process, when isolation is required for reasons of sterility or safety, when the materials involved are hard to handle, and when production rates are not high. As shown in figure 3.4, the batch (or tubular) reactor is kinetically superior to the continuous stirred tank reactor [11]. The batch reactor has a smaller reaction time and can produce the same amount of product faster than the back-mixed one.

Reactor Stability and Time Constants

The amount of heat generated by an exothermic reactor increases as the reaction temperature rises. If the reactor is operated without a temperature controller (in an open loop), an increase in the reaction temperature will also increase heat removal, because of the increase in ΔT between process and coolant temperatures. If an increase in reaction temperature results in a greater increase in heat generation than in heat

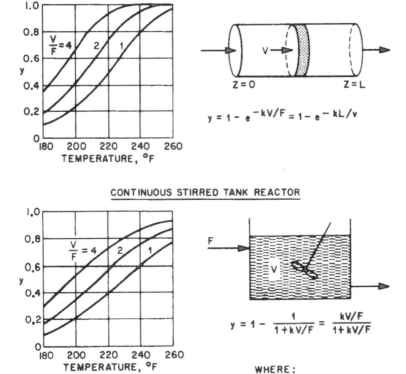

PLUG FLOW REACTOR

$$y = 1 - e^{-kV/F} = 1 - e^{-kL/v}$$

CONTINUOUS STIRRED TANK REACTOR

$$y = 1 - \frac{1}{1+kV/F} = \frac{kV/F}{1+kV/F}$$

WHERE:

F — FEED RATE (ft^3/min)
k — REACTION RATE (min^{-1})
t — TIME (min)
T — TEMPERATURE (°F)
v — VELOCITY (ft/min)
V — REACTOR VOLUME (ft^3)
y — FRACTIONAL CONVERSION
L = Z — AXIAL DISTANCE (ft)

BATCH REACTOR

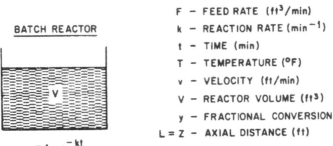

$$y = 1 - e^{-kt}$$

Fig. 3.3 Conversion equations and conversion vs. temperature characteristics differ depending on the type of reactor. (From Luyben [11] and Shinskey [15])

removal, the process is said to display "positive feedback"; as such, it is considered to be "unstable in the open loop."

The positive feedback of the open loop process can be compensated for by the negative feedback of a reactor temperature controller, which will increase the heat removal rate as the temperature rises. The addition of such a feedback controller can

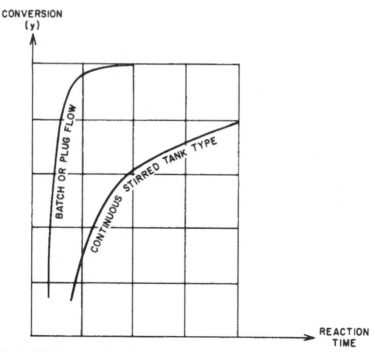

Fig. 3.4 Batch reactors have better conversion efficiencies than back-mixed reactors. (From Luyben [11])

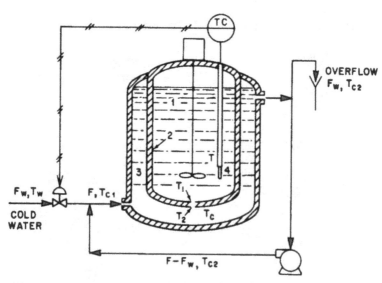

THE PROCESS GAIN: $K_p = \dfrac{T_{c2} - T_w}{F_w}$

Fig. 3.5 There are four interacting time lags in a chemical reactor process. The smaller the process gain, the narrower the proportional band and the more sensitive the control loop. (From Shinskey [15])

stabilize an open-loop unstable process only if the control loop is fast and does not contain too much "dead time." Cascade control can increase speed, and maximized coolant flow can reduce dead time. Shinskey [15] suggests that if the dead time can be kept under 35 percent of the thermal time constant of the reactor, the process can be stabilized, whereas if it approaches 100 percent the reactor will not be controllable.

A real reactor has several lags and delays, including those of measurement and heat removal, as illustrated in figure 3.5. These are listed and defined below [15]:

Thermal Time Constant $\quad \tau_1 = \dfrac{W_1C_1}{k_1A} = \dfrac{W_1C_1}{Q}(T - T_1)$

Reactor Wall Time Constant $\quad \tau_2 = \dfrac{W_2C_2\ell}{k_2A} = \dfrac{W_2C_2}{Q}(T_1 - T_2)$

Coolant Time Constant $\quad \tau_3 = \dfrac{W_3C}{k_3A} = \dfrac{W_3C}{Q}(T_2 - T_c)$

Thermal Bulb Time Constant $\tau_4 = \dfrac{W_4C_4}{k_1A_4}$

where

A = Heat-transfer area, ft^2
A_4 = Surface area of bulb, ft^2
C = Specific heat of coolant, Btu/(lb) (°F)
C_1 = Specific heat of reactants, Btu/(lb)(°F)
C_2 = Specific heat of wall, Btu/(lb)(°F)
C_4 = Specific heat of bulb, Btu/(lb)(°F)
k_1 = Heat-transfer coefficient, Btu/(h)(ft^2)(°F)
k_2 = Thermal conductivity, Btu/(h)(ft^2)(°F/in)
k_3 = Heat-transfer coefficient Btu/(h)(ft^2)(°F)
ℓ = Wall thickness, in
Q = Rate of heat evolution, Btu/h
T = Reactor temperature, °F
T_1 = Wall temperature, °F
T_2 = Outside wall temperature, °F
T_c = Average coolant temperature, °F
W_1 = Weight of reactants, lb
W_2 = Weight of wall, lb
W_3 = Weight of jacket contents, lb
W_4 = Weight of bulb, lb

Typical values of these time constants are:

τ_1 = 30 to 60 min
τ_2 = 0.5 to 1.0 min

τ_3 = 2 to 5.0 min

τ_4 = 0.1 to 0.5 min (can be minimized by the use of bare bulbs)

The total dead time in the loop is the sum of jacket transport lag, the dead time due to imperfect mixing, and miscellaneous smaller contributing factors. The dead time due to jacket displacement can be reduced by increasing the pumping rate. This should be kept under two minutes in a well-designed reactor [15]. The dead time caused by imperfect mixing can be reduced by increasing the agitator pumping capacity. In a well-designed reactor it should be held to less than 10 percent of the thermal time constant τ_1.

In the case of a typical reactor, the period of oscillation might be around 30 minutes. This period approximately equals four dead times; therefore, the dead time of such a loop is around 7.5 minutes. In the case of interacting controllers, the correct setting for such a controller would be 7.5 minutes for both integral and derivative times.

Temperature Control

The control loop features required during heat-up are substantially different from those needed during an exothermic reaction or those required during stripping or refluxing. Each will be discussed in the following paragraphs, starting with the controls of exothermic reactors.

The most frequently used controlled variable is reaction temperature, which usually needs to be maintained within 0.5°F of set point. In case of an exothermic reaction, the reaction heat can be removed, as shown in figure 3.6. This "once-through" method of cooling is undesirable, however, because the coolant temperature is not uniform. This can cause cold spots near the inlet and hot spots near the outlet. Another disadvantage of this configuration is the variable residence time of the cooling water within the jacket as the flow rate changes. This causes the dead time of the jacket to vary, which in turn necessitates the modification of the control loop tuning constants as the load varies. In addition, when the water flow is low, the Reynolds number will drop off, and with it, the heat transfer efficiency will also diminish. Low water velocity can

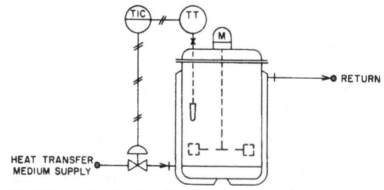

Fig. 3.6 In once-through cooling of chemical reactors, the coolant temperature is not uniform. (From Lipták [9])

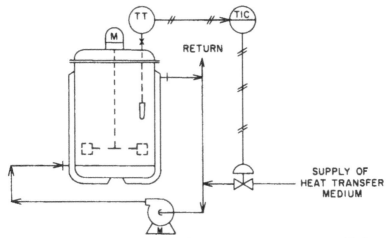

Fig. 3.7 Recirculated cooling of chemical reactors guarantees a constant and high rate of water circulation. (From Lipták [9])

also result in fouling of the heat transfer surfaces. For all the above reasons, the recirculated cooling water configuration shown in figure 3.7 is more desirable, because it guarantees a constant and high rate of water circulation [9]. This keeps the jacket dead time constant, the heat transfer coefficient high, and the jacket temperature uniform, eliminating cold and hot spots.

Cascade Control

Perhaps the most common cascade configuration is when two temperature controllers are in series, with the output of the reactor temperature controller becoming the set point of the jacket water temperature controller. It is preferred that the slave control the jacket outlet temperature rather than the inlet temperature, because this way the jacket and its dynamic response is included in the slave loop.

The purpose of the slave loop is to correct for all outside disturbances, without allowing them to affect the reaction temperature. For example, if the control valve is sticking or if the temperature or pressure of the heat transfer media changes, this would upset the reaction temperature in figures 3.6 or 3.7 but not in figure 3.8, because here the slave would notice the resulting upset at the jacket outlet and would correct for it before it had a chance to upset the master.

The period of oscillation of the master loop is usually cut in half as direct control is replaced by cascade. This might mean a reduction from 40 to 20 minutes in the period and a corresponding reduction of perhaps 30 to 15 percent in the proportional band [15]. The derivative and integral settings of an interacting controller would also be reduced from about 10 to about 5 minutes. This represents a fourfold overall loop performance improvement.

Another advantage of the cascade loop is that it removes the principal nonlinearity

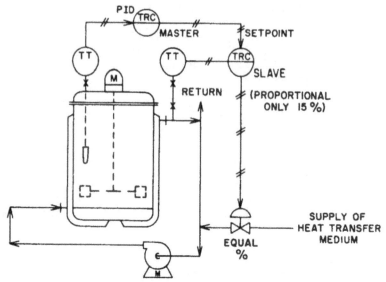

Fig. 3.8 Cascade control of a reactor with recirculation reduces the period of oscillation of the master loop. (From Lipták [9])

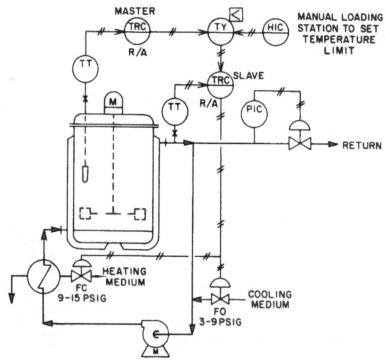

Fig. 3.9 Two-directional cascade loop with a maximum limit on jacket temperature allows heat to be added in some phases of the operation and removed in others. (From Lipták [9])

of the system from the master loop, because reaction temperature is linear with jacket outlet temperature. The nonlinear relationship between jacket outlet temperature and heat transfer media flow is now within the slave loop, where it can be compensated by an equal percentage valve, whose gain increases as the process gain drops off. In most instances, the slave will operate properly with proportional plus derivative or proportional control only, set for a proportional band of 10 to 20 percent.

If heat needs to be added in some phases of the reaction while in other phases it must be removed, the controls must be configured in a two-directional manner. Such a cascade loop is illustrated in figure 3.9, with the heating and cooling medium control valves split range controlled in such a manner that as the control signal rises from 3 to 15 PSIG, the jacket gradually moves from maximum cooling to maximum heating. This is a fail-safe arrangement, because in case of air failure the heating valve is closed and the coolant valve is opened.

Figure 3.9 also provides a maximum limit on the slave set point. This can be an important feature with temperature-sensitive products or if high wall temperatures would adversely affect the reaction.

Limitations of Cascade Control

Figure 3.10 illustrates a cascade loop consisting of three controllers in series. The master is the reaction temperature controller, the primary slave is the jacket temperature controller and the secondary slave is the valve position controller. A positioner is a position controller that detects the opening of the valve and corrects for any deviations between measurement (the mechanically detected position) and set point (the pneumatic signal received). In such a hierarchical arrangement, *only the master* can control its variable independently; the slave controller set points must be freely adjustable to satisfy the requirements of the master.

Whenever the master is prevented from modifying the set point of the slave—because of a limiter, such as in figure 3.9, or because the slave has been manually switched from remote to local set point—reset wind-up can occur. Reset wind-up is the integration of an error that the controller is prevented from eliminating. Consequently, the controller output is saturated at an extreme value. Once saturated, the controller is ineffective when control is returned to it until an equal and opposite error unsaturates it. This problem is eliminated by the external reset (ER) shown in figure 3.10. The external reset signal is used to disable the integral mode whenever the slave is not on set point. This feature eliminates the need for switching the master to manual and thereby also eliminates the need for the auto/manual station. In addition, it eliminates reset wind-up upsets due to start-ups, shutdowns, or emergency overrides.

Another limitation is that the cascade loop will be stable only if each slave is faster than its master. Otherwise, the slave cannot respond in time to the variations in the master output signal, and a cascade configuration will in fact degrade the overall quality of control. A rule of thumb is that the period of oscillation of the slave should not exceed 30 percent of the period of oscillation of the master loop. This requirement is not always satisfied. For example, in figure 3.10 it is important to select valve positioners that are

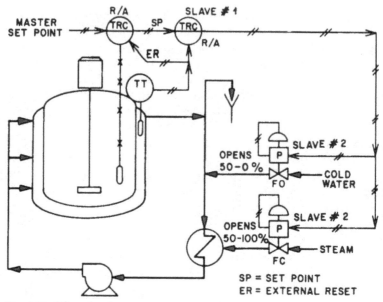

Fig. 3.10 If the valves have positioners, a cascade loop will consist of three controllers in series. All cascade masters should be provided with external reset. (From Lipták [9])

TRC Output	Steam Valve Opening	Water Valve Opening
0%	0%	100%
25	0	Throttling
50	0	0
75	Throttling	0
100	100	0

faster than the slave temperature controller on the jacket. Similarly, the jacket temperature control loop should contain less dead time than its master, which might not be possible if a once-through piping configuration (figure 3.6) is used. One possible method of reducing the dead time of the cascade slave loop is to move the measurement from the jacket outlet (figure 3.8) to the jacket inlet. This usually is not recommended, because when this is done, the slave will do much less work because the dynamics of the jacket have been transferred into the master loop.

Gas Reactions

Gas phase reactions (such as ammonia synthesis) are usually conducted at relatively high temperatures and pressures. Because of the competing reactions taking place in parallel with the main reactions, productivity and partial pressures are interrelated. For example, in ammonia synthesis a pressure increase is required to increase production when the ammonia concentration is low, and a pressure decrease is required to increase production when the ammonia concentration is high [14]. If inerts are also

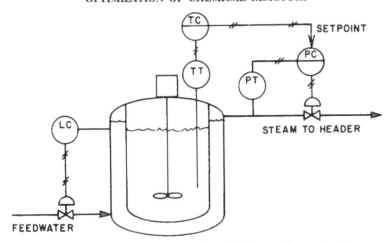

Fig. 3.11 An internal heat sink is provided by the boiling coolant in the jacket. This stabilizes exothermic reactors. (From Shinskey [14])

present, partial pressures are usually determined by detecting both the total pressure and the gas composition.

In such processes, which are both temperature and pressure sensitive, a very stable heat removal system is desired. If the reactor jacket is filled with boiling water, the rate of heat removal can vary without causing a change in the jacket temperature. Figure 3.11 illustrates such a cascade system. In order to increase the sensitivity of the loop, pressure is selected as the controlled variable for the slave controllers.

In a hydrocracking reactor (figure 3.12), product quality, catalyst life, and productivity are all a function of accurate temperature control. In this process, the reaction rate is fast, residence time is a few seconds, and reactant concentration is low. This combination would allow the reaction temperature to be controlled by manipulating the feed rate. If it is desirable to set production rate, and therefore feed rate, independently of cooling capacity, the temperature can be controlled by throttling a diluent. In figure 3.12, the diluent is hydrogen, which is admitted under separate temperature controls into each zone. The introduction of hydrogen diluent lowers the reaction rate by reducing the reactant concentration and also by cooling.

Constraint Optimization

In exothermic reactions, one of the critical safety constraints is coolant availability. That limitation is automatically configured into the control loop of a continuous reactor, illustrated in figure 3.13. Here, the optimizing controller (OIC) detects the opening of the coolant valve, and if it is less than 90-percent open, it admits more feed by increasing the set point of the FRC. Thus, production rate is always maximized, but only within the safe availability of the coolant. When the coolant valve opens beyond 90 percent, production rate is lowered so that the reactor will never be allowed to run out of coolant.

Another safety feature shown in figure 3.13 is the reaction failure alarm loop (RA).

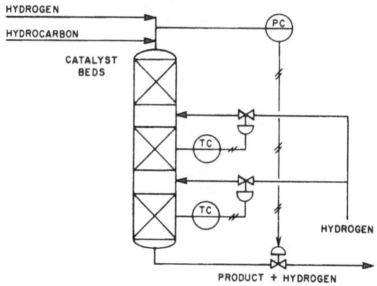

HYDROGEN

HYDROCARBON

CATALYST
BEDS

TC

TC

HYDROGEN

PRODUCT + HYDROGEN

Fig. 3.12 Dilution with cool hydrogen can also control the reaction rate. (From Shinskey [14])

This loop simply compares the reactant charging rate to the reactor with the heat removal rate from the reactor and actuates an alarm if the charging rate is in substantial excess.

In figure 3.13, the valve position controller (OIC) maintains the feed rate as high as the cooling system can handle. This results in a variable production rate, because changes in coolant temperature or variations in the amount of fouling of the heat tranfer areas will change production.

The dynamic response of the reactor temperature to changes in reactant flow is not favorable. A change in feed flow must change the reactant concentration before a change in reaction rate (and therefore in heat evolution) can change the reaction temperature. Reactant concentration adds a large secondary lag to the valve position control loop. This necessitates the use of a three-mode controller [14], which in turn limits the application to stable loops only. The slower the OIC loop (the longer its lag), the lower should be the set point of OIC in figure 3.13, to provide the required margin for stability.

Maximizing Production

In a batch reactor, production rate increases with temperature, and therefore production can be maximized by maximizing temperature. This can be accomplished by a valve position controller (VPC) that raises the batch temperature set point whenever the coolant valve is less than 90-percent open. Unfortunately, the dynamic characteristics of this loop are also undesirable. This is caused by the inverse response of the loop [14]. When the coolant valve opens to more than 90 percent, the VPC will lower

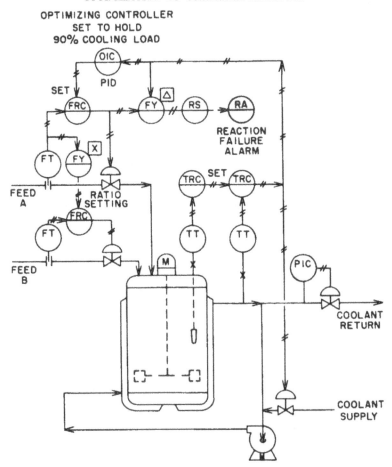

OPTIMIZING CONTROLLER
SET TO HOLD
90% COOLING LOAD

Fig. 3.13 The reaction rate in continuous reactor is matched to the capacity of the cooling system. (From Lipták [9])

the set point of the temperature controller. This will temporarily *increase* the demand for coolant, although once the excess sensitive heat is removed, it will lower it. The longer the dead time introduced by this inverse response, the larger the margin needed for safety and stability, and hence the lower the set point of the VPC.

When the concentration time constant is equal to or larger than the thermal time constant, temperature control through feed flow manipulation is no longer practical [14]. In such reactors, only the manipulation of the coolant will give stable operation. The strategy of valve position control is still useful in such applications, but in these cases it must manipulate supplementary cooling.

Supplementary cooling can be provided in a split range sequence, as shown in figure 3.14, or can be added through a separate coil, modulated by valve position control.

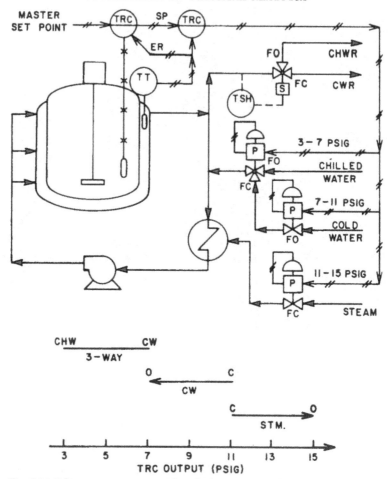

Fig. 3.14 Split-range sequencing with multiple coolants can be used to minimize cost.

Figure 3.15 describes such controls for a reactor with a separate chilled water coil. This coil is inoperative until the cold water valve approaches full opening. When the VPC detects that condition, it starts opening the chilled water valve. The resulting increased heat removal will cause the temperature cascade loop to close the cold water valve until it drops to the setting of the VPC.

This is also an optimizing strategy in the sense that the low-cost cold water has to be nearly exhausted before the high-cost chilled water is used.

Multiple Sensors

Multiple temperature sensors can be used for safety or maintenance reasons. In a fixed-bed reactor, for example, in which the location of the maximum temperature

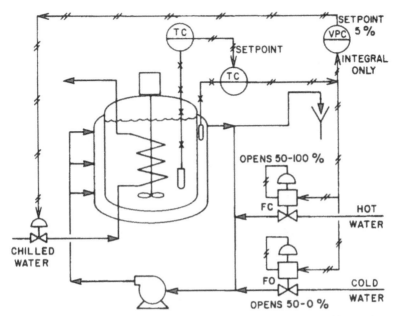

Fig. 3.15 Higher-cost chilled water is used only when the lower-cost cold water is insufficient to meet the demand for cooling. (From Shinskey [14])

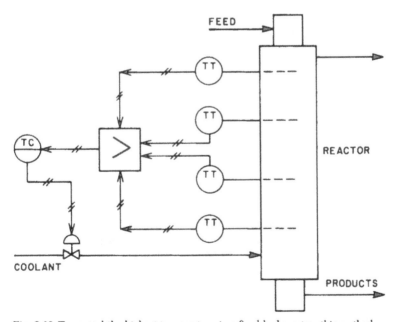

Fig. 3.16 To control the highest temperature in a fixed-bed reactor, this method can be used. (From Shinskey [15])

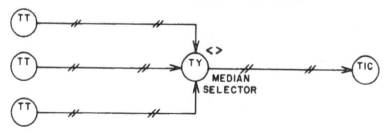

Fig. 3.17 A median selector protects from sensor error or failure.

might shift as a function of flow rate or catalyst age, multiple sensors would be installed and the highest reading selected for control, as shown in figure 3.16.

On the other hand, if the reason for the use of multiple sensors is to increase the reliability of the measurement, the use of median selectors is the proper choice (figure 3.17). A median selector rejects both the highest and the lowest signals and transmits the third. This type of redundancy protects against the consequences of sensor mis-operation, while also filtering out noise and transients that are not common to two of the signals [14].

Another method of increasing sensor reliability is the use of voting systems. These also consist of at least three sensors. Reliability is gained, as the voting system disregards any measurement that disagrees with the "majority view."

Initial Heat-Up

In most chemical reactions, a certain temperature must be reached in order to initiate the reaction. An ideal reactor temperature controller will permit rapid automatic heat-up to reaction temperature without overshoot and then will accurately maintain that reaction temperature for several hours. This is a difficult goal to accomplish, because the dynamics of the controlled process will go through a substantial change as the heat load first drops to zero and then as the cooling load gradually evolves when the reaction is started. The master temperature controller is usually a three-mode one, tuned for the exothermic phase of the reaction cycle. It might have a 30-percent proportional band and 5 minutes set for both integral and derivative. If such a controller was kept on automatic during heat-up, a substantial temperature overshoot would result (figure 3.18) because of reset wind-up. Therefore, the conventional PID controller must be supplemented by added features to provide it with the required start-up characteristics. The added feature is either the "batch unit" or the "dual mode unit," depending on the proportional band. If it is less than 50 percent, the batch unit will give good results, whereas if a proportional band wider than 50 percent is required, the dual-mode unit will be more effective.

The Batch Unit

Without the batch unit, the PI controller illustrated in figure 3.19 would receive a feedback signal (F) equaling the output (O). Therefore, whenever there is an error (E), the output signal is driven continuously by the positive feedback through I (a first-

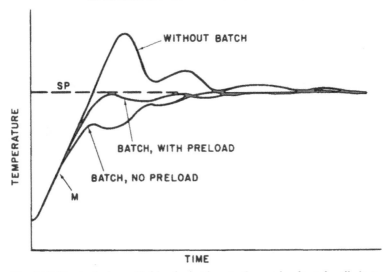

Fig. 3.18 The preload provided by the batch unit, if properly adjusted, will eliminate overshoot. (From Shinskey [15])

order lag, having a time constant I) until B reaches the saturation limit. Once in this saturated state (the reset is wound up), the output signal "O" will equal "B" even if the error has returned to zero. This is the reason that in figure 3.18 the temperature keeps rising even after it has reached set point (SP = M, E = O). Without the batch unit, therefore, control action cannot begin until an equal and opposite area of error is experienced. This is why reset wind-up always results in overshoot and why this wind-up must be eliminated by the addition of the batch unit shown in figure 3.19.

Under normal operation, the output to the valve is below the high limit. Therefore, G is positive and the amplifier drives D upward, which causes O to be less than C. In this state, the low selector selects O as the feedback signal, and the controller behaves as a conventional PI controller.

When O exceeds HL, the amplifier drives down D, C, F, and B and thereby limits O from exceeding the HL setting.

It is also necessary to provide a low limit (called preload) to the feedback signal; otherwise, the opposite of an overshoot would be experienced—an excessively sluggish approach to set point, as shown in figure 3.18 by the "no preload" curve. If the PL setting did not prevent the feedback (F) from dropping too low at times of high error (such as at the beginning of heat-up), B could saturate at the low limit, keeping output (O) below zero even when the measurement has returned to set point. With preload, the controller output "O" will equal PL when the error is zero.

It can be seen from the above that a PID-type batch controller requires a total of five adjustments, because HL and PL must be set and the three control modes must be tuned. HL should be set at the maximum allowable jacket water temperature, which would then eliminate the need for a separate limit, such as the HIC in figure 3.9.

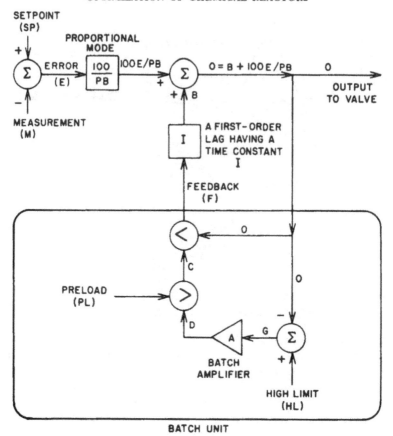

Fig. 3.19 The batch unit disables the integral mode until the error is nearly zero. (Adapted from Shinskey [15])

The correct setting for PL is the master controller output at that time when reaction has started and a steady state has been reached between the generation and the removal of the heat of reaction. If, for example, the jacket water temperature during steady state is 90°F, this value could be selected as the preload setting, which will be the output of the master (and the set point of the slave) when the reaction temperature has been reached. Actually, the PL setting should be a few degrees lower than this value, say 87°F or 88°F, to allow for the contribution of the integral action of the controller from the time the proportional band is entered to the time when the set point is reached. This is illustrated in figure 3.20.

Dual-Mode Unit

The effectiveness of the batch unit, described earlier, is lost when the reactor requires a wide proportional band, say in excess of 50 percent. With a wide band (as shown in figure 3.20) reset action would begin much earlier, which would lengthen the heat-up time. In such a case, the dual-mode unit (figure 3.21) is the proper selection.

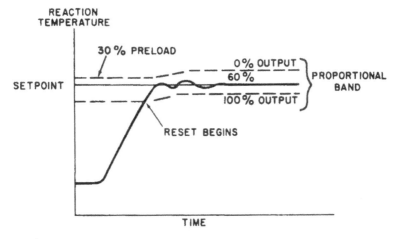

100 % OUTPUT MEANS FULL HEATING
0 % OUTPUT MEANS FULL COOLING
30 % OUTPUT MIGHT CORRESPOND TO THE NORMAL LEVEL OF COOLING
AT STEADY STATE

Fig. 3.20 If preload is correctly set, overshoot will be eliminated and heat-up
time will still be held at a minimum. (Adapted from Shinskey [15])

In the dual-mode unit, the preload is estimated as in case of the batch unit, but
it is not reduced for integral correction (not lowered from 90°F to 87 or 88°F in our
example), because in this case reset does not begin until the error is zero.

The sequence of operation is as follows:

1. Full heating is applied until the reactor is within 1 or 2 percent of its set point
   temperature. This margin is set by Em. During this state SS-1 and SS-2 are in
   position "A."
2. When E drops to Em, time delays TD-1&2 are started, and full cooling is
   applied to the reactor for a minute or so to remove the thermal inertia of the
   heat-up phase. When TD-1 times out the period required for full cooling,
   SS-1 switches to position "B" and the PID controller output is sent to the slave
   as set point. This output is fixed at the preload (PL) setting, which corresponds
   to the steady state jacket temperature (estimated in our example as 90°F).
3. When the error and its rate of change are both zero, estimated by TD-2, this
   time delay will switch SS-2 to position "B." This switching also transfers the
   PID loop from manual to automatic, with its external feedback loop closed.

If properly tuned, the dual-mode unit is the best possible controller, because by
definition, optimal switching is unmatched in the unsteady state by any other technique
[15]. On the other hand, this loop requires seven settings. Three of these—P, I, and
D—pertain only to the steady state of the process; the other four—PL, Em, TD-1,
and TD-2—will determine start-up performance. The effect of these adjustments is
self-evident:

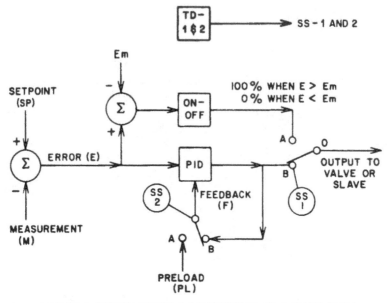

Em = A MINIMUM ERROR SETTING CORRESPONDING TO A STATE WHEN
MEASUREMENT HAS APPROACHED SETPOINT TO WITHIN 1 OR 2 %.

Fig. 3.21 The dual-mode controller sequences full heating, full cooling, and finally preloaded PID for optimum start-up of potentially unstable batch reactors.

Em should be increased in case of overshoot and lowered if undershoot is experienced.

PL has the same effect as in figure 3.18.

TD-1, if set too long, will bring the temperature down after the setpoint is reached.
TD-2 is not very critical.

From the start-up performance of the reactor, it can be determined which setting needs adjustment, as shown in figure 3.22.

Rate of Temperature Rise Constraint

In highly unstable, accident-prone reactors that have a history of runaway reactions, an added level of protection can be provided, based on the permissible rate of temperature rise during heat-up. This is usually superimposed on the previously discussed control systems on a selective basis as a safety back-up. The protection is provided by calculating the actual rate of temperature rise in units of °F/minute and sending that signal as the measurement to RRIC in figure 3.23. The set point of this controller is programmed as a function of the heat-up state. Toward the end of the heat-up period, when the reaction temperature has been almost reached (the error is near zero), the value of the permissible rate of rise is set to be much lower than at the beginning of

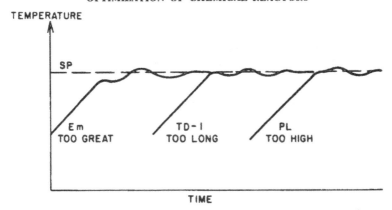

Fig. 3.22 It is possible to diagnose the dual-mode loop tuning constants based on start-up performance. (From Shinskey [15])

heat-up. Doing so prevents the process from building up a thermal inertia as it approaches the region of potential instability.

Heat Release Determination and Control

By multiplying the jacket circulation rate by the difference in temperature between inlet and outlet, it is possible to determine the amount of heat released or taken up by the reaction. The reaction is in an endothermic state when the value of Q in figure 3.24 is negative, and it is in an exothermic phase when Q is positive.

Under steady state conditions, at high circulation rates the difference between To and Ti is about 5°F or less. The method of reading this ΔT is critical, because during

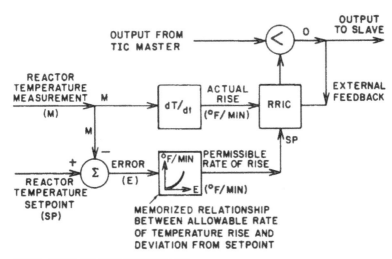

Fig. 3.23 The rate of temperature rise constraint can be applied as added protection.

load changes, dynamic compensation is needed. As more cooling is requested, the jacket inlet temperature will drop, and the jacket outlet temperature will stay unaffected until the jacket contents are displaced. If Q is to be used for control purposes, such dynamic errors must be removed by compensation. This is accomplished by delaying the jacket inlet temperature into the ΔT transmitter by a time equal to the delay through the jacket [15]. In figure 3.24, this is accomplished by simulating the jacket using a length of tubing whose dead time is adjustable by changing the flow rate. A similar result can be obtained by placing an inverse derivative relay, set for a jacket displacement time of, say, 30 seconds, into the transmitter output signal; the desired goal can also be reached by electronically simulating the transportation lag in digital systems.

The correct selection of the ΔT transmitter is very important, because in order to be able to detect Q within a ± 1-percent error, the ΔT must be detected within $\pm 0.02°F$. This can be accomplished only with the best RTD type transmitters (Rosemount model 444RD9/788542 or equal), with spans as narrow as -5 to 0 to $+5°F$. When the absolute value of the jacket temperature varies from, say, $50°F$ to $250°F$ throughout the reaction cycle, it will also be necessary to correct the calibration of the ΔT transmitter as the absolute temperature changes.

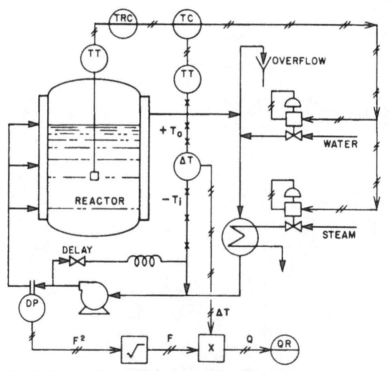

Fig. 3.24 To correctly determine heat release, the jacket inlet temperature reading must be delayed by the residence time of the jacket, so that the ΔT transmitter will be dynamically compensated. (From Shinskey [15])

Once the value of Q is accurately determined, it can be used for many purposes. The instantaneous value of Q signals the rate of heat removal or addition, and the time integral of Q gives the total heat that has been added or removed. A change in the value of ΔT under standard conditions can signal fouling and the need for cleaning the reactor heat transfer surfaces.

Stripping, Refluxing or Percentage Conversion

During stripping or refluxing phases, the reactor might be controlled on the basis of heat input (Q), because refluxing tends to be done at constant temperature and the increase in temperature during stripping is usually too small for control purposes. Therefore, the system would be switched automatically from temperature control (figure 3.24) to heat input control (figure 3.25) whenever the reactor enters a stripping or

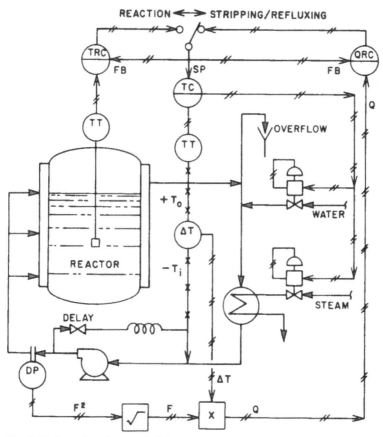

Fig. 3.25 Control can be switched from temperature to heat input as the reaction cycle enters a stripping or refluxing phase.

refluxing phase. The heat input during refluxing is usually set to be sufficient to maintain a state of slow boiling. During stripping, the heat input is usually set to complete the stripping in some empirically established time period, for example, an hour or two.

The time integral of Q represents the total reaction heat, which is an indicator of product concentration or percentage conversion. It can be used to introduce additives at predetermined conversions and to determine reaction end point. Through these automated steps, the need for taking grab samples can be eliminated, resulting in reduced overall cycle time and, therefore, increased production rate.

Pressure Compensated Temperature Control

In large polymerization reactors having low heat transfer coefficients and large changes in heat evolution, the conventional temperature cascade loop is not fast enough. On the other hand, pressure measurement gives an almost instantaneous indication of changes in temperature.

Figure 3.26 illustrates the application of a pressure compensated temperature control system to a reactor with both jacket and overhead condenser cooling (U.S. patent no. 3,708658 of January 2, 1973, assigned to Byrd Hopkins of the Monsanto Company). This same approach can also be applied to reactors with jacket cooling only. Under steady state conditions the reactor temperature (Tm) is on set point, set by the cam programmer, and therefore the calculated temperature $Tc = Tm$. When an upset occurs, the pressure transmitter (PT) will detect it first, causing the calculated temperature (Tc) to change as the AP part of the expression is changed. This will make the measurement of TC-1 much faster than it otherwise would have been; it also allows the overhead condenser to start removing the excess heat even before the temperature transmitter (TT-1) is able to detect it. After each dynamic upset, the PI controller slowly returns the calculated temperature (Tc) to equal the measured temperature (Tm). This then automatically reestablishes the correct pressure-temperature relationship as the composition in the reactor changes.

The net result is the ability to operate the reactor at a much higher reaction rate, thereby obtaining higher productivity than was possible with temperature control alone.

TC-1 in figure 3.26 would normally be provided with proportional and derivative control modes only. When TC-1 is on set point, the output signal returns to a value set by an adjustable internal bias. If the preferred means of cooling is through the jacket, then this bias will be set to a low value, but not to zero.

If there is only a single manipulated variable, the control system would be configured as in figure 3.27. The control modes are distributed between the measured variables so that integral action will act on the slowly responding temperature while derivative action is applied to the more sensitive pressure [14]. The derivative time setting will be much shorter than the integral, because pressure responds faster than temperature. Integral action cannot be applied to the pressure measurement, because the pressure can vary even under constant temperature conditions (as a result of variation in feed composition or catalyst activity), and the intent here is to use the pressure loop *only in the unsteady state*. Naturally, integral action is applied to the temperature

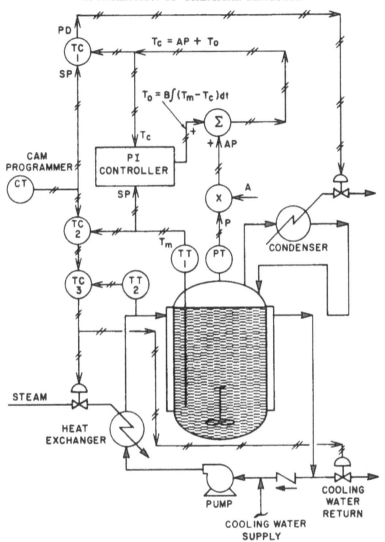

Fig. 3.26 Pressure compensated temperature control can be used for improved speed of response. (Source: Adapted from Byrd Hopkins, U.S. patent no. 3,708658)

measurement signal, because it is steady state temperature that determines product quality, and integral action will assure its return to set point.

Pressure Control

In gas phase reactions, in oxidation and hydrogenation reactions, or in high-pressure polymerization, the reaction rate is also a function of pressure. If, in a batch reaction,

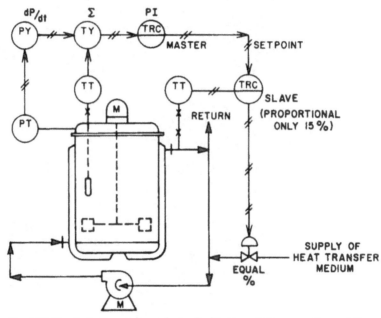

Fig. 3.27 The derivative mode can be applied to the quickly responding variable, and the integral mode can be applied to the slowly responding variable.

the process gas is completely absorbed, the controls in figure 3.28 would apply. Here, the concentration of process gas in the reactants is related to the partial pressure of the process gas in the vapor space. Therefore, pressure control results in the control of reaction rate. This loop is fast and easily controlled.

Certain reactions not only absorb the process gas feed but also generate by-product gases. Such a process might involve the formation of carbon dioxide in an oxidation

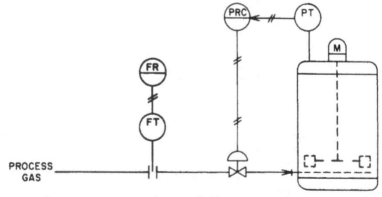

Fig. 3.28 Reactor pressure can be controlled by gas make-up. (From Lipták [9])

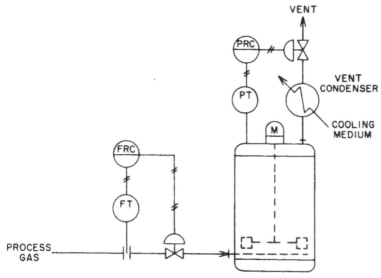

Fig. 3.29 Reactor pressure can be controlled by vent throttling.

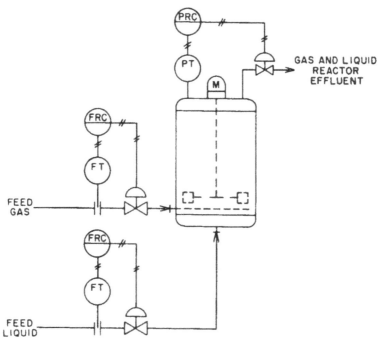

Fig. 3.30 Arrangements such as this can be used to control pressure in continuous reactors. (From Lipták [9])

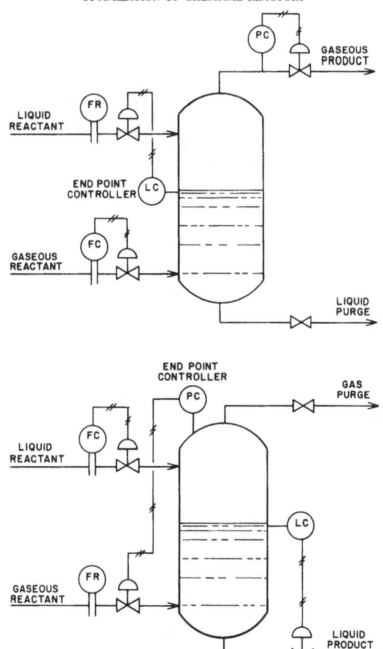

Fig. 3.31 Inventory control may be used with multiphase reactants. (From Shinskey [15])

reaction. Figure 3.29 illustrates the corresponding pressure control system. Here, the process gas feed to the reactor is on flow control, and reactor pressure is maintained by throttling a gas vent line. This particular illustration also shows a vent condenser, which is used to minimize the loss of reactor products through the vent.

In case of continuous reactors, a system such as shown in figure 3.30 is often employed. Here the reactor is liquid full, and both the reactor liquid and any unreacted or resultant by-product gases are relieved through the same outlet line. Reactor pressure is sensed, and the overflow from the reactor is throttled to maintain the desired operating pressure. Process gas feed and process liquid feed streams are on flow control.

All these illustrations are simplified. For instance, it may be desirable to place one of the flow controllers on ratio in order to maintain a constant relationship between feed streams. If the reaction is hazardous, and there is the possibility of an explosion in the reactor (for example, oxidation of hydrocarbons), it may be desirable to add safety devices, such as a high-pressure switch, to stop the feed to the reactor automatically.

If one of the reactants differs in phase from both the other reactants and the product, inventory control can be applied (figure 3.31). In the case of gaseous products, reactor level is controlled by modulating the liquid reactant; with liquid products, the gaseous reactant is modulated to keep reactor pressure constant. A purge is needed in both cases to rid the reactors of inerts either in the liquid or in the gas phase.

Vacuum Optimization

Some reactions must be conducted under vacuum. The vacuum source is frequently a steam jet type ejector. Such units are essentially venturi nozzles with very little turndown. This constant capacity vacuum source is frequently matched to the variable capacity reactor by wasting the excess capacity of the ejector. This is done by creating an artificial load through the admission of ambient air. When the vapor generation rate in the reactor is low, the steam jet is still operating at full capacity, sucking in and ejecting ambient air. The corresponding waste of steam can be substantially reduced if the system shown in figure 3.32 is applied. Here, a low load condition is detected by the fact that the air valve is more than 50-percent open, and this automatically transfers the system to the "small" jet. A later increase in load is detected as the air valve closes (at around 5 percent), at which point the system is switched back to the high capacity jet.

Product Quality Control

Measuring the composition of the contents of a chemical reactor is a method that can help maintain product quality in continuous reactors and can detect end points in batch reactors. The sensors for composition measurement, called analyzers, are limited by sampling difficulties, by the intermittent nature of some models, and by incomplete mixing within the reactor. Because sampling and sample preparation result in dead time, sampling time should be minimized or eliminated altogether; this can be done by placing the sensor directly into the reactor, which results in tight control and fast response.

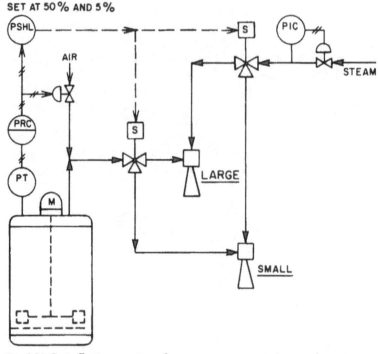

Fig. 3.32 Cost-effective operation of vacuum pressure controls can reduce steam waste substantially. (From Lipták [9])

Because analyzers in general are low-reliability and high-maintenance devices, most users are reluctant to close an automatic loop around them. The main concern is that the failure of the analyzer might cause a hazardous condition by driving the reactor into an unsafe state. As shown in figure 3.33, this concern can be alleviated by the use of redundant sensors through a high signal selector [15]. In case of a "downscale" failure of one of the analyzers, the back-up unit takes over automatically. An "upscale" failure

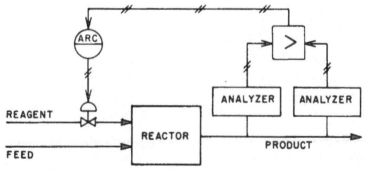

Fig. 3.33 The reactor must be protected from the consequences of analyzer failure. (From Shinskey [15])

in this arrangement would shut down the reactor, providing safety at the cost of production. If neither an accident nor an interruption of production can be tolerated, three analyzers are required, arranged in a "voting" configuration. In such an arrangement, if one of the sensors disagrees with the others, it is automatically disregarded and the "majority view" is used for a closed loop control. Maintenance and recalibration is then initiated for the defective sensor. Another solution is to use a median selector in combination with the three analyzers.

Residence Time Control

In continuous reactors it is usually desirable to provide a constant residence time for the reaction. If the production rate (inflow to the reactor) varies, the most convenient way of keeping the residence time constant is to maintain a constant ratio between reactor volume (V) and reactor outflow (F). The ratio V/F is the residence time, where the volume (V) can be approximated by a level measurement, and F can be measured as outflow. Figure 3.34 illustrates a residence time control system using this approach.

End Point Control

In batch reactors, if the completion of the reaction cannot be reliably detected, it is common practice to terminate reactant flow on the basis of total charge. This can be based on flow or weight measurement. In other reactions, the end point is signaled by the fall or rise of reaction pressure, which triggers a pressure switch or pressure controller to shut off the reactant flow (figure 3.31). In still other reactions, the total reaction heat signals the end point, as is illustrated in figures 3.24 and 3.25.

If some analytical property, for example, pH, is used in detecting the end point, it is important to realize that during most of the reaction the measurement will be away from set point. Therefore, integral action must not be used, because it would saturate and overshoot would result. This is one of the few processes in which the proper selection is a proportional plus derivative controller with zero bias, modulating an equal percentage reagent valve. If the derivative time is set correctly, the valve will shut quickly as the desired end-point pH is reached. With excessively long derivative time setting, the valve will close prematurely (figure 3.35), but this is not harmful because it will reopen as long as the pH is away from the set point.

If one reactant is charged to the reactor in substantial excess to the others, this will usually guarantee the complete conversion of the other reactants, and end point control is not required, unless used on the recycle stream.

If the reactants are fed to the reactor in the same proportion in which they react, end point control is required. If the reactants and product are all in the same phase (all liquids or all gases) the end point is usually determined through the use of an analyzer, such as the pH unit, which was described in connection with figure 3.35.

On the other hand, when the reactants are in different phases, the end point can be controlled on the basis of level or pressure. As was shown in figure 3.31, if the product is gas, the end point is maintained by the level controller, whereas if the product is liquid, the end point control is based on pressure. When this pressure rises, that is an indication of excess gaseous reagent accumulation in the reactor. In response,

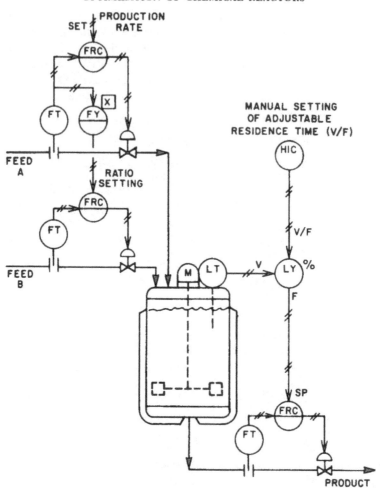

Fig. 3.34 Residence time control is used in chemical reactors to maintain a constant ratio between reactor volume and reactor outflow.

the pressure controller (PC) will increase the liquid reagent concentration in the reactor by increasing its flow. This in turn will increase gas consumption, and the balance will be reestablished and the correct end point maintained.

All reactions with a liquid product can benefit from residence time control. This requires the combination of the features in the control system shown in figure 3.31 and 3.34, resulting in figure 3.36.

Analyzers Used for End Point Detection

Probe-type analyzers do not require sampling systems, because the analyzer itself is inserted directly into the process. In the case of chemical reactors, analyzers can be

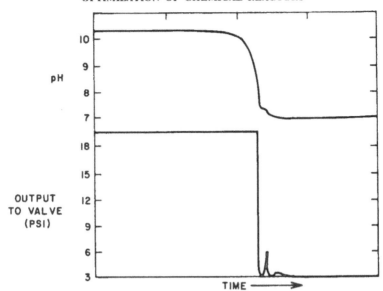

Fig. 3.35 The end point of a batch reaction is controlled with a proportional plus derivative controller and an equal percentage valve. (From Shinskey [15])

inserted either into the reactor vessel or into the discharge piping. Regardless of their locations, probes will give accurate readings only when clean. Therefore, automatic cleaning of such on-stream analyzer probes is a good idea. Figure 3.37 illustrates a flow-through, brush-type probe cleaner assembly that is provided with a sight glass allowing visual observation of the cleanliness of the probe. Such units are also available with spray nozzle attachments for water or chemical cleaning of the electrode.

One common method of end point detection is the measurement of viscosity either by monitoring the torque on the agitator or by using viscosity sensors [8]. Such viscometers can be inserted in pipelines or directly into the reactor (see figure 3.38). It is essential that they be located in a representative area and be temperature-compensated.

In reactions in which the end point can be correlated to density, some of the more sensitive and less maintenance prone densitometers [8] can also be considered. The radiation-type design tends to meet these requirements if pipeline installation is acceptable. These units can operate with a minimum span of 0.05 specific gravity unit and an error of ± 1 percent of that span, or an absolute error of ± 0.0005 Sp. G.

In a number of reactions, refractometers [8] have been found to be acceptable solutions to end point detection. In many installations, they are used as laboratory devices: a grab sample is brought to the bench analyzer. In other installations, they are used "on-stream." When the RI analyzer is on-stream, human error tends to be reduced and analysis time is shortened, with a corresponding increase in productivity. On the other hand, an on-stream installation requires more maintenance attention,

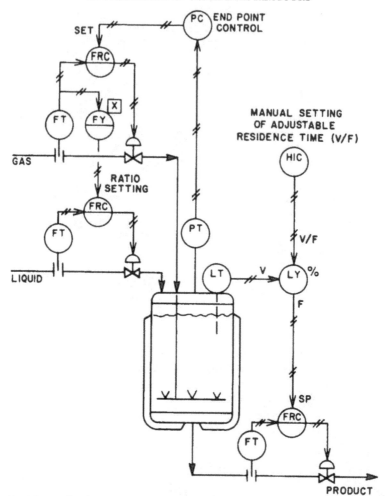

Fig. 3.36 End point control is based on pressure, whereas residence time control is based on level, when the product is in the liquid phase.

because the RI measurement must be made to high precision, such as ±0.0005 RI units (figure 3.39). This requires accurate temperature compensation, plus protection through cooling of the heat-sensitive photocells, which are time-consuming and expensive to replace when ruined by high temperature.

If the end point of a reaction is detectable by such measurements as a change in color, opacity, or the concentration of suspended solids [8], the self-cleaning probe shown in figure 3.40 can be considered for use. Here the reciprocating piston in the inner cavity of the probe not only serves to clean the optical surfaces through its wiping seals but also guarantees the replacement of the sample in the sample chamber by fresh

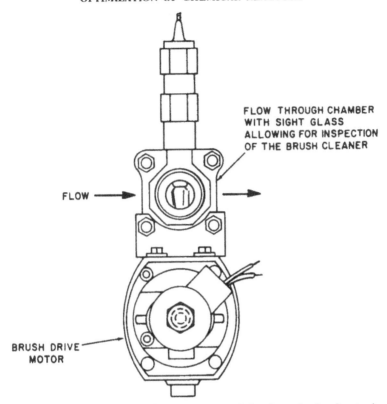

FLOW THROUGH CHAMBER
WITH SIGHT GLASS
ALLOWING FOR INSPECTION
OF THE BRUSH CLEANER

FLOW

BRUSH DRIVE
MOTOR

Fig. 3.37 The flow-through probe cleaner is provided with a sight-glass for visual observation. (Courtesy of Amico)

material upon each stroking of the piston. The temperature considerations and limitations mentioned in connection with the RI analyzer also apply here.

If the end point of the reaction is detectable by infrared beam attenuation, such analyzers can also be put on-stream. The crystal probe illustrated in figure 3.41 can be inserted in pipelines and will detect the attenuated total reflectance of the IR beam. It is particularly suitable for the detection of water in hydrocarbons.

Reactor Safety

Safety problems can arise in chemical reactors as a result of many causes, including equipment failure, human error, loss of utilities, or instrument failures. Depending on the nature of the problem, the proper response can be to "hold" the reaction sequence until the problem has been cleared or to initiate an orderly emergency "shutdown" sequence. Such actions can be taken manually or automatically and might consist of the types of steps described in figure 3.42.

After a "hold" or "emergency" condition has been cleared, an orderly sequence of

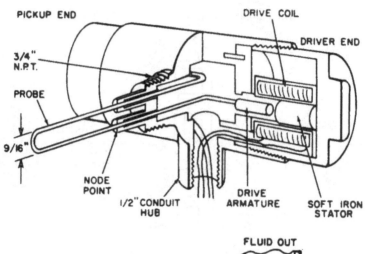

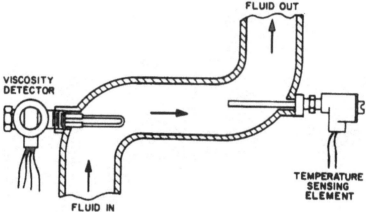

Fig. 3.38 This temperature-compensated, vibrating-reed type of viscosity detector can be inserted in pipelines or directly into the reactor. (Courtesy of Automation Products)

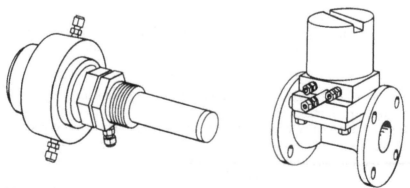

Fig. 3.39 On-stream refractometers can be used in both probe and flow-through configurations. (Courtesy of ANACON, Inc.)

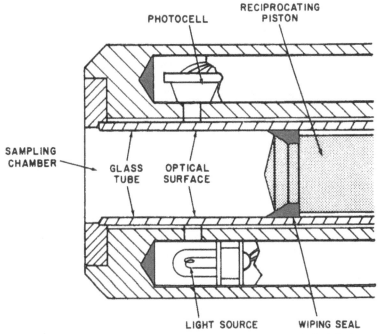

Fig. 3.40 Cross-sectional view of a self-cleaning opacity or suspended solids detector probe. (Courtesy of Biospherics, Inc.)

transitional logic is required to return the reactor to normal operation (figure 3.43). It makes no difference whether the emergency was caused by a pump or valve failure or whether the reactor was put on hold to allow for manual sampling and laboratory analysis of the product: a return sequence is still required. The reentry logic determines the process state when the interruption occurred and then decides whether to return to that process state or the previous one in order to reestablish the conditions that existed at the time of the interruption.

A reactor control system should provide the following features [2]:

- The ability to maximize production
- The ability to minimize shutdowns
- The ability to maximize on-line availability of the reactor
- The ability to minimize the variations in utility and raw material demand
- The ability to provide smooth operation in terms of constant conversion, yield, and product distribution
- Easy start-up and shutdown

However, the overriding, primary design objective is that the reactor must be safe for both the operating personnel and the environment.

Safety is guaranteed by monitoring both the potential causes of emergencies and

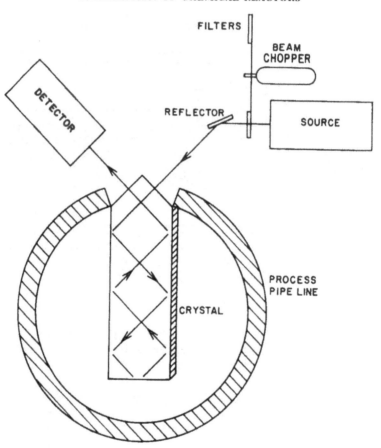

Fig. 3.41 A probe-type IR detector can be used on-stream if the end point of the reaction is detectable by infrared beam attenuation. (Courtesy of Sieger Ltd.)

their consequences. If the cause is detected and responded to in time, the symptoms of an emergency will never develop; in that sense, the monitoring of potential causes of safety hazards is preferred. Some of the potential causes of emergencies and the methods used to sense them are listed below:

Causes	Methods of Detection (Lipták [8])
Electric power failure	Volt or watt meters
Instrument air failure	Pressure switch
Coolant failure	Pressure or flow switches
Reactant flow failure	Pressure or flow switches
Catalyst flow failure	Pressure or flow switches
Loss of vacuum	Vacuum switch

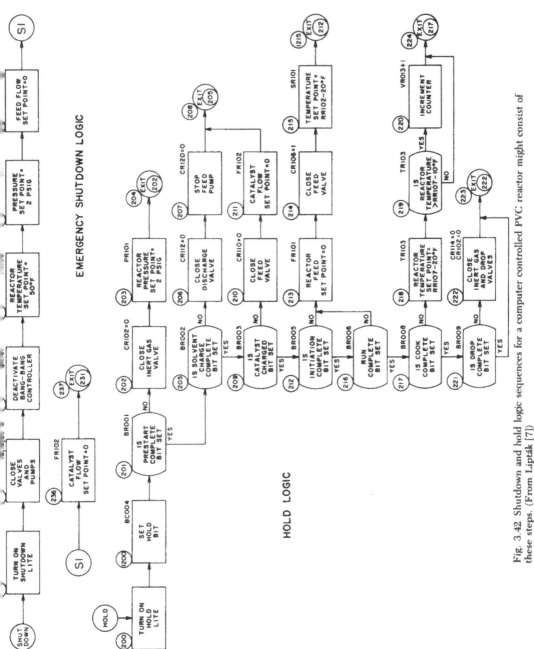

Fig. 3.42 Shutdown and hold logic sequences for a computer controlled PVC reactor might consist of these steps. (From Lipták [7])

133

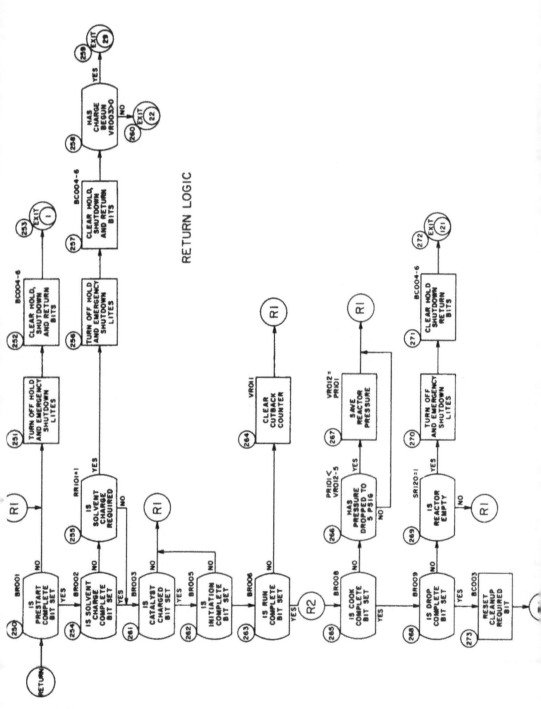

RETURN LOGIC

134

Causes	*Methods of Detection*
Agitator failure	Torque or RPM detector
Pump failure	Pressure or flow switch
Valve failure or mispositioning	Position sensing limit switches
Measuring device failures	Validity checks

Each of the above failures can actuate an annunciator point to warn the operator and can also initiate a "hold" or "shutdown" sequence, if the condition cannot be corrected automatically. Automatic correction will require different responses in each case.

For example, in case of electric power failure, a diesel generator or steam turbine might be started. If instrument air failure is detected, an alternate compressor might be started, or an alternate gas, such as dry nitrogen, might be used. In addition, the safe failure positions for all control valves must be predetermined, as in figure 3.10. In case of loss of coolant, an alternate source might be used, or the type of automatic constraint control that was shown in figure 3.13 might be applied.

If reactant or catalyst flow fails, automatic ratio loops (figure 3.13) can cut back the flow rates of related streams; alternatively, if there are alternate storage tanks from which the reactant can be drawn, these tanks might be accessed. A loss of vacuum can also be corrected by switching to an alternate vacuum source if such redundancy is economically justifiable. Critical pumps and valves can be provided with spare back-up units, and agitators can be furnished with multiple drives. Failed sensors and defective instruments can be automatically replaced by redundant and voting systems, as was explained in connection with figures 3.16, 3.17, and 3.33.

If the cause of an emergency goes undetected, eventually it will affect the operating conditions of the reactor. These symptoms might involve temperature, pressure, composition, or other conditions as listed below:

Symptom	*Detectors*
High reactor temperature	High temperature switch
High jacket temperature	High temperature switch
Low jacket temperature	Low temperature switch
High reaction rate	High heat release detector
Reaction failure	Low heat release detector
High rate of temperature rise	Rate of rise detector
High pressure	High pressure switch and relief valve
Abnormal level	Level and weight switches
Abnormal composition	On-stream analyzers

The control system will respond to the symptom of an abnormal process variable much as it did to the previously discussed failure causes, but here the probability of an emergency shutdown is higher. For example, if, in a potentially unstable reactor, the temperature or pressure has exceeded its high limit, it is possible to make sure that the reactor is on full cooling and that the reactant feeds are shut off, but after that little else can be done but to take drastic action to prevent a runaway reaction. Such drastic action might include depressurization or even the transferring of the reactor contents into a blowdown tank.

The consequences of high jacket temperature are less serious, and the limit shown in figure 3.9 is usually sufficient to prevent the formation of hot spots. In some processes, the critical consideration is low jacket temperature [2], because that can cause "frosting" or "freezing" of the reactants onto the inner wall of the reactor. As the frozen layer thickens, heat transfer is blocked and the reactor must be shut down. If used on such a process, the control system in figure 3.9 would also be provided with a low limit on the slave set point.

When the reaction rate suddenly rises or when it approaches the capacity of the cooling system, it is necessary either to increase the heat removal capacity quickly (figures 3.15 and 3.23) or to cut back on the reactant flows (figure 3.13). It is also necessary to detect if the reaction has failed by comparing the reactant feed rate with the heat removal rate (figure 3.13) or by monitoring heat release (figure 3.24). In yet other processes, it is desirable to limit the rate of temperature rise (figure 3.23).

Instrument Reliability

A risk analysis requires data on the failure rate for the components of the control system. Such data has been collected from user reports; the findings of one such survey are shown in table 3.1. When the MTBF of the individual instrument components has been estimated, the MTBF of the various control loops can be established [13]. An example of this technique is illustrated in figure 3.44. Once the MTBF of each loop has been established, the reliability of the whole reactor control system can be increased by increasing the MTBF of the highest risk loops. This usually is done through self-diagnostics; preventive maintenance; the use of back-up devices; and, in critical instances, the technique of "voting," described in connection with figure 3.17.

While risk analysis is complex and time-consuming, a plant can be designed for a particular level of safety, just as it can be designed for a particular level of production. The reliability of the result is as good as the data used in the analysis. For this reason data collected by users, testing laboratories, or insurance companies should be used, not manufacturers' estimates. For instrument reliability and performance data, one good source is the International Association of Instrument Users, SIREP-WIB.

Sequencing Logic Controls

A chemical reactor control system is a combination of continuous PID-type analog and sequential logic control functions. The various analog loops used in reactor control have already been discussed. The logic controls serve to initiate the various process states in logical sequence, as shown in figure 3.45. A process state is a distinct phase of operations in a batch cycle, such as charge, reaction initiation, or run. When a system is returning to the normal control state, the reentry logic determines what process state the system was in at the time it went into hold or shutdown. It may be necessary to

Table 3.1
FAILURE RATE FOR CONTROL SYSTEM
COMPONENTS

Variable	Instrument	Mean Time Between Failures
Level	Bubbler	1–2 yrs.
	d/p transmitter	1–5 yrs.
	Float & cable	0.2–2 yrs.
	Optical	0.1–5 yrs.
Flow	Flume & weir	0.5–5 yrs.
	Venturi, etc.	2 mo.–5 yrs.
	Propellers	1 mo.–1 yr.
	Positive displacement	1 mo.–1 yr.
	Magnetic	0.5–10 yrs.
Density	Nuclear	1–3 yrs.
	Mechanical	1–6 mos.
Analysis	pH and ORP	1–4 mos.
	Dissolved O_2	1–9 mos.
	Turbidity	1–6 mos.
	Conductivity	1–4 mos.
	Chlorine gas	0.5–1 yr.
	Explosive gas	0.2–1 yr.
	TOC	0.1–1 mo.
Miscellaneous	Temperature	0.5–2 yrs.
	Pressure	0.1–5 yrs.
	Speed	0.6–5 yrs.
	Weight	0.6–2 yrs.
	Position	0.1–1 yr.
	Sampling	0.1–1 yr.

Adapted from Molvar [12].

return to an earlier process state to reestablish the conditions that existed at the time of interruption. For instance, if the system was in hold and the reactor contents were allowed to cool below the run temperature, the system must return to the heat-up state and repeat a series of steps designed to achieve run conditions before reinitializing the reaction.

The sequence logic can be reduced to a time-ordered combination of a few basic actions, which are the components of the process and control states described in figure 3.45. These actions are listed below:

1. Operate on-off devices (pumps, valves, etc.)
2. Activate or deactivate control loops
3. Open and close cascade loops
4. Supply control loop parameters, such as set point, ramp rate, alarm limits, tuning parameters, and flow integrals

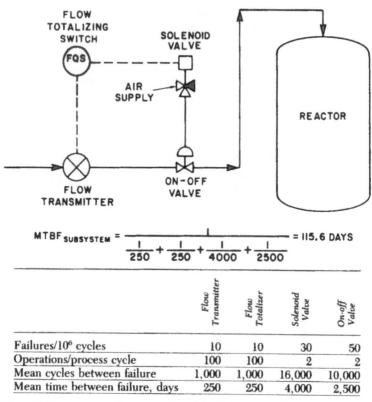

$$MTBF_{SUBSYSTEM} = \cfrac{1}{\cfrac{1}{250} + \cfrac{1}{250} + \cfrac{1}{4000} + \cfrac{1}{2500}} = 115.6 \text{ DAYS}$$

	Flow Transmitter	Flow Totalizer	Solenoid Valve	On-off Valve
Failures/10^6 cycles	10	10	30	50
Operations/process cycle	100	100	2	2
Mean cycles between failure	1,000	1,000	16,000	10,000
Mean time between failure, days	250	250	4,000	2,500

Fig. 3.44 The mean time between failure (MTBF) for a control loop can be calculated based on the MTBF of its components. (From Scott [13])

5. Integrate flows
6. Initiate times or delays between processing steps
7. Perform calculations
8. Compare values, measurements, and test indicators and branch according to the results of the comparisons
9. Initiate alarm status or operator messages
10. Release control if nothing more can be done until the next sequencing interval

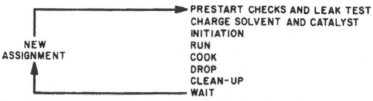

Fig. 3.45 Logic controls initiate the sequential states in a chemical reaction. (From Lipták [7])

It is important to note that the data acquisition and loop control functions are not included as part of the sequence logic. Rather, the sequence logic plays a supervisory role, communicating changes to the control loops as required, in the same way that an operator would adjust set points and alarm limits during the course of a batch.

Figure 3.46 gives a specific example of the above generalized description. The upper portion of the illustration shows the physical equipment involved, including the

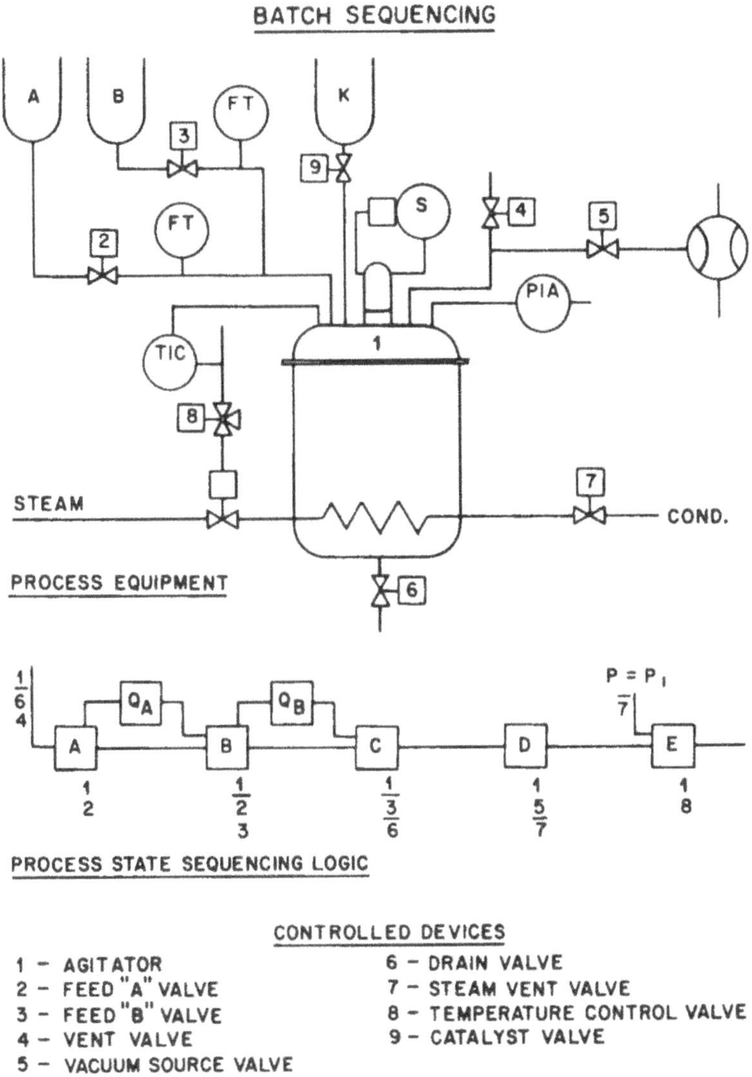

BATCH SEQUENCING

CONTROLLED DEVICES

1 — AGITATOR
2 — FEED "A" VALVE
3 — FEED "B" VALVE
4 — VENT VALVE
5 — VACUUM SOURCE VALVE

6 — DRAIN VALVE
7 — STEAM VENT VALVE
8 — TEMPERATURE CONTROL VALVE
9 — CATALYST VALVE

Fig. 3.46 In this batch reactor example of sequencing logic, the upper portion shows the physical equipment involved, and the lower portion shows the sequenced process states.

nine controlled devices whose states are being logically sequenced. The lower portion of the illustration identifies the sequenced process states. During process state "A" the required quantity of raw material "A" is being charged. The prerequisites for the initiation of state "A" are that 1 and 4 be on and 6 be off. This means that the agitator must be on, the vent valve must be open, and the drain valve must be closed. When these prerequisites are satisfied, state "A" is initiated by turning 2 on. This means that the charge valve of reactant "A" is opened and a control loop, such as that shown in figure 3.44, is activated. The required quantity of this ingredient is set by the recipe as Qa. When this target is reached, process state "A" is terminated and process state "B" is initiated, which then performs its own tasks defined by the sequencing logic. In a more sophisticated reactor control system, the number of controlled devices and the number of logical sequence steps is much greater, but the basic concept is the same.

Figure 3.47 shows process state "A," which is the logic for charging raw material "A" (solvent), in an actual PVC plant.

Time-Sequence Diagrams

When microprocessor-based distributed controls are used, the control system design follows the step-by-step approach described in figure 3.48. Of the six steps shown, step 2 is the most important. In step 2, all individual continuous (regulatory) and discrete functions are defined using adequate symbology or by developing control documents such as logic diagrams (ladder diagrams), boolean equations, and time sequence diagram(s), which must be referenced on the P&IC-D. All necessary calculations, such as material and heat balances and recipe corrections based on quality control data, should also be defined.

The time-sequence diagram shows the sequential changes in control actions and their time relationships. The sequence of events can be followed along dotted lines for vertical time coincidence and along the solid lines in a horizontal direction for sequence control.

The diamond symbol is used to indicate a trigger event, and a vertical dotted line indicates the time coincidence of trigger events. When two or more diamonds occur in the same relative time line, an "AND" logic condition is assumed.

The format of the time-sequence diagram proceeds from left to right in discrete time steps, with relative time being the horizontal coordinate.

Figure 3.49 illustrates a time-sequence diagram for a flow loop, such as the one shown in figure 3.44. First, the valve XV-1 is opened and the controller FIC-101 is placed in automatic (t_0). When XV-1 is open and FIC-101 is in "auto," the continuous task of flow integration is started (t_1). When the desired total is reached (t_2), XV-1 is closed. When XV-1 is closed (t_3), the integration of flow continues to check for leakage. If, at the end of a preset time interval, the total flow is still zero, FIC-101 is switched back to manual (t_4).

Figure 3.50 describes a control system in both ladder and time-sequence terminology. Here, when the start button is closed and XV-1 is open (its contact is closed), motor coil "M" is energized. When the coil is energized, its auxiliary "M-contact" closes. Once the "M-contact" is closed, it keeps the coil energized even if the start button contact reopens. If at any time the XV-1 valve would fail closed, its limit switch contact will open to deenergize the M coil. Once the coil is deenergized, the auxiliary contact "M" reopens.

In batch process units, the batch sequence is subdivided into *process states*. Each state is given a unique name, such as CHARGE, REACT, HEAT, COOL, HOLD, DISCHARGE, WASH, and EMPTY. Within each state, discrete control functions, continuous-regulatory control functions, and safety and permissive interlocks are performed. For example, figure 3.51 describes a time sequence diagram for the states of PREPARATION, PREMIX, CHARGING, HEAT, COOL, and DUMP.

In the area of computer simulation and control, substantial progress has been made in the last few years. Some of these advances are reported in references 4, 5, 6, 10, and 16.

Batch Process Control Definitions

Progress has also been made in developing a standardized language for batch process control. The International Purdue Workshop on Industrial Computer Systems has proposed the following terminology [1] for the precise definition, discussion, and solution of control problems:

Activity: The actual production of a batch in progress, resulting from a particular recipe, set of unit descriptors, and set of operator-entered data.

Batch: The product that is produced by one execution of a recipe.

Campaign: The total production run of one product, for example, for a single order or season, consisting of one or more lots.

Formula: The necessary data and procedural operations that define the distinct control requirements of a particular type or grade of product. For example, a formula might take the form of a list of parameters for control but could include modifiers to the procedure or its subordinate operations. (This definition allows the procedure for a recipe to be distinguished from its data.)

Lot: A collection of batches prepared using the same recipe. Typically, all batches of a lot are prepared from the same homogeneous source of raw material.

Operation: A major programmed processing action or set of related actions, normally consisting of one or more phases. Operations are naturally related to a distinct regime of production: for example, all processing carried out in one batch unit of a multiunit production line.

Phase: A sequence or collection of steps or actions associated with a state of processing with natural event boundaries. Normally, the phase boundaries represent points of process transition, hold, or activity. The boundaries define major milestones and possible points of safe intervention.

Procedure: A user-defined set of instructions whose purpose is to specify the order of execution of operations and related control functions necessary to carry out the plan for making a specific class of end products (as modified by the formula and unit descriptors). Typically, a procedure consists of a sequence of operations, but it may also include other computations.

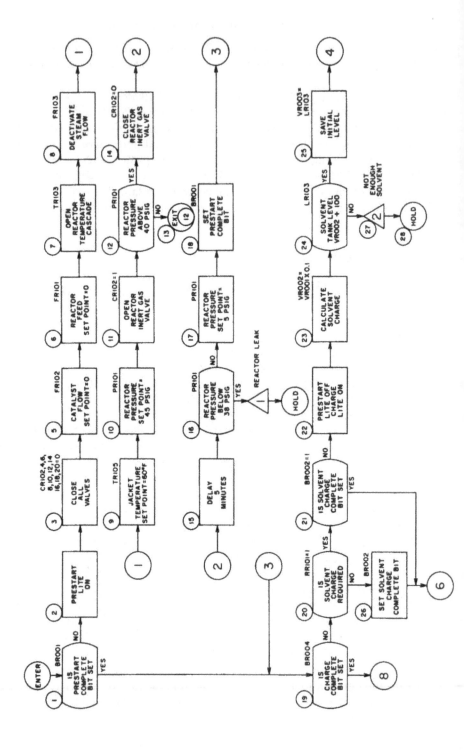

142

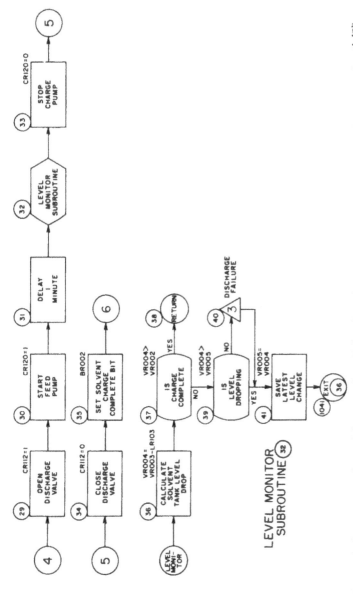

Fig. 3.47 The sequencing logic of charging solvent "A" is shown here as it occurs in an actual PVC plant. (From Lipták [7])

143

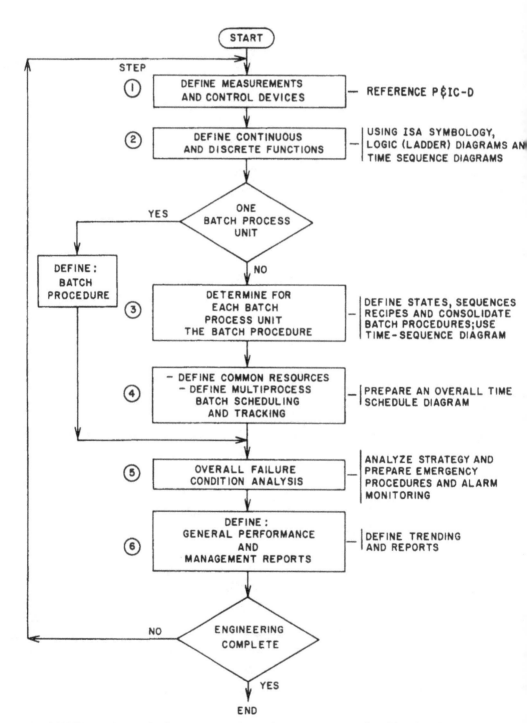

Fig. 3.48 This step-by-step batch process control is used in micro-processor-based distributed control systems. (From Lipták [9])

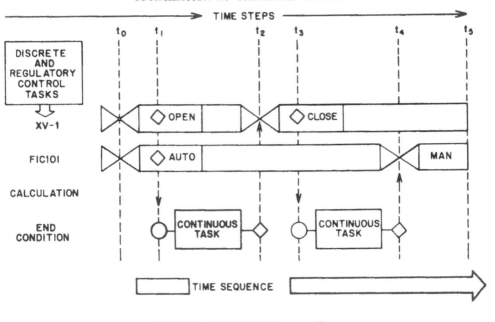

to   XV-1 IS OPENED AND FIC101 IS PLACED IN AUTO

t₁   IF XV-1 IS OPEN AND FIC101 IS IN AUTO THEN
      START END CONDITION CONTINUOUS TASK

t₂   XV-1 CLOSES AFTER END CONDITION MET

t₃   IF XV-1 CLOSED THEN START END CONDITION

t₄   FIC101 IS PUT IN MANUAL AFTER END CON-
      DITION MET

t₅   TIME SEQUENCE ENDS

◯   CONTINUOUS TASK START

◇   TRIGGER EVENT

Fig. 3.49 The time-sequence diagram proceeds from left to right in discrete time steps. (From Lipták [9])

Recipe: The complete set of data and operations that defines the control requirements of a particular type or grade of product. Specifically, the combination of procedure and formula.

Unit descriptor: A set of parameters related only to the equipment that particularize the quantities and identify the specific points or loops referenced generally in the procedure. The initial and final equipment states and values are included.

Step: The lowest-level term that describes an operator-observable event or action.

Batch Charging

Reactants and catalysts can be charged into the reactors either by weight or by volume, using flow meters, pumps, feeders, and weighing systems, as shown in table 3.2.

The main considerations in making the selection are accuracy and reliability of the

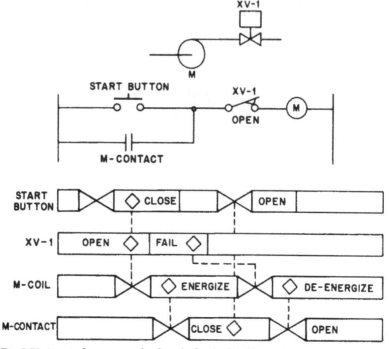

Fig. 3.50 A control system can be described in both ladder and time-sequence diagramming. (From Lipták [9])

data obtained. Because material balances are always based on the weight (not the volume) of the reactants, all things being equal, the mass or weight type sensors would be preferable; using such sensors would eliminate the need for density or temperature compensation. If they cannot be considered for some reason, the solid state volumetric sensors (vortex) would be preferred over the ones with moving parts (turbines, pumps, positive displacement meters), because sensors with moving parts tend to be high-maintenance devices, with frequent need for recalibration.

Therefore, in a new installation with no restrictions on hardware selection, it might be advisable to place the reactor on load cells and charge all ingredients that weigh more than 10 percent of the total batch by weight. Reactants and additives that are needed in lesser quantities can be added under mass flow meter control with the load cells providing only back-up. Catalysts or ingredients needed in extremely small quantities, representing less than 1 percent of the total batch, can be added by metering pumps or specialized feeders.

Control Equipment, Present and Future

In the last decade, manufacturers of shared, distributed controllers (which started out as microprocessor-based multiple PID controllers) have added more and more logic

BATCH PROCEDURES– TIME SEQUENCE DIAGRAM

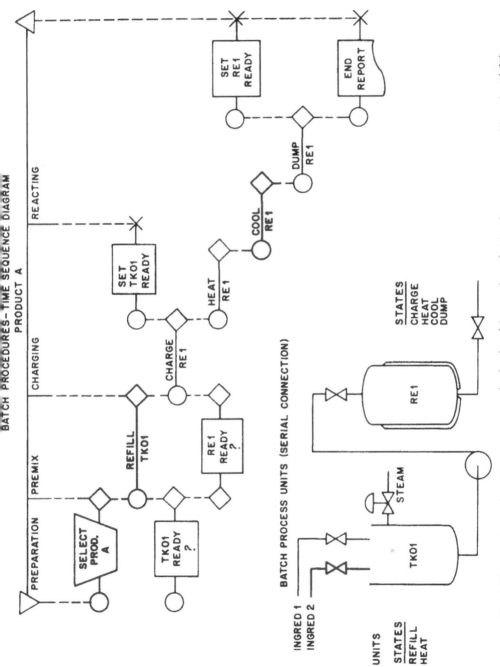

Fig. 3.51 A batch procedure time sequence can be developed for a multiproduct/single-stream process. (From Lipták [9])

147

Table 3.2

PERFORMANCE CAPABILITIES OF BATCH CHARGING INSTRUMENTS

Detector	Type	Min. Error	Rangeability
Load cell weighing	Mass	±0.1%FS	NL
Turbine type flow meters	Volumetric	±0.25%AF	10:1
Positive displacement flow meters	Volumetric	±0.25%AF	10:1
Metering pumps	Volumetric	±0.5%FS	20:1
Mass flow meters	Mass	±0.5%AF	20:1
Vortex shedding flow meters	Volumetric	±0.75%AF	10:1
Level	Volumetric	±0.25 inch	NL

AF – Actual flow
FS – Full scale
NL – No limitation

From Lipták [8].

capability to their products. At the same time, manufacturers of PLC's (which started out as microprocessor-based programmable logic controllers) added more and more PID capability to their products. The result today is the merging of these products into devices that are capable of handling both analog and logic sequencing control tasks. Although the control industry does not yet speak in these terms, a new era of process control has arrived: the era of the "unit controller."

The unit controller is capable of controlling a unit operation in the plant. That unit operation can be a reactor, a distillation tower, a compressor or any other subsystem. This represents a major step forward because it makes it necessary to stop thinking in terms of controlling single loops—such as pressures, flows, or temperatures—and to start thinking in multivariable terms, controlling the overall unit operation. This is a prudent and logical direction to take, because plants do not produce and sell pressures, flows, or temperatures: they sell a product. Therefore, the truly relevant control variable is maximum productivity. To move from the age of single to multivariable mentality, the unit controller is required. The microprocessor is needed to memorize the complex nature of the unit process in order for control to take place on this new and higher level.

The chemical reactor is obviously a natural candidate for control by unit controller. The number and types of input and output signals required and the types of control actions to be performed on them are within the capabilities of many microprocessor-based multiloop products on today's market. Therefore, the user can implement right now all that was discussed in this chapter by programming one of these multiloop controllers—that is, if the user is capable of generating the software required to configure the necessary analog loops (see figures 3.19 to 3.27) or logic sequencing loops (see figures 3.42 to 3.51). In other words, most of the multiloop controllers on today's market provide only the hardware and the capability for it to be programmed by the user. Depending on the manufacturer, these "empty boxes" are more or less "friendly" to the user, but they are "empty" in the same sense as the Atari video game is "uneducated" until a program for playing chess, tennis, etc. is plugged in.

From this perspective, today's process control industry is software-limited. A large variety of well-designed "empty" boxes are available, which are provided with a library of "low-level" software subroutines for PID, logic, and other functions. Unfortunately, with few exceptions, it is up to the user to educate this box to become a reactor controller or some other unit operation controller.

High-level software packages will be developed in the future to give "personality" to the previously empty boxes. One plug-in cassette or floppy disk might educate an empty box to become a distillation tower controller, and other disks could give it the personality of a dryer, evaporator, reactor, or other controller. It goes without saying that these high-level programs would also require adjustment to fit the universal unit controller to the particular process, but this "custom fit" would occur at a much higher level. The required user inputs could all be provided by process engineers dealing with recipes, temperature profiles, equipment or piping configuration, etc. Based on this process description, the "educated" or "full" box of the future will sequentially display the proposed control system, giving the user the flexibility to make modifications at every step. If the user accepts the canned "expert" solutions proposed, the resulting controls will represent the state of the art of reactor control at that time. If the user modifies the recommended solution—say, by replacing dual-mode control with rate-of-rise control—the user's selection will be implemented using the best rate-of-rise control algorithm available at that time. Such high-level unit controllers would also be provided with subroutines for scaling, automatic loop tuning, validity checks, decoupling, and all those other features that today involve substantial development risk and expense.

As these educated unit controllers become available (and some early models already are), the design of the control systems of chemical plants will involve much less "rediscovery of the wheel" activity at the single-loop level and much more activity concerned with interconnecting a plant-wide network of distributed unit controllers and then optimizing plant productivity by controlling these unit operations.

To a reader who today is unable to get a single control loop properly tuned, all this might sound like "pie in the sky," but it should be realized that the main bottleneck in our industrial society is knowledge. The reason why loops are cycling or are in manual is that there is nobody in that plant who knows how to tune those loops. The unit controller, with its canned intelligence, overcomes this education gap by providing all users with the knowledge of the process control specialists who programmed the unit. The difference is similar to the difference between the task of playing chess and that of programming a computer to do so.

Conclusions

The various goals of chemical reactor optimization include the following:

- Precise recipe charging
- Minimum heat-up time without temperature overshoot
- Maximized reaction rate within limits of coolant availability
- Accurate and fast end point determination

- Elimination of bad batches
- Designing safety and reliability into the system
- Optimized scheduling and inventory controls

If the potentials of all of the above optimization strategies were fully exploited, the resulting productivity increase would be around 25 percent.

REFERENCES

1. Bristol, E. H., "A Design Tool Kit for Batch Process Control," *InTech*, October 1985.
2. Buckley, P. S., "Protective Controls for a Chemical Reactor," *Chemical Engineering*, April 20, 1970 and April 4, 1975.
3. *Control Roundup*, vol. 4, no. 1, January 1984, published by The Foxboro Company.
4. Hodson, A.J.F., and D. W. Clarke, "Self-Tuning Applied to Batch Reactors," *Proceedings of the IEE Vacation School on Industrial Digital Control Systems* (London: IEE, September 1984).
5. Horn, H. E., "Feasibility Study of the Application of Self-Tuning Controllers to Chemical Batch Reactors," Report OUEL-1248/78 (Oxford: Engineering Lab, Oxford University, 1978).
6. Jutan, A., and A. Uppal, "Combined Feedforward-Feedback Servo Control for an Exothermic Batch Reactor," *Industrial and Engineering Chemistry, Process Design and Development*, July 1984, pp. 597–602.
7. Lipták, B. G., ed., *Instrument Engineer's Handbook*, vol. 2 (Radnor, PA: Chilton Book Co., 1970), p. 1309.
8. Lipták, B. G., ed., *Instrument Engineer's Handbook, Process Measurement* (Radnor, PA: Chilton Book Co., 1982).
9. Lipták, B. G., ed., "Reactor Control and Optimization," in *Instrument Engineer's Handbook, Process Control* (Radnor, PA: Chilton Book Co., 1985).
10. Longwell, E. J., "Endo/Exo Batch Reaction Requires Complex Temperature Control," *InTech*, August 1986.
11. Luyben, W. L., "Batch Reactor Control," *Instrumentation Technology*, August 1975.
12. Molvar, A. C., "Instrumentation and Automation Experiences," U.S. Environmentation Protection Agency publication no. EPA-600/2–76–198, October 1976.
13. Scott, D. H., "A Systems Approach to Batch Control," *Instrumentation Technology*, August 1970.
14. Shinskey, F. G., *Controlling Multivariable Processes* (Instrument Society of America, 1981).
15. Shinskey, F. G., *Process Control Systems*, 2nd ed. (New York: McGraw-Hill, 1979).
16. Tolfo, F., "Numerical Simulation and DDC of an Exothermic Batch Reactor," in *Computerized Control and Operation of Chemical Plants* (Vienna, Austria: Verein Osterreichischer Chemiker, September 1981), pp. 380–85.

Chapter 4

OPTIMIZATION OF CHILLERS

HEAT PUMPS serve to transport heat from a lower to a higher temperature level. They do not make heat; they just transport it. This is similar to a regular pump transporting water from a lower to a higher elevation. The required energy input to a regular pump is a function not so much of the amount of water to be transported but of the elevation to which it is to be lifted. Similarly, the required energy input of a heat pump is a function not only of the amount of cooling it has to do but also of the temperature elevation against which it is pumping. The reduction of this temperature difference is the goal of optimization.

Each °F reduction in the ΔT will lower the yearly operating cost by 1.5 percent (each °C by 2.7 percent). There are not many unit operations whose efficiency can be doubled through optimization; chiller optimization is one of these few. It is possible to cut the operating cost of a chiller in half through the reduction of its ΔT by 33°F (18.5°C). Such results from optimization are not unprecedented [8].

In this review of the state of the art of chiller controls, the thermodynamic aspects of the refrigeration process will be discussed first, after which conventional chiller controls will be discussed briefly. Finally, a detailed treatment of various aspects of chiller optimization will be provided.

Thermodynamic Aspects

Figure 4.1 illustrates how heat pumps can transport heat from a lower to a higher elevation and thereby provide cooling of the low temperature process. The heat pump removes $Q1$ amount of heat from the chilled water and, at the cost of investing W amount of work, delivers Qh quantity of heat to the warmer cooling tower (or condenser) water.

In the lower part of this illustration, the idealized temperature-entropy cycle is shown for the refrigerator. The cycle consists of two isothermal and two isentropic (adiabatic) processes:

1→2:Adiabatic process through expansion valve
2→3:Isothermal process through evaporator
3→4:Adiabatic process through compressor
4→1:Isothermal process through condenser

The isothermal processes in this cycle are also isobaric (constant pressure). The efficiency of a refrigerator is defined as the ratio between the heat removed from the process (Ql) and the work required to achieve this heat removal (W).

$$\beta = \frac{Ql}{W} = \frac{Tl}{Th - Tl}$$

Because the chiller efficiency is much more than 100 percent, it is usually called the *coefficient of performance* (COP).

If a chiller requires 1.0 kWh (3412 BTU/h) to provide a ton of refrigeration (12,000 BTU/h), its coefficient of performance is said to be 3.5. This means that each unit of energy introduced at the compressor will pump 3.5 units of heat energy to the cooling tower water. Conventionally controlled chillers operate with COPs in the range of 2.5 to 3.5. Optimization can double the COP by increasing Tl, by decreasing Th, and through other methods that will be described later.

Figure 4.1 shows the four principal pieces of equipment that make up a refrigeration machine. In path 1→2 (through the expansion valve) the high-pressure subcooled refrigerant liquid becomes a low-pressure, liquid-vapor mixture. In path 2→3 (through the evaporator), this becomes a superheated low-pressure vapor stream, whereas in the compressor (path 3→4) the pressure of the refrigerant vapor is increased. Finally, in path 4→1 this vapor is condensed at constant pressure. The liquid leaving the condenser is usually subcooled, whereas the vapors leaving the evaporator are usually superheated by controlled amounts.

The unit most frequently used in describing refrigeration loads is the ton. Because several types of tons are referred to in the literature, it is important to distinguish among them:

Standard ton	200 BTU/min (3520W)
British ton	237.6 BTU/min (4182 W)
European ton (Frigorie)	50 BTU/min (880 W)

Refrigerants and Heat Transfer Fluids

The fluid that carries the heat from a low to a high temperature level is referred to as the *refrigerant*. Table 4.1 provides a summary of the more frequently used refrigerants.

The data are presented on the assumption that the evaporator will operate at 5°F (−15°C) and that the temperature of the cooling water supply for the condenser will allow it to maintain 86°F (30°C). Other temperature levels would have illustrated the relative characteristics of the various refrigerants equally well. It is generally desirable to avoid operation under vacuum in any part of the cycle because of sealing problems. At the same time, very high condensing pressures are also undesirable because of the resulting structural strength requirements. From this point of view, the refrigerants between propane and methyl chloride in table 4.1 display favorable characteristics. An exception to this reasoning is when very low temperatures are required. For such

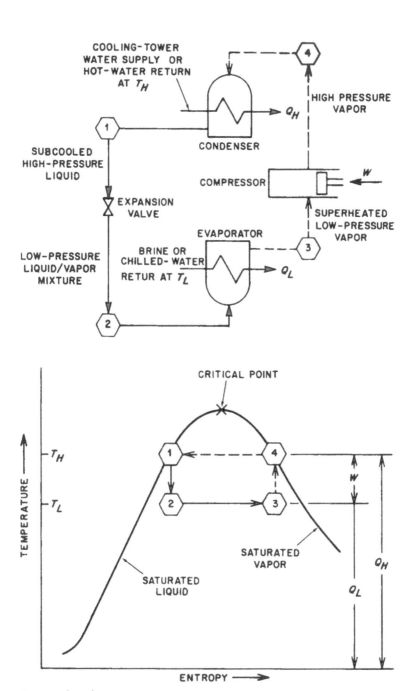

Fig. 4.1 The refrigeration cycle consists of two isothermal and two isentropic processes. (From Lipták [5])

service, ethane can be the proper selection in spite of the resulting high system design pressure.

Another consideration is the latent heat of the refrigerant. The higher it is, the more heat can be carried by the same amount of working fluid and, therefore, the smaller the corresponding equipment size can be. This feature has caused many users in the past to compromise with the undesirable characteristics of ammonia.

One of the most important considerations is safety. In industrial installations, the desirability of nontoxic, nonirritating, nonflammable refrigerants cannot be overemphasized. It is similarly important that the working fluid be compatible with the compressor lubricating oil. Corrosive refrigerants are undesirable for the obvious reasons of higher initial cost and increased maintenance.

Most of the working fluids listed in table 4.1 are compatible with reciprocating compressors. Only the last four fluids in the table, which have high volume-to-mass ratios and low compressor discharge pressures, can justify the consideration of rotary or centrifugal machines.

When all the advantages and drawbacks of the many refrigerants are considered, Freon-12 is found to be suited for the largest number of applications.

In the majority of industrial installations, the refrigerant evaporator is not used directly to cool the process. More frequently, the evaporator cools a circulated fluid, which is then piped to cool the process.

For temperatures below the point at which water can be used as a coolant, brine is frequently used. It is important to remember that weak brines may freeze and that strong brines, if they are not true solutions, may plug the evaporator tubes. For operation around 0°F (−18°C), the sodium brines (NaCl) are recommended; for services down to −45°F (−43°C), the calcium brines (CaCl) are best.

Care must be exercised in handling brines, because they are corrosive if they are not kept at a pH of 7 or if oxygen is present. In addition, brine will initiate galvanic corrosion between dissimilar metals.

Conventional Refrigeration Units

The refrigeration unit shown in figure 4.2, although far from a "standard" system, does contain some of the features typical of conventional industrial units in the 500-ton (1760-kw) and larger sizes. These features include the capability for continuous load adjustment as contrasted with stepwise unloading, the application of the economizer expansion valve system, and the use of hot gas bypass to increase rangeability.

The unit illustrated provides refrigerated water at 40°F (4.4°C) through the circulating header system of an industrial plant. The flow rate is fairly constant, and therefore process load changes are reflected by the temperature of the returning refrigerated water. Under normal load conditions, this return water temperature is 51°F (10.6°C). As process load decreases, the return water temperature drops correspondingly. With the reduced load on the evaporator, TIC-1 gradually closes the suction damper or the prerotation vane of the compressor. By throttling the suction vane, a

Table 4.1
REFRIGERANT CHARACTERISTICS

Refrigerant		Applicable compressor (R = Reciprocating, RO = Rotary, C = Centrifugal)	Boiling point in °F at atmospheric pressure	Evaporator pressure in PSIA† if operating temperature is 5°F (−15°C)	Condenser pressure in PSIA† if operating temperature is 86°F (30°C)	Latent heat in BTU/lbm‡ at 18°F (−7.8°C)	Toxic (T), Flammable (F), Irritating (I)	Mixes and/or compatible with the lubricating oil	Chemically inert and noncorrosive	Remarks
Ethane	C_2H_6	R	−127	236	675	148	T&F	No	Yes	For low-temperature service
Carbon dioxide	CO_2	R	−108	334	1039	116	No	Yes	Yes	Low-efficiency refrigerant
Propane	C_3H_8	R	−48	42	155	132	T&F	No	Yes	For low-temperature service
Freon-22	$CHClF_2$	R	−41	43	175	92	No	(1)	Yes	High-efficiency refrigerant
Ammonia	NH_3	R	−28	34	169	555	T&F	No	(2)	Most recommended
Freon-12	CCl_2F_2	R	−22	26	108	67	No	Yes	Yes	Expansion valve may freeze if water is present
Methyl chloride	CH_3Cl	R	−11	21	95	178	(3)	Yes	(4)	Common to these refrigerants:
Sulphur dioxide	SO_2	R	+14	12	66	166	T&I	No	(4)	a. Evaporator under vacuum
Freon-21	$CHCl_2F$	RO	+48	5	31	108	No	Yes	Yes	b. Low compressor discharge pressure
Ethyl chloride	C_2H_5Cl	RO	+54	5	27	175	F&I	No	(5)	c. High volume-to-mass ratio across compressor
Freon-11	CCl_3F	C	+75	3	18	83	No	Yes	Yes	
Dichloro methane	CH_2Cl_2	C	+105	1	10	155	No	Yes	Yes	

Source: From Lipták [4].

$°C = \dfrac{°F - 32}{1.8}$

†PSIA = 6.9 kPa
‡BTU/lbm = 232.6 J/kg

(1) Oil floats on it at low temperature
(2) Corrosive to copper-bearing alloys
(3) Anesthetic
(4) Corrosive in the presence of water
(5) Attacks rubber compounds

10:1 turndown ratio can be accomplished. If the load drops below this ratio, the hot gas bypass system has to be activated.

The hot gas bypass is automatically controlled by TIC-2. Its purpose is to keep the constant speed compressor out of surge: when the load drops to levels sufficiently low to approach surge, this bypass valve is opened. If the chilled water flow rate is constant, the difference between chilled water supply and return temperatures is an indication of the load. If full load corresponds to a 15°F (8.3°C) difference on the chilled water side of the evaporator and the chilled water supply temperature is controlled by TIC-1 at 40°F (4.4°C), then the return water temperature detected by TIC-2 is also an indication of load.

If surge occurs at 10-percent load, this would correspond to a return water temperature of 41.5°F (5.3°C). In order to stay safely away from surge, TIC-2 in figure 4.2 is set at 42°F (5.6°C), corresponding to approximately 13-percent load. When the temperature drops to 42°F (5.6°C), this valve starts to open, and its opening can be proportional to the load detected. This means that the valve is fully closed at 42°F (5.6°C), fully open at 40°F (4.4°C), and throttled in between. This throttling action is accomplished by a plain proportional controller that has a 2°F throttling range, which, on a span of 0 – 100°F, corresponds to a proportional band of 2 percent or a gain of 50.

The hot gas bypass makes it theoretically possible to achieve a very high turndown ratio by temporarily running the machine on close to zero process load. This operation can be visualized as a heat pump, transferring heat energy from the refrigerant itself to the cooling water. In the process, some of the refrigerant vapors are condensed, resulting in an overall lowering of operating pressures on the refrigerant side.

The main advantage of a hot gas bypass, therefore, is that it allows the chiller to operate at low loads without going into surge. The price of this operational flexibility is an increase in operating costs, because the work introduced by the compressor is wasted as friction drop (friction between gas molecules and valve) through the hot gas bypass valve. As will be shown later, optimized control systems eliminate this waste through the use of variable speed compressors, which will respond to a reduction in load by lowering their speed instead of throwing away the unnecessarily introduced energy in TCV-1 and TCV-2.

Instead of controlling the hot gas bypass on the basis of return water temperature (as in the upper section of figure 4.2), other conventional packages control it on the basis of prerotation vane opening. This is illustrated in the lower portion of figure 4.2. The problem with opening the hot gas bypass when the prerotation vane has closed to some fixed point is that this control technique disregards condensing temperature. This causes the hot gas bypass to open sooner and to a greater extent than needed. This is illustrated in the lower section of figure 4.2, where under normal loads the compressor operates at point "O." As the load drops off at constant discharge pressure (constant condenser temperature), the prerotation vane is gradually closed, until at point "1" it opens the hot gas bypass, to protect the compressor from going into surge. If the actual load drops to "1S," the HGBP will furnish the differential flow between "1S" and "1."

Now, if the cooling water temperature is reduced, normal operation falls at point

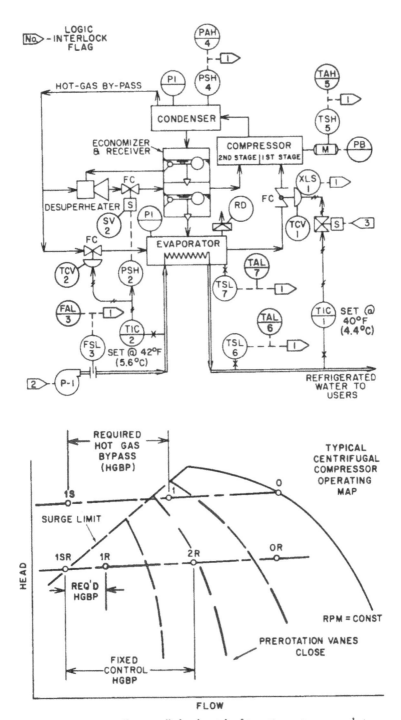

Fig. 4.2 A conventionally controlled industrial refrigeration system uses a hot gas bypass to increase rangeability. (Top: From Lipták [4]; bottom: from Cooper [2])

159

"OR." As the load drops, the fixed setting on the prerotation vane is reached at point "2R." This is much sooner than necessary. It is possible to obtain optimized controls that recognize the impact of cooling water temperature variations and that, in this case, will open the HGBP only when load drops to point "1R."

The economizer shown in figure 4.2 can increase the efficiency of operation by 5 to 10 percent. This is achieved by the reduction of space requirements, savings on compressor power consumption, reduction of condenser and evaporator surfaces, and other effects. The economizer shown in figure 4.2 is a two-stage expansion valve with condensate collection chambers. When the load is above 10 percent, the hot gas bypass system is inactive. Condensate is collected in the upper chamber of the economizer, and it is drained under float level control, driven by the condenser pressure. The pressure in the lower chamber floats off the second stage of the compressor, and it, too, is drained into the evaporator under float level control, driven by the pressure of the compressor second stage. Economy is achieved as a result of the vaporization in the lower chamber by precooling the liquid that enters the evaporator and at the same time desuperheating the vapors that are sent to the compressor second stage.

When the load is below 10 percent, the hot gas bypass is in operation, and the solenoid valve SV-2, which is actuated by the high pressure switch PSH-2, opens. Some of the hot gas goes through the evaporator and is cooled by contact with the liquid refrigerant, and some of the hot gas flows through the open solenoid. This second portion is desuperheated by the injection of liquid refrigerant upstream of the solenoid, which protects against overheating the compressor.

Safety Interlocks

Operating safety in the system illustrated in Figure 4.2 is guaranteed by a number of interlocks. The first interlock system prevents the compressor motor from being started if one or more of the following conditions exist, and it also stops the compressor if any except the first condition listed occurs while the compressor is running.

- Suction vane is open, detected by limit switch XLS-1
- Refrigerated water temperature is dangerously low, approaching freezing as sensed by TSL-6
- Refrigerated water flow is low, measured by FSL-3
- Evaporator temperature has dropped near the freezing point, as detected by TSL-7
- Compressor discharge pressure (and, therefore, pressure in the condenser) is high, indicated by PSH-4
- Temperature of motor bearing or winding is high, detected by TSH-5
- Lubricating oil pressure is low (not shown in figure 4.2)

The second interlock system guarantees that the following pieces of equipment are started or are already running upon starting of the compressor:

- Refrigerated water pump (P-1)

- Lubricating oil pump (not shown)
- Water to lubricating oil cooler, if such exists (not shown)

The third interlock usually ensures that the suction vane is completely closed when the compressor is stopped.

Nomenclature

The symbols and abbreviations used in figure 4.2 and in the rest of this chapter are listed below for reference purposes:

CHWP: Chilled-water pump
CTWP: Cooling-tower water pump
FAH, FAL: Flow alarm, H = high, L = low
FC: Fail closed
FO: Fail open
FSH, FSL, FSHL: Flow switch, H = high, L = low, HL = high-low
HLL: High-low limit switch
HWP: Hot-water pump
LCV: Level-control valve
M_{1-5}: Motor that drives equipment or "place" where energy is consumed, 1 = cooling-tower fan(s), 2 = cooling-water pump(s), 3 = compressor(s), 4 = chilled-water pump(s), 5 = hot-water pump(s)
P-1: Pump
P, P_{1-3}: Pressures
PAH: Pressure alarm, H = high
PB: Pushbutton
PCV: Pressure-control valve
PDIC: Pressure-differential indicating controller
PI: Pressure indicator
PSH, PSL: Pressure switch, H = high, L = low
Q_H: Amount of heat delivered to cooling-tower water
Q_L: Amount of heat removed from chilled water
RD: Rupture disk
S: Solenoid
SC: Speed controller
SIC: Speed-indicating controller
SP: Setpoint
SV: Solenoid valve
T_c: Temperature of refrigerant in condenser inlet
T_e: Temperature of refrigerant in evaporator inlet
T_H: Temperature of cooling water at condenser exit, absolute
T_L: Temperature of chilled water at evaporator exit, absolute
T_{chwr}: Temperature of chilled-water return
T_{chws}: Temperature of chilled-water supply
T_{ctwr}: Temperature of cooling-tower water return

T_{ctws}: Temperature of cooling-tower water supply
T_{hwr}: Temperature of hot-water return
T_{hws}: Temperature of hot-water supply
T_{wb}: Temperature of wet bulb
TAH, TAL: Temperature alarm, H = high, L = low
TCV: Temperature control valve
TIC: Temperature-indicating controller
TSH, TSL: Temperature switch, H = high, L = low
TT: Temperature transmitter
TY: Temperature relay
VPC: Valve position controller
W: Work
XLS: Limit switch
XSCV: Superheat control valve

Optimizing the Total System

The layout of a typical cooling system is illustrated in figure 4.3. In order to develop a completely generalized method for optimizing such systems, the duplication of equipment will be disregarded, and all systems will be treated as in figure 4.4. Here any number of cooling towers, pumps, or chillers are represented by single symbols, because variations in their numbers will not affect the optimization strategy.

In this generalized cooling system, the cooling load from the process is carried by the chilled water to the evaporator, where it is transferred to the freon refrigerant. The freon takes the heat to the condenser, where it is passed on to the cooling tower water, so that it might finally be rejected to the ambient air. This heat pump operation involves four heat transfer substances (chilled water, freon, cooling tower water, air) and four heat exchanger devices (process heat exchanger, evaporator, condenser, cooling tower). The total system operating cost is the sum of the costs of circulating the four heat transfer substances (M1, M2, M3, and M4).

In the traditional (unoptimized) control systems, such as the one illustrated in figure 4.2, each of these four systems were operated independently in an uncoordinated manner. In addition, conventional control systems did not vary the speed of the four transportation devices (M1 to M4). By operating them at constant speeds, they introduced more energy than was needed for the circulation of freon, air, or water and therefore had to waste that excess energy.

Load-following optimization eliminates this waste while operating the aforementioned four systems as a coordinated single process, with the goal of control being to maintain the cost of operation at a minimum. The controlled variables are the supply and return temperatures of chilled and cooling tower waters, and the manipulated variables are the flow rates of chilled water, freon, cooling tower water, and air. If water temperatures are allowed to float in response to load and ambient temperature variations, the waste associated with keeping them at arbitrarily selected fixed values is eliminated and the chiller operating cost is drastically reduced.

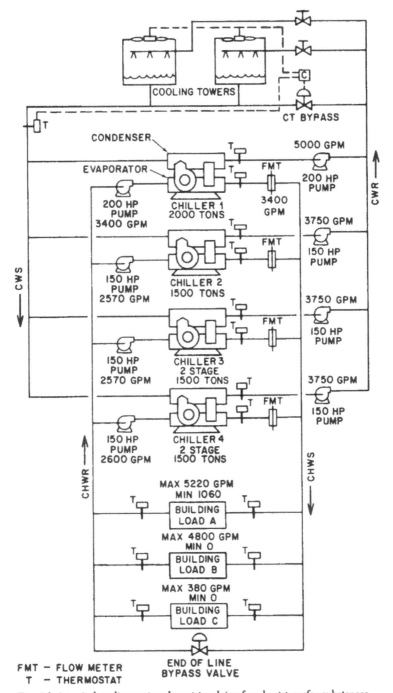

Fig. 4.3 A typical cooling system layout involving four heat transfer substances and four heat exchange devices. (From Hallanger [3])

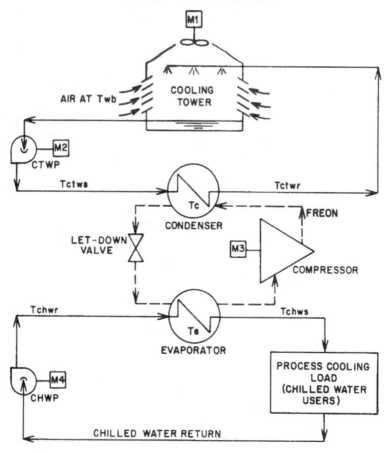

ABBREVIATIONS:

Twb   : WET BULB TEMP.
Tctws : COOLING TOWER WATER SUPPLY TEMP.
Tctwr : COOLING TOWER WATER RETURN TEMP.
Tchws : CHILLED WATER SUPPLY TEMP.
Tchwr : CHILLED WATER RETURN TEMP.
CTWP : COOLING TOWER WATER PUMP
CHWP : CHILLED WATER PUMP
Tc   : FREON TEMPERATURE IN CONDENSER
Te   : FREON TEMPERATURE IN EVAPORATOR

Fig. 4.4 Optimized control of refrigeration machines can be accomplished by treating the process as an integrated system. (From Lipták [5])

Evaporator-Side Cost Model

The yearly cost of operating the total cooling system can typically be broken down as follows:

M1 (Fans)	10%
M2 (CTWP)	15
M3 (Compr.)	60
M4 (CHWP)	15
	100%

The overall optimization model can be looked at in two steps, by discussing first the lower and then the upper half of figure 4.4. The lower (evaporator) portion is easily comprehended, because the sum of M3 and M4 will be minimum when both Tchws and Tchwr are maximized. This is so because at any load condition the compressor operating cost (M3) will be reduced by an increase in either the evaporator inlet or outlet temperature. The chilled water pumping cost (M4) depends only on the ΔT across the users. The higher this ΔT, the less water needs to be pumped to transport the required amount of heat. As the chilled water supply temperature is set by the load, ΔT can be maximized only by maximizing the chilled water return temperature (Tchwr). The increasing of this temperature therefore has the following benefits:

- M3 is lowered by about 1.5 percent per °F rise
- The heat transfer efficiency of the evaporator is improved by increasing the ΔT through it
- Assuming that on the average the ΔT is 15°F, increasing it by 1°F will lower M4 by about 6%

As M3 is about 60 percent and M4 about 15 percent of the total operating cost, a 1°F increase in ΔT will lower the yearly cost of operation by about $(1 \times 60)/100 + (6 \times 15)/100 = 1.5$ percent.

Condenser-Side Cost Model

The condenser portion of the cost model is a little more complicated. Here, for any set of load and ambient conditions, the minimum operating cost is at the minimum point of a three-dimensional surface. The x, y, and z coordinates of that space are defined as follows:

z = Total cost of operation (M1 + M2 + M3)
x = Approach (Tctws-Twb)
y = Condenser ΔT (Tctwr-Tctws), which is also called the *range* of the cooling
tower.

In actual operation, when x and y are at particular values, the minimum cost point can be found by looking up the minimum, first on the x–z curve (figure 4.5) and then on the y–z curve (figure 4.6). For the x–z curve the value of y is treated as a measured

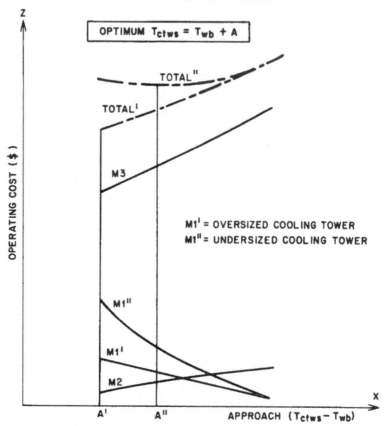

Fig. 4.5 At each particular load, Twb and Tctwr, an empirically determined cost-versus-approach curve can have a minimum.

constant; for the y–z curve, the value of x. If either x or y is found to be away from their minimum, the corresponding set points are reset, and after a "hold" period, the system is reevaluated. Naturally, as load changes, the x–y–z surface and the x–z and y–z curves will change with it. Therefore, at the end of the "hold" period, the new search for the minimum cost point will use different curves.

The curves in figures 4.5 and 4.6 can be based on the measurement of the actual total operating cost, on a projected cost based on past performance, or on some combination of the two. The continuous storing and updating of operating cost history as a function of load, ambient conditions, and equipment configuration not only will provide the curves required for optimization but can also be used to signal the need for maintenance.

In figure 4.5, the optimum Tctws is found by summing M1, M2, and M3. In most installations, the major cost element is M3, which increases by about 1.5 percent of compressor operating cost (or 1 percent of total cooling system cost) for each 1°F

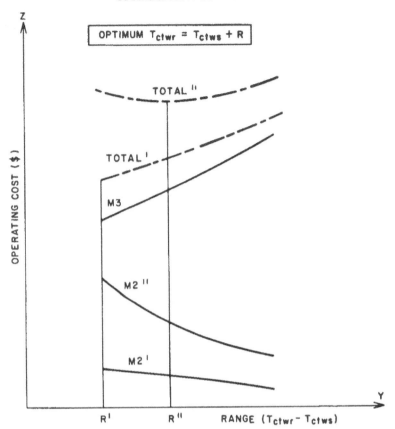

Fig. 4.6 At each particular load, an empirically determined cost-versus-range curve can have a minimum.

reduction in approach. The cost-versus-approach curve is not necessarily a smooth one, because if constant-speed machines are stopped and started or if positive displacement compressors are loaded and unloaded, there will be steps in the M3 curve. This also is true for constant speed fans (M1) or pumps (M2). The pumping cost also increases as approach rises, but M2 is usually only a fraction of M3, unless the piping is very long or undersized.

The fan cost M1 tends to drop off as approach rises. If the load is low relative to the size of the cooling towers, the rate of increase in M1 with a reduction in approach will also be small. In such cases (M1') the optimum Tctws is the safe lowest temperature that the tower can generate. If, on the other hand, the load is high, the fan cost of lowering approach will also increase (M1"). In that case, the optimum approach (A") is at some intermediate value.

In figure 4.6, the optimum Tctwr can be found by determining the optimum condenser ΔT (or cooling tower range) for a particular load and Tctws. As the range increases (Tctwr rises), the cost of operating the compressor also rises, and the cost of pumping drops. If the relative cost of pumping is low (M2'), the optimum range will be the minimum one (R'), because it will allow the compressor to work against the lowest head. If there is a point below which the rise in pumping costs is more than the drop in compressor operating cost, then the optimum operating range ("R") is at that value.

Chilled Water Supply Temperature Optimization

The yearly operating cost of a chiller is reduced by 1.5 percent to 2 percent for each 1°F (0.6°C) reduction in the temperature difference across this heat pump (Th-Tl in figure 4.1). In order to minimize this difference, Th must be minimized and Tl maximized. Therefore, the optimum value of Tchws is the maximum temperature that will still satisfy all the loads. Figure 4.7 illustrates a method of continuously finding and maintaining this value.

It should be noted that an energy-efficient refrigeration system cannot be guaranteed by instrumentation alone, because equipment sizing and selection also play an important part in the overall result. For example, the evaporator heat transfer area should be maximized so that Tl is as close as possible to the average chilled water temperature in the evaporator. Similarly, at the compressor the freon flow should be made to match the load by motor speed control rather than by throttling. TCV-1 and TCV-2 in figure 4.2 represent sources of energy waste, whereas figure 4.7 shows the energy-efficient technique of motor speed control. The variable-speed operation can be achieved by using variable-speed drives on electric motors or by using steam turbine drives. If several constant-speed motors are used, then all compressors except one should be driven to the load at which they are most efficient (TCV-1 and TCV-2 in figure 4.2 fully open); the remaining compressor should be used to match the required load by throttling.

Figure 4.7 shows the proper technique of maximizing the chilled water supply temperature in a load-following, floating manner. The optimization control loop guarantees that all users of refrigeration in the plant will always be satisfied while the chilled water temperature is maximized. This is done by selecting (with TY-1) the most open chilled water valve in the plant and comparing that signal with the 90-percent set point of the valve position controller, VPC-1. If even the most open valve is less than 90-percent open, the set point of TIC-1 is increased; if the valve opening exceeds 90 percent, the TIC-1 set point is decreased. This technique allows all users to obtain more cooling (by further opening their supply valves) if needed, while the header temperature is continuously maximized. The VPC-1 set point of 90 percent is adjustable. Lowering it gives a wider safety margin, which might be required if some of the cooling processes served are very critical. Increasing the set point maximizes energy conservation at the expense of the safety margin.

An additional benefit of this load-following optimization strategy is that because all chilled water valves in the plant are opened as Tchws is maximized, valve cycling is reduced and pumping costs are lowered because when all chilled water valves are

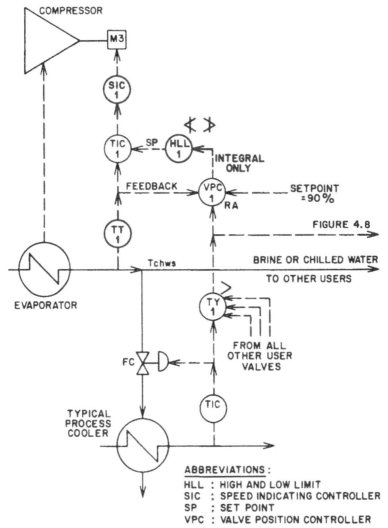

Fig. 4.7 Load-following floating control of chilled water supply temperature is one method of continuously finding and maintaining the optimum value of Tchws. (From Lipták [4])

opened, they require less pressure drop. Valve cycling is eliminated when the valve opening is moved away from the unstable region near the closed position [5].

In order for the control system in figure 4.7 to be stable, it is necessary to use an integral-only controller for VPC-1, with an integral time that is tenfold that of the integral setting of TIC-1 (usually several minutes). This mode selection is needed to allow the optimization loop to be stable when the valve opening signal selected by TY-1 is either cycling or noisy. The high/low limit settings on the set point signal (HLL-1) to TIC-1 guarantee that VPC-1 will not drive the chilled water temperature to unsafe or undesirable levels. Because these limits can block the VPC-1 output from affecting Tchws, it is necessary to protect against reset windup in VPC-1. This is done through the external feedback signal shown in figure 4.7.

Chilled Water Return Temperature Optimization

The combined cost of operating the chilled water pumps and the chiller itself is a function of the temperature drop across the evaporator (Tchwr-Tchws). Because an increase in this ΔT decreases compressor operating costs while also decreasing pumping costs, the aim of this optimization strategy is to maximize this ΔT.

This ΔT will be the maximum when the chilled water flow rate across the chilled water users is the minimum. As Tchws is already controlled, this ΔT will be maximized when Tchwr is maximum. This goal can be reached by looking at the opening of the most open chilled water valve in figure 4.7. If even the most open chilled water valve is not yet fully open, the chilled water supply temperature (set by TIC-1 in figure 4.7) can be increased, or the temperature rise across the users can be increased by lowering the ΔP (set by PDIC-1 in figure 4.8); both methods may be used. Increasing the chilled water supply temperature reduces the yearly compressor operating cost (M3) by approximately 1.5 percent for each °F of temperature increase, whereas lowering the ΔP reduces the yearly pump operating cost (M4) by approximately 50 cents/GPM for each PSID (or $11.5/m^3$/s for each Pa).

The set points of the two valve position controllers (VPC-1s in figures 4.7 and 4.8) will determine if these adjustments are to occur in sequence or simultaneously. If both set points are the same, simultaneous action will result; if one adjustment is economically more advantageous than the other, the set point of the corresponding VPC will be positioned lower than the other.

This will result in sequencing, which means that the more cost-effective correction will be fully exploited before the less effective one is started. In figures 4.7 and 4.8, it was assumed that increasing Tchws is the more cost-effective step. This is not always the case, and even when it is, there might be process reasons that make it undesirable to float Tchws up or down. If the settings are as shown in figures 4.7 and 4.8, the system will function as follows: If the most open valve is less than 90-percent open, VPC-1 in figure 4.8 will lower the set point of PDIC-1, and VPC-1 in figure 4.7 will increase the set point of TIC-1. When the opening of the most open valve reaches 90 percent VPC-1 in figure 4.7 will slowly start lowering the TIC set point. If it is not lowered enough (or if the lowering is not fast enough), and the most open valve continues to open further, at 95 percent the VPC-1 in figure 4.8 will take fast corrective action, quickly raising the set point of PDIC-1. Thus, no user valve will ever be allowed to open fully and go out of control as long as the pumps and chillers are capable of meeting the load.

VPC-1 in figure 4.8 is the cascade master of PDIC-1, which guarantees that the pressure difference between the chilled water supply and return is always high enough to motivate water flow through the users but never so high as to exceed their pressure

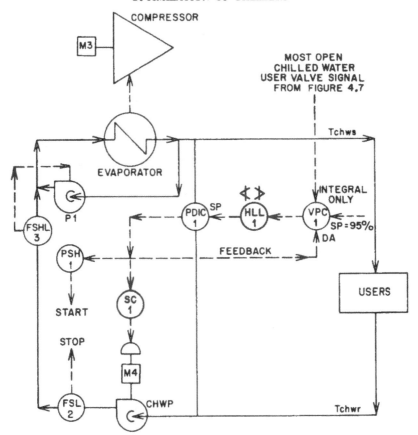

Fig. 4.8 Floating of chilled water flow rate keeps evaporator ΔT always at maximum (optimum) values. (From Lipták [4])

ratings. The high and low limits are set on HLL-1, and VPC-1 is free to float this set point within these limits to keep the operating cost at a minimum.

In order to protect against reset wind-up when the output of VPC-1 reaches one of these limits, an external feedback is provided from the PDIC-1 output signal to the pump speed controller SC-1. The VPC is an integral-only controller, which is tuned to be much more responsive than the VPC in figure 4.7

When the chilled water pump station consists of several pumps, only one of which is variable-speed, additional pump increments are started when PSH-1 signals that the pump speed controller set point is at its maximum. When the load is dropping, the excess pump increments are stopped on the basis of flow, detected by FSL-2. In order to eliminate cycling, the excess pump increment is turned off only when the actual total flow corresponds to less than 90 percent of the capacity of the *remaining* pumps.

This load-following optimization loop will float the total chilled water flow to achieve maximum overall economy. In order to maintain efficient heat transfer and appropriate turbulence within the evaporator, a small local circulating pump (P1) is provided at the evaporator. This pump is started and stopped by FSHL-3, guaranteeing that the water velocity in the evaporator tubes will never drop below the adjustable limit of about 4 ftps (1.2m/s).

Cooling Tower Supply Temperature Optimization

Minimizing the temperature of the cooling tower water is one of the most effective ways to contribute to chiller optimization. Conventional control systems of the past were operated with constant cooling tower temperatures of 75°F (23.9°C) or higher. A constant utility condition is an enemy of efficiency and therefore of optimization. Each 10°F (5.6°C) reduction in the cooling tower water temperature will reduce the yearly operating cost of the compressor by approximately 15 percent. For example, if a compressor is operating at 50°F (10°C) condenser water instead of 85°F (29.4°C) it will meet the same load while consuming half as much power. Operation at condenser water temperatures of less than 50°F (10°C) is quite practical during the winter months. Savings exceeding 50 percent have been reported [8].

As shown in figure 4.9, an optimization control loop is required in order to maintain the cooling tower water supply continuously at an economical minimum temperature. This minimum temperature is a function of the wet bulb temperature of the atmospheric air. The cooling tower cannot generate a water temperature that is as low as the ambient wet bulb, but it can approach it. The temperature difference between Tctws and Twb is called the *approach*, as was explained in connection with figure 4.5.

Figure 4.9 illustrates the fact that as the approach increases, the cost of operating the cooling tower fans drops and the costs of pumping and of compressor operation increase. Therefore, the total operating cost curve has a minimum point that identifies the optimum approach that will allow operation at an overall minimum cost. This ΔT automatically becomes the set point of TDIC-1. This optimum approach increases if the load on the cooling tower increases or if the ambient wet bulb decreases.

If the cooling tower fans are centrifugal units or if the blade pitch is variable, the optimum approach is maintained by continuous throttling. If the tower fans are two-speed or single-speed units, the output of TDIC-1 will start and stop the fan units incrementally in order to maintain the optimum approach. In cases in which a large number of cooling tower cells constitutes the total system, it is also desirable to balance the water flows to the various cells automatically as a function of the operation of the associated fan. In other words, the water flows to all cells whose fans are at high speed should be controlled at equal high rates; cells with fans operating at low speeds should receive water at equal low flow rates, and cells with their fans off should be supplied with water at equal minimum flow rates. Subjects such as these and the cooling towers in general are discussed in more detail in chapter 7.

Cooling Tower Return Temperature Optimization

As was depicted in figure 4.6, the combined cost of operating the cooling tower pumps and the chiller compressor is a function of the temperature rise across the

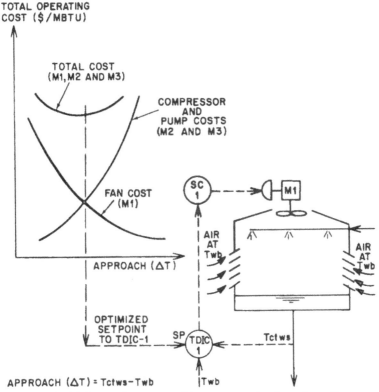

Fig. 4.9 A load-following floating control loop is required to keep the cooling tower water supply at an economical minimum temperature. (From Lipták [4])

condenser. Because an increase in this ΔT increases compression costs while decreasing pumping costs, the combined total curve can have a minimum cost point. The ΔT corresponding to this minimum automatically becomes the set point of TDIC-1 in the optimized control loop shown in figure 4.10. This controller is the cascade master of PDIC-1, which guarantees that the pressure difference between the supply and return cooling tower water flows is always high enough to provide flow through the users but never so high as to cause damage. The high and low limits are set on HLL-1. TDIC-1 freely floats this set point within these ΔP limits, to keep the operating cost at a minimum.

In order to protect against reset wind-up (when the output of TDIC-1 reaches one of these limits), an external feedback is provided from the PDIC-1 output signal to the pump speed controller SC-1.

When the cooling tower water pump station consists of several pumps, only one of which is variable-speed, additional pump increments are started when PSH-1 signals that the pump speed controller set point is at its maximum. When the load is dropping, the excess pump increments are stopped on the basis of *flow*, detected by FSL-2. In

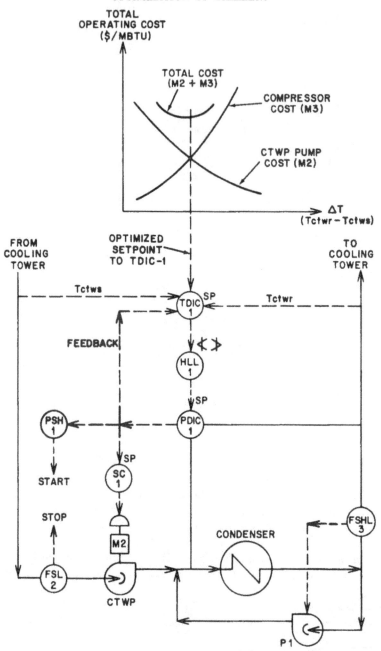

Fig. 4.10 The cooling tower water flow rate can be floated to keep condenser ΔT always at optimum values. (From Lipták [4])

order to eliminate cycling, the excess pump increment is turned off only when the actual total flow corresponds to less than 90 percent of the capacity of the *remaining* pumps.

This load-following optimization loop will float the total cooling tower water flow to achieve maximum overall economy. In order to maintain efficient heat transfer and appropriate turbulence within the condenser, a small local circulating pump (P1) is provided at the condenser. This pump is started and stopped by FSHL-3, guaranteeing that the water velocity in the condenser tubes will never drop below the adjustable limit of, for example, 4fps.

Heat Recovery Optimization

Figure 4.11 depicts the required optimizing control loop when the heat pumped by the chiller is recovered in the form of hot water.

Like the control system shown for chilled water temperature floating in figure 4.7, the hot water temperature can also be continuously optimized in a load-following floating manner. If, at a particular load level, it is sufficient to operate with 100°F (37.8°C) instead of 120°F (48.9°C) hot water, this technique will allow the same tonnage of refrigeration to be met by the chiller at 30 percent lower cost. The reason is that the compressor discharge pressure is determined by the hot water temperature in the split condenser.

The optimization control loop in figure 4.11 guarantees that all hot water users in the plant will always obtain enough heat while the water temperature is minimized. TY-1 selects the most open hot water valve in the plant, and VPC-1 compares that transmitter signal with a 90-percent set point. If even the most open valve is less than 90-percent open, the set point of TIC-1 is decreased; if the opening exceeds 90 percent, the set point is increased. This allows all users to obtain more heat (by further opening their supply valves) if needed, while the header temperature is continuously optimized.

Figure 4.11 also shows that an increasing demand for heat will cause the TIC-1 output signal to rise as its measurement drops below its set point. An increase in heat load will cause a decrease in the heat spill to the cooling tower, since the control valve TCV-1A is closed between 3 and 9 PSIG (0.2 and 0.6 bar). At a 9-PSIG (0.6 bar) output signal, all the available cooling load is being recovered and TCV-1A is fully closed. If the heat load continues to rise (TIC-1 output signal rises over 9 PSIG, or 0.6 bar), this will result in the partial opening of the "pay heat" valve, TCV-1B. In this mode of operation, the steam heat is used to supplement the freely available recovered heat to meet the prevailing heat load.

A local circulating pump, P1, is started whenever flow velocity is low. This prevents the formation of deposit in the tubes. P1 is a small 10- to 15-hp pump operating only when the flow is low. The main cooling tower pump (usually larger than 100 hp) is stopped when TCV-1A is closed.

Optimized Operating Mode Selection

The cost-effectiveness of heat recovery is a function of the outdoor temperature, the unit cost of energy from the alternative heat source, and the percentage of the

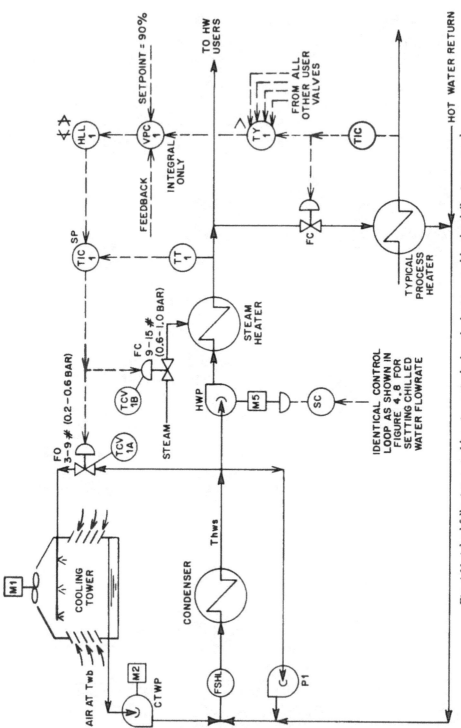

Fig. 4.11 A load-following control loop is required when the heat pumped by the chiller is recovered as hot water. (From Lipták [4])

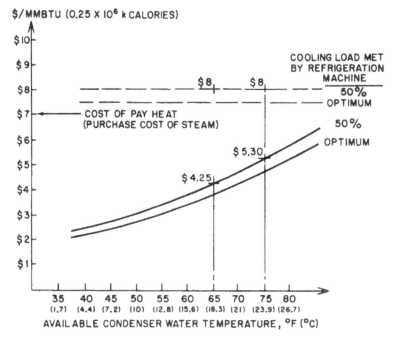

— — OPERATING COSTS OF HEAT RECOVERY CHILLERS GENERATING 105 °F
(40.6 °C) HOT WATER WITH 50 % AND "OPTIMUM" LOADING OF CHILLER.

—— OPERATING COSTS OF CONVENTIONAL MECHANICAL CHILLER AT VARIOUS
CONDENSER WATER TEMPERATURES AND WITH 50 % AND "OPTIMUM"
LOADING OF THE CHILLER.

Example*

Temperature of Condenser Water	Cost Components	Mechanical Refrigeration Mode	Heat Recovery Mode
	Cost of Cooling	$4.25	$8.00
65°F (18.3°C)	Cost of Heating	(0.5) (7.0) = $3.50	$0.00
	Total	$7.75	$8.00
	Cost of Cooling	$5.30	$8.00
75°F (23.9°C)	Cost of Heating	(0.5) (7.0) = $3.50	$0.00
	Total	$8.80	$8.00

*This example is based on the following assumptions.
  a. The actual cooling load is 50% of chiller capacity. (CL = 0.5 CAP)
  b. The heating load (the demand for hot water) is 50% of cooling load. (HL = 0.5 CL)

Fig. 4.12 The most cost-effective mode of refrigeration can be selected automatically.
(From Lipták [4])

cooling load that can be used as recovered heat. According to figure 4.12, if steam is available at $7/MMBTU and only half of the cooling load is needed in the form of hot water, it is more cost-effective to operate the chiller on cooling tower water and use steam as the heat source when the outside air is below 65°F (18°C). Conversely, when the outdoor temperature is above 75°F (23.9°C), the penalty for operating the split condenser at hot water temperatures is no longer excessive; therefore, the plant should automatically switch back to recovered heat operation. This cost-benefit analysis is a simple and continually used element of the overall optimization scheme.

In plants in locations such as the southern states, where there is no alternative heat source, another problem can arise because all the heating needs of the plant must be met by recovered heat from the heat pump. It is possible that during cold winter days there might not be enough recovered heat to meet this load. Whenever the heat load exceeds the cooling load and there is no alternative heat source available, an artificial cooling load must be placed on the heat pump.

This artificial heat source can in some cases be the cooling tower water itself. A direct heat exchanger between cooling tower and chilled water streams is also advantageous when there is no heat load but when there is a small cooling load during the winter. At such times, the chiller can be stopped, and the small cooling load can be met by direct cooling from the cooling tower water that is at a winter temperature (Figure 4.13).

Because coolant can be provided from many sources in a typical plant, another approach to optimization is to reconfigure the system in response to changes in loads, ambient conditions, and utility costs. For example, during some operating and ambient conditions, the cooling tower water may be cold enough to meet the demands of the

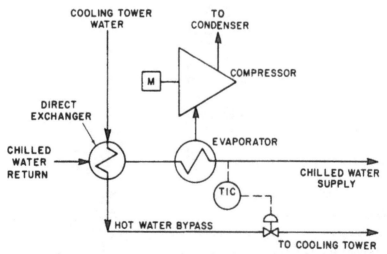

Fig. 4.13 The cooling tower water can be used either as a heat source (with the compressor on) or as a means of direct cooling (with the compressor off). (From Lipták [6])

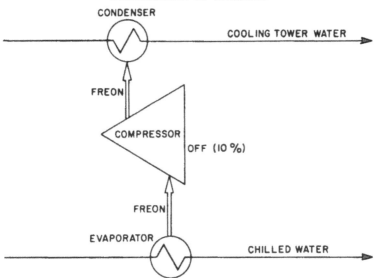

Fig. 4.14 In the thermosiphon mode of operation, the compressor is off, and freon vaporizes in the evaporator and liquifies in the condenser because the temperature gradient is reversed. (From Lipták [6])

process directly. Alternatively, if the cooling tower water temperature is below the temperature required by the process, the chillers can be operated in a thermosiphon mode. Freon circulation is then driven by the temperature differential rather than by a compressor (figure 4.14). In the thermosiphon mode of operation, the chiller capacity drops to about 10 percent of its rating.

Optimized Use of Storage

If storage tanks are available, it is cost-effective to generate the daily brine or chilled water needs of the plant at night, when it is the least expensive to do so, because ambient temperatures are low and night-time electricity is less expensive in some areas [7].

When demand is low, it may be possible to lower operating costs by operating the chillers part of the time at peak efficiency rather than continuously at partial loading. Efficiency tends to be low at partial loads because of losses caused by friction drop across suction dampers, prerotation vanes, or steam governors. Cycling is practical if the storage capacity of the distribution headers is enough to avoid frequent stops and starts. When operation is to be intermittent, data such as the heat to be removed and the characteristics of the available chillers are needed to determine the most economical operating strategy.

When chiller cycling is used, the thermal capacity of the chilled water distribution system is used to absorb the load while the chiller is off. For example, if the pipe distribution network has a volume of 100,000 gallons (378,000 l), this represents a thermal capacity of approximately 1 million BTUs for each degree F of temperature rise (1.9 10⁹ J/°C). So, if one can allow the chilled water temperature to float 5°F (2.8°C) (for example, from 40°F to 45°F or from 4.4°C to 7.2°C) before the chiller is restarted, this represents the equivalent of approximately 400 tons (1405 kW) of thermal capacity. If the load happens to be 200 tons (704 kW), the chiller can be turned off for two hours at a time. If the load is 1,000 tons (3,514 kW), the chiller will be off for only 24 minutes. This illustrates the natural load-following, time-proportioning nature of this scheme.

If the chiller needs a longer period of rest than the thermal capacity of the distribution system can provide, three options are available:

- Tankage can be added to increase the water volume
- A second chiller can be started (not the one that was just stopped)
- The load can be distributed among chillers of different sizes by keeping some in continuous operation while cycling others

Load Allocation

Continuous measurement of the actual efficiency ($/ton) of each chiller can enable all loads to be met through the operation of the most efficient combination of machines for the load. In plants with multiple refrigerant sources, the cost per ton of cooling can be calculated from direct measurements and used to establish the most efficient combination of units to meet present or anticipated loads. As with boilers, this calculation accounts for differences among units as well as for the efficiency-versus-load characteristics of the individual coolant sources, as shown in figure 4.15.

In simple load allocation systems, only the starting and stopping of the chillers is optimized. In such systems, when the load is increasing, the most efficient idle chiller is started; when the load is dropping, the least efficient one is stopped. In more sophisticated systems, the load distribution between operating chillers is also optimized. In such systems, a computer is used to calculate the real-time efficiency of each chiller—that is, to calculate the incremental cost for the next load change for each chiller.

If the load increases, the incremental increase is sent to the set point of the most cost-effective chiller. If the load decreases, the incremental decrease is sent to the least cost-effective chiller (figure 4.16). The required software packages with proved capabilities for continuous load balancing through the predictions of costs and efficiencies are readily available [10]. With the strategy described in figure 4.16, the most efficient chiller will either reach its maximum loading or will enter a region of decreasing efficiency and will no longer be the most efficient. When the loading limit is reached on one chiller, or when a chiller is put on manual, the computer will select another as the most efficient unit for future load increases.

The least efficient chiller will accept all decreasing load signals until its minimum limit is reached. Its load will not be increased unless all other chillers are at their maximum load or in manual. As shown in figure 4.15, some chillers can have high efficiency at normal load while being less efficient than the others at low load. Such

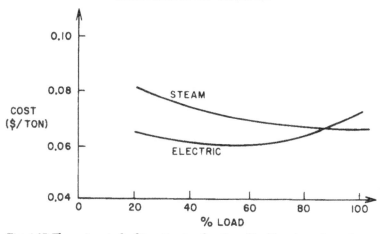

Fig. 4.15 The unit cost of refrigeration is a function of load but depends on the characteristics of each chiller. The indicated cost figures are for representative installed equipment. (From Lipták [7])

units are usually not allowed to shut down, but are given a greater share of the load by a special subroutine.

If all chillers are identical, some will be driven to maximum capacity while others will be shut down by this strategy, and only one chiller will be placed at an intermediate load.

In starting up chiller stations, the optimization system "knows" how many BTUs need to be removed before start-up and the size and efficiency of the available chillers. Therefore, the length of the "pull-down" period can be minimized and the energy cost of this operation optimized.

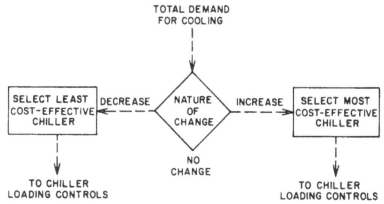

Fig. 4.16 Computer-based load allocation directs load increases to the most cost effective chiller and sends load decreases to the least cost effective chiller.

Multistage Refrigeration Units

It is not practical to obtain a compression ratio outside the range of 3:1 to 8:1 with the compressors used in the process industry. This places a limitation on the minimum temperature that a single-stage refrigeration unit can achieve. For example, in order to maintain the evaporator at $-80°F$ ($-62°C$) and the condenser at $86°F$ ($30°C$), the required compression ratio would be as follows, if Freon-12 was the refrigerant:

$$\text{Compression ratio} = \frac{\text{Refrigerant pressure at } 86°F}{\text{Refrigerant pressure at } -80°F} = \frac{108}{2.9} = 37$$

Such a compression ratio is obviously not practical; therefore, a multistage system is required.

Figure 4.17 illustrates a cascade refrigeration system, with ethylene as the lower and propane as the higher temperature refrigerant. In order to minimize the total operating cost, the loading of the two stages must be coordinated to balance the total work between the stages. The higher the interstage temperature (evaporator #2), the more work is done by stage #1 and the less remains for stage #2. If the compression ratios (and therefore the temperature differences) of the two stages were the same, the work would be nearly equally distributed. If true equality is to be achieved, the differences in the properties of the refrigerants and the added load on the higher stage caused by the work introduced by the lower stage compressor must also be considered.

Another goal of optimization is to maximize the rangeability of the multistage unit. The rangeability is maximum when the two stages approach their surge limits together. This, too, can be guaranteed by redistributing the load between the stages so that they maintain equal distance from their respective surge lines as the load drops.

In figure 4.17, TIC-01, the temperature controller of the chilled process stream, sets the speed of the first stage compressor by modifying the set point of its suction pressure controller. The speed of compressor #2 is set in ratio to #1 by PY-01. This speed ratio set point is optimized by a multivariable control envelope, which considers both cost and rangeability in moving the system to its optimum. As long as both stages are far from surge, the algorithm redistributes the load to minimize total operating cost. When one of the stages is nearing surge, the envelope algorithm increases its loading to keep the system out of surge. Therefore, at high loads the goal of load distribution between the stages is to minimize the total operating cost, and at low loads the goal is to keep the system out of surge.

As ambient temperatures drop, the cascaded stages will operate against a lower total temperature difference; therefore, each of the stages will require less work. As the condenser temperature drops, compressor #2 will lower its suction pressure. This in turn will cause the temperature in evaporator #2 to drop, which will cause compressor #1 to also lower its suction pressure. As the suction pressures drop, PIC-01 and PIC-02 will reduce the speed of their respective compressors, thereby reducing the total work by the heat pumps. Using these procedures, this cascade system responds to ambient variations in a flexible and efficient manner.

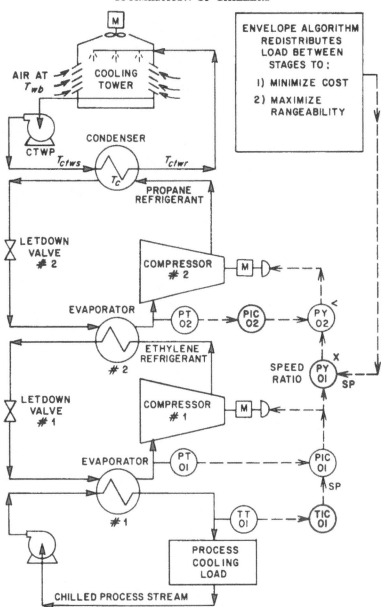

Fig. 4.17 Multistage refrigeration units can be optimized to balance the work between the stages. (Adapted from Shinskey [9])

Control of Multiple Users

When evaporators operate in parallel, off the same compressor, the load can be followed by varying either the evaporator temperature or the refrigerant level in the evaporators. Figure 4.18 shows the controls when the evaporator levels (their wetted heat transfer areas) are kept constant and load variations result in modifications in the evaporator temperature. TY-03 selects the most open TCV and keeps it from opening

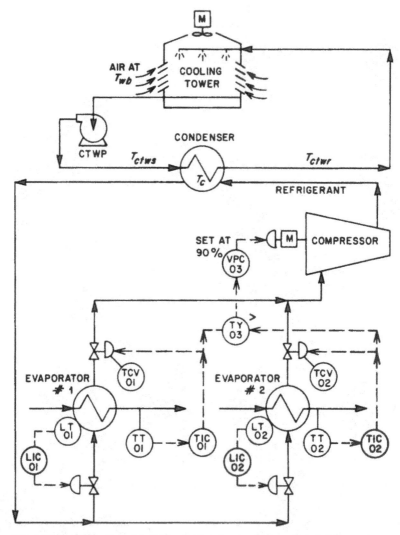

Fig. 4.18 Load-following control of multiple users can be set up so that the most demanding user sets the speed of the compressor.

to more than 90 percent by speeding up the compressor when the valve opening reaches 90 percent. The resulting reduction in compressor suction pressure lowers the evaporator temperature and increases the rate of heat transfer. Thus, no user is ever allowed to run out of coolant, and compressor operating costs are kept at a minimum. This type of load-following optimization system, where compressor speed varies with the requirements of the most heavily loaded user, is very sensitive to dynamic upsets. Therefore, VPC-03 must be tuned for slow adjustment of the compressor speed to avoid upsetting the other users, which are not selected for control.

If one of the user valves is consistently more open than the others, it might be possible to repipe the other loads to the interstage of the compressor. This might provide a better balance between users.

Figure 4.19 illustrates a control system in which the vapor-side control valves have been eliminated. This advantage must be balanced against a number of disadvantages. In this system, heat transfer is modulated by varying the tube surface area exposed to boiling. The user with the highest load will have the highest refrigerant level. LY-03 selects the most flooded evaporator for control by LC-03, which increases the speed of the compressor when the level exceeds 90 percent and lowers it when the level is below.

This method of load following has some drawbacks. One problem is that the heat transfer surfaces within the evaporators are not fully utilized. Another difficulty is that the relationship between level and heat transfer area can be confused by foaming, which keeps the tubes wet even when the level has dropped [9]. For these reasons, tuning of such control system can be difficult, necessitating the sacrifice of responsiveness for stability. Therefore, if responsiveness is critical, it is better to keep the refrigerant level constant and manipulate the evaporating pressure, as shown in figure 4.18.

Precise and Responsive Control

Feedforward anticipation can increase the responsiveness of the control system while also providing more precise temperature control. Such precision is desired when, for example, the goal of optimization is to maximize chilling without freezing.

When the heat transfer area is fixed (constant level), the rate of heat transfer (load) is a function of the temperature difference between the process fluid and the boiling refrigerant. Therefore, the boiling temperature (and pressure) must change as the load varies.

Figure 4.20 shows the main components of this control system. The instantaneous load is calculated by multiplying (FY-05) the flow rate of the cooled process fluid by its desired drop in temperature (TY-04). Based on this load and the desired process temperature, FY-06 determines the desired refrigerant temperature. "K" represents heat transfer area and coefficient. The value of "K" in this summing device is adjustable to match the actual slope of load versus refrigerant temperature. PY-07 serves to convert boiling temperature to the corresponding vapor pressure set point.

Dynamic compensation is provided by FY-03 to match the response of process temperature to refrigerant pressure [9]. A valve position control system, similar to the

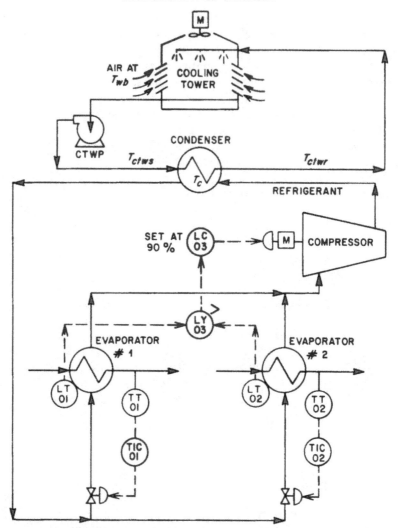

Fig. 4.19 In this alternative method for controlling multiple users, the vapor-side control valves have been eliminated. (Adapted from Shinskey [9])

one shown in figure 4.18, can also be used here to minimize the operating cost of the compressor.

Retrofit Optimization

In new plants it is easy to install variable-speed pumps, to provide the thermal capacity required for chiller cycling, or to locate chillers and cooling towers at the same elevation and near each other so that the cooling tower water pumping costs will be

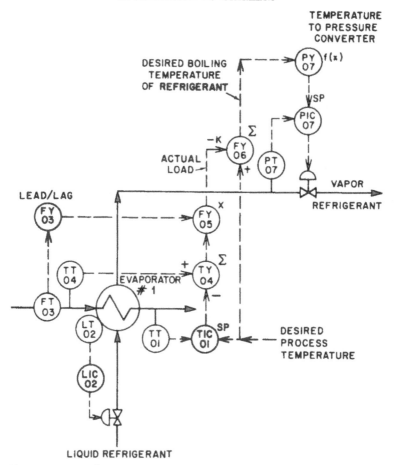

Fig. 4.20 Increased responsiveness can be achieved through the use of feedforward setting of refrigerant pressure. (Adapted from Shinskey [9])

minimized. In existing plants, the inherent design limitations must be accepted, and the system must be optimized without changes to the equipment. This is quite practical to do and can still produce savings up to 50 percent [8], but certain precautions are needed.

In optimizing existing chillers, it is important to give careful consideration to the constraints of low evaporator temperature, economizer flooding, steam governor rangeability, surge, and piping/valving limitations.

A detailed discussion of surge controls is given in the chapter devoted to compressor optimization. Therefore, here it should suffice to just mention that surge occurs at low loads when not enough freon is circulated. A surge condition can cause violent vibration and damage.

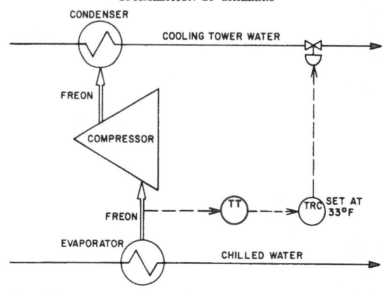

Fig. 4.21 Protection against evaporator freeze-up can be provided by temperature control. (From Lipták [6])

Old chillers usually do not have automatic surge controls and have only vibration sensors for shutdown. If the chillers will operate at low loads, it is necessary to add an antisurge control loop. Surge protection is always provided at the expense of efficiency. To bring the machine out of surge, the freon flow must be artificially increased if there is no real load on the machine. The only way to provide this increase in flow is to add artificial and wasteful loads (for example, hot gas or hot water bypasses). Therefore, it is much more economical either to cycle a large chiller or to operate a small one than to meet low load conditions by running the machine near its surge limit.

Low temperatures can occur in the evaporator when an old chiller is optimized—one that has been designed for operation at 75°F (23.9°C) condenser water and is run at 45 or 50°F (7.2 or10°C) condenser water (in the winter). This phenomenon is exactly the opposite of surge, because it occurs when freon is being vaporized at an excessively high rate. Such vaporization occurs because the chiller is able to pump twice the tonnage for which it was designed as a result of the low compressor discharge pressure. In such a situation, the evaporator heat transfer area becomes the limiting factor; furthermore, the only way to increase heat flow is to increase the temperature differential across the evaporator tubes. This shows up as a gradual lowering of freon temperature in the evaporator until it reaches 32°F (0°C), at which point the machine shuts down to protect against ice formation.

There are two ways to prevent this phenomenon from occurring in existing chillers. The first is to increase the evaporator heat transfer area (a major equipment modification). The second is to prevent the freon temperature in the evaporator from dropping

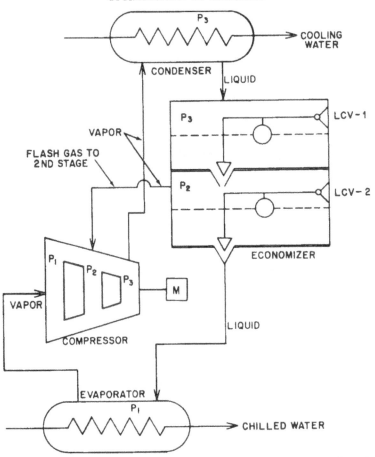

Fig. 4.22 Economizer flooding can be solved by supplementing LCV 1 and 2 with larger external valves. (From Lipták [4])

below 33°F (0.6°C) by not allowing the cooling tower water to cool the condenser to its own temperature. The latter solution requires only the addition of a temperature control loop (figure 4.21). This prevents the chiller from taking full advantage of the available cold water from the cooling tower by throttling its flow rate, thereby causing its temperature to increase.

Retrofitting the Economizer and the Steam Governor

On existing chillers, the economizer control valves (LCV-1 and LCV-2 in figure 4.22) are often sized on the assumption that the freon vapor pressure in the condenser (P3) is constant and corresponds to a condenser water temperature of 75°F or 85°F (23.9°C or 29.4°C). Naturally, when such units are operated with 45°F or 50°F (7.2°C

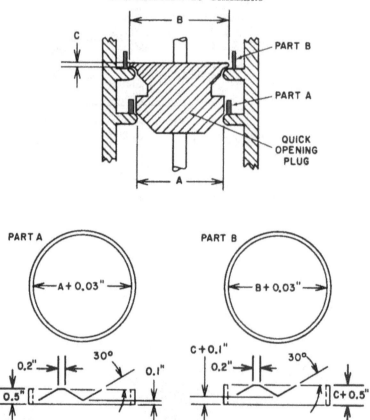

Fig. 4.23 A double-seated steam governor valve can be rebuilt for optimized variable-speed service. The notched rings provide the necessary rangeability. (From
Lipták [6])

or 10°C) condenser water, P3 is much reduced, as is the pressure differential across
LCV-1 and LCV-2. If this occurs—when the freon circulation rate is high—as it easily
can—the control valves will be unable to provide the necessary flow rate, and flooding
of the economizer will occur (the flow is higher and the ΔP is lower than was the basis
of valve sizing). The solution is to install larger valves, preferably external and with
proportional and integral control modes. Proportional controllers alone cannot maintain
the set point as load changes; the addition of the integral mode eliminates this offset.
This is important in machines that were not designed originally for optimized, low-
temperature condenser water operation, because otherwise the compressor can be
damaged by the liquid freon that overflows into it from the flooded flash evaporator.

In order for a steam turbine-driven compressor to be optimized, its rotational
velocity must be modulated over a reasonably wide range. This is not possible with

old, existing machines if they are provided with quick-opening steam governor valves. With such governors a slight increase in lift from the fully closed position results in a substantial steam flow and therefore a substantial rotational velocity. If this steam flow is throttled, the valves become unstable and noisy.

The valve characteristics can be changed from quick-opening to linear at minimal cost. The desired wide rangeability can be obtained by welding two rings with V-notches to the seats of the existing steam governor valves, as shown in figure 4.23.

Retrofitting the Water Distribution Network

Figure 4.7 illustrates the ideal water distribution system, in which the individual users are served by two-way valves and the optimum supply temperature is found by keeping the most open valve at 90-percent opening. Although this is a very effective control configuration, it is not always used in existing installations. If the number of user valves is very high and distributed over a large area or if the loops are not properly tuned and the valves have a tendency of cycling from fully closed to fully open, this control strategy is not practical. In such situations, the controls shown in figure 4.7 can still be applied by selecting a few representative user valves that are not cycling and by basing the optimization on those.

It is also possible to group the user valves into high and low priority categories. Figure 4.24 illustrates a control system in which the high priority users are treated the same way as they were in figure 4.7. The low priority users, on the other hand, are grouped together, and their demand is detected through the measurement of total flow. If all low priority valves are the two-way type, and if supply temperature is constant, flow will vary directly with cooling load. If valve-position-based optimization was applied, it would attempt to maximize valve opening, which in turn would maximize flow by raising Tchws to the highest acceptable level. This same goal is achieved by FIC-2 in figure 4.24. As the flow drops off as a result of a reduction in demand for cooling, FIC-2 will raise the supply temperature, which in turn will increase the total flow, as the valves open up. Therefore, the total flow is kept constant, and load variations result in supply temperature variations. VPC-1 and FIC-2 should be integral-only controllers, with time constants of several minutes.

The control of total flow in figure 4.24 can satisfy the average user, but not all the users. Therefore, the critical users cannot be treated in this manner, but must be protected by valve position control. Such control is provided by VPC-1, which overrides the flow control signal whenever a high priority valve reaches 90-percent opening.

In many existing installations—particularly in air-conditioning applications— three-way control valves are used. This is a highly undesirable practice, because it unnecessarily increases the required amount of pumping and lowers return water temperature. Yet, if the system is already in operation and if it consists of hundreds of air-conditioning type users, it would be better not to attempt to redesign it but to optimize it in its existing form. A control system that will do that is shown in figure 4.25.

In this system, the total flow is relatively constant, and it is the return water temperature that reflects the variations in cooling load. The control system in figure

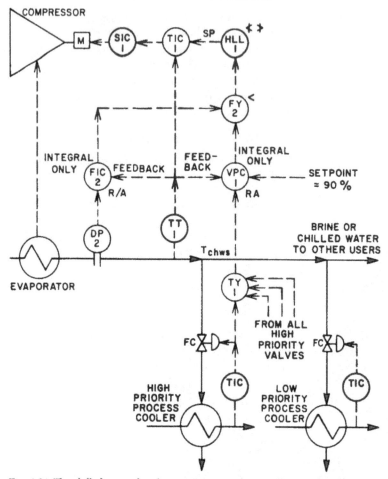

Fig. 4.24 The chilled water distribution system can be retrofitted for supply temperature optimization. (Adapted from Lipták [5])

4.25 determines the set point for TIC-1 by relating the temperature rise (Tchws-Tchwr) and the desired space temperature set points (To) as follows [9]:

$$\text{TIC-1 Set Point} = \text{To} - \text{G (Tchwr} - \text{Tchws)}$$

If the thermostats are set at To = 80°F (27°C) and if, at full load, Tchws = 50°F (10°C) and Tchwr = 65°F (18°C), the value of G can be determined as follows:

$$G = (\text{To} - \text{Set Point}) / (\text{Tchwr} - \text{Tchws}) = (80 - 50) / (65 - 50) = 2$$

TDY-2 in figure 4.25 therefore calculates the desired set point for TIC-1 at any load. For example, if the load drops to 50 percent, the space temperature To will begin to fall, and the thermostats will divert more coolant into the return line. This will reduce

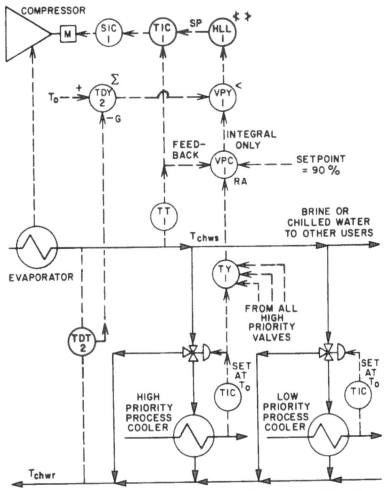

Fig. 4.25 Chilled water users with 3-way valves can also be retrofitted for optimization. (Adapted from Lipták [6])

Tchwr from 65°F (18°C) to 57.5°F (14°C). In conventional systems, this then would become the new steady state. With the control system shown in figure 4.25, TDY-2 will revise the TIC-1 set point as follows:

$$\text{Set Point} = 80 - 2\,(57.5 - 50) = 65°F$$

As Tchws is slowly increasing, the space thermostats will divert less and less water until they come to rest at the same percentage of water diverted as at full load. The benefit of this optimization strategy is in reducing the temperature differential across the chiller, which in turn lowers the operating cost by 15 percent for each 10°F (5.5°C).

The setting of G determines the percentage of coolant diverted by the average valve. If G is set too low, some valves will be 100-percent open to the chilled water supply and therefore out of control. If G is set too high, the chilled water supply temperature will be lower than needed, and compressor energy will be wasted.

In order to protect the high-priority users from running out of coolant when G is set low, a valve-position-based override is used in figure 4.25. Whenever a high-priority valve is more than 90-percent open, VPC-1 is going to override the set point of TIC-1 and lower it until all high-priority valves are less than 90-percent open.

Conclusions

The various goals of chiller optimization include the following:

- Minimizing the temperature difference across the chiller by minimizing cooling tower water temperature and maximizing chilled water temperature
- Minimizing pumping costs by transporting only as much water as is required to meet the load and by using variable speed pumping

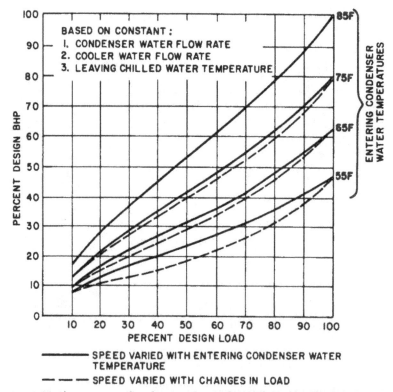

Fig. 4.26 Chant gives combined savings from reducing the condenser water temperature and from using variable speed compressors. (Adapted from Carrier Corporation [1])

- Minimizing cooling tower operating costs by approach optimization
- Operating chillers at the load at which their efficiency is the highest, by meeting partial loads with part-time operation at maximum efficiency and making coolant when least expensive, if night storage is available
- Maximizing the efficiency of load allocation by always using the most efficient chiller combination and using efficiency information to initiate maintenance

The effects of reducing the condenser water temperature and using variable speed compressors are shown in figure 4.26. For example, at 50-percent design load and 55°F condenser water temperature the power consumption will be less than 20 percent. This compares to 53 percent if guide vanes are used on a constant speed machine. If the potentials of all of the above optimization strategies are fully exploited, the operating costs of chiller stations which operate throughout the year and are located in the northern states can be cut in half [8]. The corresponding payback period on investment for optimization controls is usually under a year [7].

With the availability of inexpensive solid state sensors and microprocessors, it is possible to incorporate all the optimization strategies in a single refrigeration unit controller. Such multivariable controllers make it possible to replace the uncoordinated control of flows and temperatures with load following by floating parameters that always correspond to the lowest possible cost of operation.

REFERENCES

1. Carrier Corporation, "Centrifugal Liquid Chillers," *Application Data* 17–IXA, 1975.
2. Cooper, K. W., "Saving Energy with Refrigeration," *ASHRAE Journal*, December 1978.
3. Hallanger, E. C., "Operating and Controlling Chillers," *ASHRAE Journal*, September 1981.
4. Lipták, B. G., "Chiller Controls, *Instrument Engineer's Handbook, Process Control* (Radnor, PA: Chilton Book Co., 1985), section 8.4.
5. Lipták, B. G., "Optimizing Controls for Chillers and Heat Pumps," *Chemical Engineering*, October 17, 1983.
6. Lipták, B. G., "Optimizing Plant Chiller Systems," *InTech*, September 1977.
7. Lipták, B. G., "Save Energy by Optimizing Your Boilers, Chillers, and Pumps," *InTech*, March 1981.
8. Romita, E., "A Direct Digital Control for Refrigeration Plant Optimization," *ASHRAE Transactions*, vol. 83, part 1, 1977.
9. Shinskey, F. G., *Energy Conservation Through Control*, Academic Press, 1978.
10. Zimmer, H., "Chiller Control Using On-line Allocation for Energy Conservation," paper no. 76–522, *Advances in Instrumentation*, part 1, ISA/76 International Conference, Houston, Texas.

Chapter 5

OPTIMIZATION OF CLEAN-ROOMS

CLEAN-ROOMS are used for experimental studies of a wide variety. Testing and analysis laboratories, used in the medical, military, and processing industries, are all designed to operate as clean-rooms. The optimization of clean-rooms, therefore, is an important subject to many industries. The production of semiconductors is one of several processes done in a clean-room environment. Optimization of the clean-room control system will drastically reduce both the energy cost of operation and the cost of manufacturing of off-spec products [5].

The Semiconductor Plant

The process, or production, area of a typical semiconductor manufacturing laboratory is between 100,000 and 200,000 square feet (9,290 and 18,580 square meters). The market value of the daily production from this relatively small area is over one million dollars. Plant productivity is increased if the following conditions are maintained in the process area:

- No drafts (+0.02″ H_2O ± 0.005″ H_2O or 5 Pa ± 1.3 Pa)
- No temperature gradients (72°F ± 1°F or 22°C ± 0.6°C)
- No humidity gradients (35% RH ± 3%)
- No air flow variations (60 air changes/hr ± 5%)

Therefore, the primary goal of optimization is to maximize plant productivity continuously through the accurate control of these parameters. The secondary goal is to conserve energy. The main control elements and control loop configurations required for arriving at a high productivity semiconductor manufacturing plant will be described in this chapter.

In order to prevent contamination through air infiltration from the surrounding spaces, the clean-room pressure must be higher than that of the rest of the building. As shown in figure 5.1, the clean-room is surrounded by a perimeter corridor. The pressure within the clean-room is higher than that in the surrounding corridor to protect against the leakage of dirty air. The clean-room is made up of rows of work stations. These rows are also referred to as *zones*. A typical semiconductor producing plant has approximately 200 work stations, also called *subzones*. Each subzone faces the clean work aisle, and behind it is the service core.

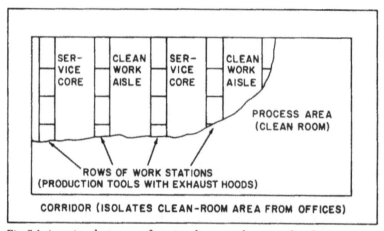

Fig. 5.1 A semiconductor manufacturing clean-room has rows of workstations. Each row is also called a *zone*, and each workstation is also called a *subzone*. (From Lipták [6])

Subzone Optimization

As shown in figure 5.2, air is supplied to the subzone through filters located in the ceiling in front of the work station. The total flow and the temperature of this air supply are both controlled. The evacuated exhaust air header is connected to the lower hood section of the work station. It pulls in all the air and chemical fumes that are generated in the work station and safely exhausts them into the atmosphere. In order to make sure that none of the toxic fumes will spill into the work aisle, clean air must enter the work station at a velocity of about 75 fpm (0.38 m/s). The air that is not pulled in by the exhaust system enters the service core, where some of it is recirculated back into the workstation by a local fan. The rest of the air is returned from the service core by the return air header. A damper in this header (RAR-1) is modulated to control the pressure (DPS-1) in the work aisle.

Pressure Controls

The pressure in the corridor is the reference for the clean-room pressure controller DPS-1 in figure 5.2. This controller is set to maintain a few hundredths of an inch of positive pressure relative to the corridor. The better the quality of building construction, the higher this set point can be, but even with the lowest quality buildings, a setting of approximately 0.02″ H_2O (5Pa) can be maintained easily. Because at such near-atmospheric pressures, air behaves as if it were incompressible, the pressure control loop shown in figure 5.2 is both fast and stable. When the loop is energized, DPS-1 quickly rotates the return air control damper (RAR-1) until the preset differential is reached. At that point, the electric motor stops rotating the damper, and it stays at its

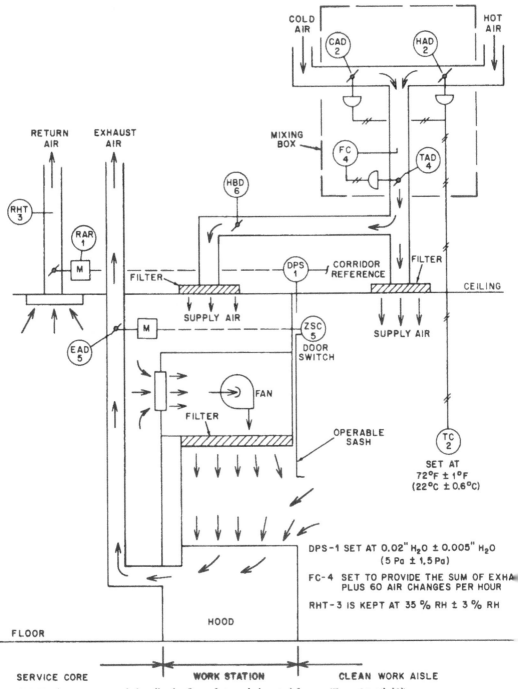

Fig. 5.2 Work station controls handle the flow of air and chemical fumes. (From Lipták [6])

201

last opening. This position will remain unaltered as long as the air balance in the area remains the same:

Return Air Flow = Supply Air Flow − (Exhaust Air Flow + Pressurization Loss)

When this air flow balance is altered (for example, as a result of a change in exhaust air flow), it will cause a change in the space pressure, and DPS-1 will respond by modifying the opening of RAR-1.

Elimination of Drafts

Plant productivity is maximized if drafts are eliminated in the clean work aisle. Drafts would stir up the dust in this area, which in turn would settle on the product and cause production losses. In order to eliminate drafts, the pressure at each work station must be controlled at the same value. Doing so eliminates the pressure differentials between stations and therefore prevents drafts. When a DPS-1 unit is provided to control the pressure at each work station, the result is a uniform pressure profile throughout the clean room.

PDS-1 usually controls the pressure in the clean work aisle on the "process" side of the work station. Yet, it is important to make sure that *all* points, including the service aisle, are under positive pressure. Because the local circulating fan within the work station draws the air in from the service core and discharges it into the clean work aisle, the pressure in the service core will always be lower than that on the process side (see figure 5.2). Therefore, it is possible for localized vacuum zones to evolve in the service core, which could cause contamination by allowing air infiltration. To prevent this from happening, several solutions have been proposed.

One possibility is to have DPS-1 to control the service core pressure. This is not recommended, because a draft-free process area can be guaranteed only if there are no pressure gradients on the work aisle side; this can be achieved only by locating the DPS-1 units on the work aisle side of the work stations.

Another possibility is to leave the pressure controls on the work aisle side but raise the set point of DPS-1 until all the service cores in the plant are also at a positive pressure. This solution cannot be universally recommended either, because the quality of building construction might not be high enough to allow operation at elevated space pressures. The pressurization loss in badly sealed buildings can make it impossible to reach the elevated space pressure. Yet another solution is to install a second DPS controller, which would maintain the service core pressure by throttling the damper HBD-6. This solution will give satisfactory performance but will also increase the cost of the control system by the addition of a few hundred control loops (one per work station).

A more economical solution is shown in figure 5.2. Here, a hand-operated bypass damper (HBD-6) is manually set during initial balancing. This solution is reasonable for most applications, because the effect of the work station fan is *not* a variable, and therefore a constant setting of HBD-6 should compensate for it.

Temperature Controls

The temperature at each work station is controlled by a separate thermostat, TC-2 in figure 5.2. This temperature controller adjusts the ratio of cold air to hot air within the supply air mixing box to maintain the space temperature.

Unfortunately, conventional thermostats cannot be used if the temperature gradients within the clean-room are to be kept within ± 1°F of 72°F (±0.6°C of 22°C). Conventional thermostats cannot meet this requirement with regard to measurement accuracy or control quality. Even after individual calibration, it is unreasonable to expect less than a ±2°F or 3°F (±1°C or 1.7°C) error in overall loop performance if HVAC-quality thermostats are used. Part of the reason for this is the fact that the "offset" cannot be eliminated in plain proportional controllers, such as thermostats [7].

Thus, in order for the thermostat to move its output from the midscale value (50%-50% mixing of cold and hot air) an error in room temperature must first develop. This error is the permanent offset. The size of this offset error for TC-2 in figure 5.2 for the condition of maximum cooling can be determined as follows:

$$\text{Offset Error} = \frac{\text{Spring Range of CAD}}{2 \text{ (Thermostat Gain)}}$$

Assuming that CAD-2 has an 8 to 13 PSIG (55 to 90 kPa) spring (a spring range of 5 PSI, or 34.5 kPa) and that TC-2 is provided with a maximum gain of 2.5 PSI/°F (31 kPa/°C), the offset error is 1°F (0.6°C). Under these conditions, the space temperature must permanently rise to 73°F (22.8°C) before CAD-2 can be fully opened. The offset error will increase as the spring range increases or as the thermostat gain is reduced. Sensor and set point dial error are always additional to the offset error.

Therefore, in order to control the clean-room temperature within ± 1°F (±0.6°C), it is necessary to use an RTD-type or a semiconductor transistor-type temperature sensor and a proportional-plus-integral controller, which will eliminate the offset error. This can be most economically accomplished through the use of microprocessor-based shared controllers that communicate with the sensors over a pair of telephone wires that serve as a data highway.

Humidity Controls

The relative humidity sensors are located in the return air stream (RHT-3). In order to keep the relative humidity in the clean-room within 35% RH ± 3% RH, it is important to select a sensor with a lower error than ± 3% RH. The repeatability of most human hair element sensors is approximately ± 1% RH. These units can be used for clean-room control applications if they are *individually calibrated* for operation at or around 35% RH. Without such individual calibration, they will not perform satisfactorily, because their off-the-shelf inaccuracy, or error, is approximately ± 5% RH. The controller associated with RHT-3 is not shown in figure 5.2, because relative humidity is not controlled at the work station (subzone) level but at the zone level (as mentioned earlier, a zone is a row of work stations). The control action is based on the relative humidity reading in the combined return air stream from all work stations within that zone.

Flow Controls

The proper selection of the mixing box serving each work station is of critical importance. Each mixing box serves the dual purpose of providing accurate control of the total air supply to the subzone and modulating the ratio of "cold" and "hot" air to satisfy the requirements of the space thermostat TC-2.

The total air supply flow to the subzone should equal 60 air changes per hour *plus* the exhaust rate from that subzone. This total air supply rate must be controlled within ± 5% of actual flow by FC-4, over a flow range of 3:1. The rangeability of 3:1 is required because as processes change, their associated exhaust requirements will also change substantially. FC-4 in figure 5.2 can be set manually, but this setting must change every time a new tool is added to or removed from the subzone. The setting of FC-4 must be done by individual in-place calibration against a portable hot wire anemometer reference. Settings based on the adjustments of the mixing box alone (without an anemometer reference) will not provide the required accuracy.

Some of the mixing box designs available on the market are not acceptable for this application. Unacceptable designs include the following:

- "Pressure-dependent" designs, in which the total flow will change if air supply pressures vary. Only "pressure-independent" designs can be considered, because both the cold and the hot air supply pressures to the mixing box will vary over some controlled minimum.
- Low rangeability designs. A 3:1 rangeability with an accuracy of ± 5% of *actual* flow is required.
- Selector or override designs. In these designs either flow or temperature is controlling on a selective basis. Such override designs will periodically disregard the requirements of TC-2 and will thus induce upsets, cycling, or both.

If the mixing box is selected to meet the aforementioned criteria, both air flow and space temperature can be accurately controlled.

Optimization of the Zones

Each row of work stations shown in figure 5.1 is called a zone, and each zone is served by a cold deck (CD), a hot deck (HD), and a return air (RA) subheader. These subheaders are frequently referred to as *fingers*. The control devices serving the individual work stations (subzones) will be able to perform their assigned control tasks if the "zone finger" conditions make it possible for them to do so.

For example, RAR-1 in figure 5.2 will be able to control the subzone pressure as long as the ΔP across the damper is high enough to remove all the return air without requiring the damper to open fully. As long as the dampers are throttling (neither fully open nor completely closed), DPS-1 in figure 5.2 is in control.

PIC-7 in figure 5.3 is provided to control the vacuum in the RA finger and thereby to maintain the required ΔP across RAR-1. PIC-7 is a non-linear controller with a fairly wide neutral band. This protects the CD finger temperature (TIC-6 set point) from being changed until a sustained and substantial change takes place in the detected RA pressure.

Similarly, the mixing box in figure 5.2 will be able to control subzone supply flow and temperature as long as its dampers are not forced to take up extreme positions. Once a damper is fully open, the associated control loop is out of control. Therefore, the purpose of the damper position controllers (DPCs) in figure 5.3 is to prevent the corresponding dampers from having to open fully.

Lastly, the relative humidity in the return air must also be controlled within acceptable limits.

Therefore, at the zone level there are five controlled, or limit, variables but only one manipulated variable:

Limit or Controlled Variables	*Manipulated Variable*
RA pressure	TIC-6 set point
RA relative humidity	
Max. CAD opening	
Max. HAD opening	
Max. TIC-6 set point	

Figure 5.3 shows the control loop configuration required to accomplish the aforementioned goals. It should be noted that dynamic lead/lag elements are not shown and that this loop can be implemented in either hardware or software.

Envelope Optimization

Whenever the number of control variables exceeds the number of available manipulated variables, it is necessary to apply multivariable envelope control. This means that the available manipulated variable (TIC-6 set point) is not assigned to serve a single task but is selectively controlled to keep many variables within acceptable limits— within the "control envelope" shown in figure 5.4.

By adjusting the set point of TIC-6, it is possible to change the cooling capacity represented by each unit of CD air. Because the same cooling can be accomplished by using less air at lower temperatures or by using more air at higher temperatures, the overall material balance can be maintained by manipulating the set point of the heat balance controls. Return air humidity can similarly be affected by modulating the TIC-6 set point, because when this set point is increased, more CD air will be needed to accomplish the required cooling. Increasing the ratio of humidity-controlled CD air in the zone supply (HD humidity is uncontrolled) also brings the zone closer to the desired 35%RH set point.

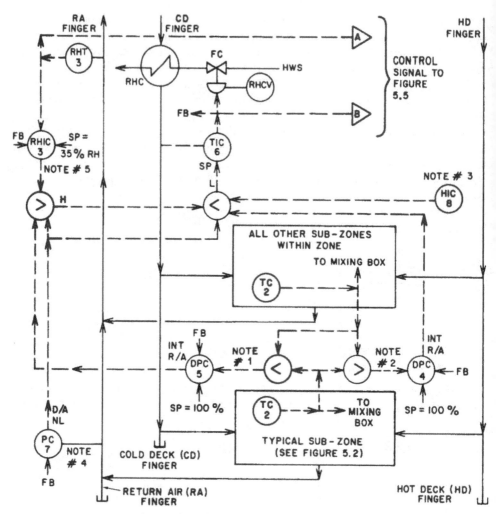

Fig. 5.3 Typical zone control system with envelope control. (From Lipták [6])

Note 1: This is the opening of the most open hot air damper. HAD-2 (shown in figure 5.2) is fully open when the TC-2 output is 3 PSIG (21 kPa).

Note 2: This is the opening of the most open cold air damper. CAD-2 (shown in figure 5.2) is fully open when the TC-2 output is 15 PSIG (104 kPa).

Note 3: HIC-8 sets the maximum limit for the CD temperature at approximately 70°F (21°C).

Note 4: As vacuum increases (and pressure decreases), the output of PIC-7 also increases.

Note 5: D/A in summer, R/A in winter.

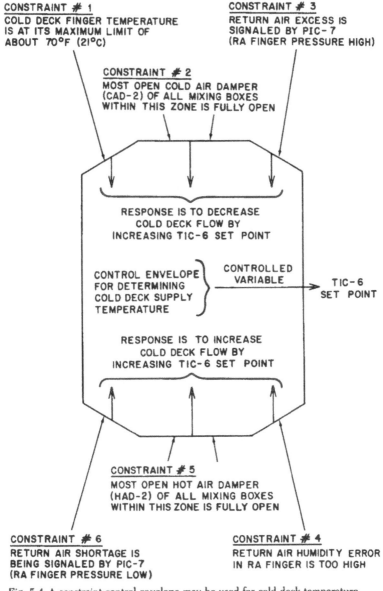

CONSTRAINT # 1
COLD DECK FINGER TEMPERATURE
IS AT ITS MAXIMUM LIMIT OF
ABOUT 70°F (21°C)

CONSTRAINT # 3
RETURN AIR EXCESS IS
SIGNALED BY PIC-7
(RA FINGER PRESSURE HIGH)

CONSTRAINT # 2
MOST OPEN COLD AIR DAMPER
(CAD-2) OF ALL MIXING BOXES
WITHIN THIS ZONE IS FULLY OPEN

RESPONSE IS TO DECREASE
COLD DECK FLOW BY
INCREASING TIC-6 SET POINT

CONTROL ENVELOPE
FOR DETERMINING
COLD DECK SUPPLY
TEMPERATURE

CONTROLLED
VARIABLE

TIC-6
SET POINT

RESPONSE IS TO INCREASE
COLD DECK FLOW BY
INCREASING TIC-6 SET POINT

CONSTRAINT # 5
MOST OPEN HOT AIR DAMPER
(HAD-2) OF ALL MIXING BOXES
WITHIN THIS ZONE IS FULLY OPEN

CONSTRAINT # 6
RETURN AIR SHORTAGE IS
BEING SIGNALED BY PIC-7
(RA FINGER PRESSURE LOW)

CONSTRAINT # 4
RETURN AIR HUMIDITY ERROR
IN RA FINGER IS TOO HIGH

Fig. 5.4 A constraint control envelope may be used for cold deck temperature
optimization. (From Lipták [6])

In this envelope control system, the following conditions will cause an increase in the TIC-6 set point:

- Return air shortage detected by a drop in the RA finger pressure (increase in the detected vacuum) measured by PIC-7. An increase in the TIC-6 set point will increase the CD demand, which in turn lowers the HD and, therefore, the RA demand.
- Hot air damper in mixing box (HAD-2 in figure 5.2) is near to being fully open. This too necessitates the aforementioned action to reduce HD demand.
- RA humidity does not match the RHIC-3 set point. This condition also necessitates an increase in the proportion of the CD air in the zone supply. Because the moisture content of the CD air is controlled, an increase in its proportion in the total supply will help control the RH in the zone.

On the other hand, the following conditions will require a decrease in the TIC-6 set point:

- Return air excess detected by a rise in RA finger pressure (drop in the detected vacuum) measured by PIC-7.
- Cold air damper in mixing box (CAD-2 in figure 5.2) is fully open.
- CD finger temperature exceeds 70°F (21°C). This limit is needed to keep the CD always cooler than the HD.

Plant-Wide Optimization

In conventionally controlled semiconductor plants, both the temperature and the humidity of the main CD supply air are fixed. This can severely restrict the performance of such systems, because as soon as a damper or a valve is fully open (or closed), the conventional system is out of control. The optimized control system described here does not suffer from such limitations, because it automatically adjusts both the main CD supply temperature and the humidity so as to follow the load smoothly and continuously, never allowing the valves or dampers to lose control by fully opening or closing. The net result is not only increased productivity but also reduced operating costs.

A semiconductor production plant might consist of two dozen zones. Each of the zones can be controlled as shown in figure 5.3. The total load represented by all the zones is followed by the controls shown in figure 5.5. The overall control system is hierarchical in its structure: the subzone controls in figure 5.2 are assisted by the zone controls in figure 5.3, and the plant-wide controls in figure 5.5 guarantee that the zone controllers can perform their tasks. The interconnection among the levels of the hierarchy is established through the various valve and damper position controllers (VPCs and DPCs). These guarantee that no throttling device, such as mixing boxes or RHC valves, anywhere in this plant will ever be allowed to reach an extreme position. Whenever a control valve or a damper is approaching the point of losing control (nearing full opening), the load-following control system at the next higher level modifies the air or water supply conditions so that the valve or damper in question will not need to open fully. This hierarchical control scheme is depicted in figure 5.6.

The overall plant-wide control system shown in figure 5.5 can be viewed as a flexible combination of material balance and heat/humidity balance controls. Through the load-following optimization of the set points of TIC-1 and TIC-11, the heat balance controls are used to assist in maintaining the material balance around the plant. For example, if the material balance requires an increase in air flow and the heat balance requires a reduction in the heat input to the space, both requirements will be met by admitting more air at a lower temperature.

Material Balance Controls

The plant-wide material balance is based on pressure control. PC-9 and PC-10 modulate the variable volume fans so as to maintain a minimum supply pressure in each of the CD and HD fingers. The suction side of the CD supply fan station is open to the outside and will draw as much outside air as the load demands (figure 5.5).

The suction pressure of the HD supply fan station is an indication of the balance between RA availability and HD demand. This balance is maintained by PIC-7 in figure 5.3 at the zone level. Because this controller manipulates a heat transfer system (RHCV), it must be tuned for slow, gradual action. This being the case, it will not be capable of responding to sudden upsets or to emergency conditions, such as the need to purge smoke or chemicals. Such sudden upsets in material balance are corrected by PC-6 and PC-7 in figure 5.5. PC-6 will open a relief damper if the suction pressure is high, and PC-7 will open a make-up damper if it is low. In between these limits, both dampers will remain closed and the suction pressure will be allowed to float.

Heat Balance Controls

Heat balance also requires a multivariable envelope control system (similar to that shown in figure 5.4), because the number of controlled variables exceed the number of available manipulated variables. Therefore, the plant-wide air and water supply temperature set points (TIC-1 and TIC-11) are not adjusted as a function of a single consideration but are selectively modulated to keep several variables within acceptable limits inside the control envelope (figure 5.5).

By adjusting the set points of TIC-1 and TIC-11, it is possible to adjust the cooling capacity represented by each unit of CD supply air and the heating capacity of each unit of hot water. This then provides not only for load following control but also for minimization of the operating costs. Cost reductions are accomplished by minimizing the HWS temperature and by minimizing the amount of simultaneous cooling and reheating of the CD air.

This control envelope is so configured that the following conditions will decrease the TIC set points (both TIC-1 and TIC-11):

- The least open RHCV is approaching full closure
- CD supply temperature is at its maximum limit

On the other hand, the following conditions will require and increase in the TIC set points:

- The most open RHCV is approaching full opening
- The CD supply temperature is at its minimum limit

The set point of VPC-5 is produced by a function generator, f(x). The dual purposes of this loop are to prevent any of the reheat coil control valves (RHCVs) from fully opening and thereby losing control and to force the least open RHCV toward a minimum opening (to minimize wasteful overlap between cooling and reheat) but to keep it from full closure as long as possible. These purposes are accomplished by f(x), which keeps the least open valve at approximately 12-percent opening as long as the most open valve is less than 90 percent open. If this 90-percent opening is exceeded, f(x) prevents the most open valve from fully opening. It does so by lowering the set point of VPC-5, which in turn will increase the set points of the TICs. Increasing the set point of TIC-1 reduces the load on the most open RHCV, whereas increasing of the TIC-11 set point increases the heating capacity at the reheat coil. Through this method of load following and through the modulation of main supply air and water temperatures, all zones in the plant will be kept under stable control if the loads are similar in each zone.

Mechanical Design Limitations

All control systems—including this one—will lose control when the design limits of the associated mechanical equipment are reached or if the dissimilarity in load distribution is greater than was anticipated by the mechanical design. Therefore, if a condition ever arises in which one zone requires large amounts of cooling (associated RHCV closed) while some other zone at the same time requires its finger reheat coil valve (RHCV) to be fully open, then the control system must decide which condition it is to correct, because it cannot correct both. The control system in figure 5.5 is so configured that it will give first priority to preventing the RHCV valve from fully opening. Therefore, if as a result of mechanical design errors or misoperation of the plant there is no CD supply temperature that can keep all valves from fully closing or fully opening, then the control system will allow some RHCV valves to close while keeping all of them less than 100-percent open. Whenever such a condition is ap-

Fig. 5.5 The total load represented by all zones is handled by the plant-wide optimization control system. (From Lipták [6])
Symbols: DPtIC = Dew point controller. PID = Controller with proportional, integral, and derivative control modes. AMC = Controller with positions for automatic-manual-cascade modes of operation. VPC = Valve position controller. INT = Controller with integral control action only. Its setting is to be ten times the integral time setting of the associated process controller(s).
Note 1: This pressure controller keeps the lowest of all finger pressures at a value above some minimum. It modulates the fan volume or speed and starts extra fan units to meet the load. When the load drops, the unnecessary fans are stopped by flow (not pressure) control.
Note 2: If the pressure rises above −1"H$_2$O (−0.25 kPa), the relief damper starts to open, and if the pressure drops below −3"H$_2$O (−0.75 kPa), the make-up damper opens. This opening is limited by TC-8, which prevents the HD temperature from dropping below 72° (22°C). Between the settings of PC-6 and PC-7, both dampers are tightly closed and the suction pressure floats.

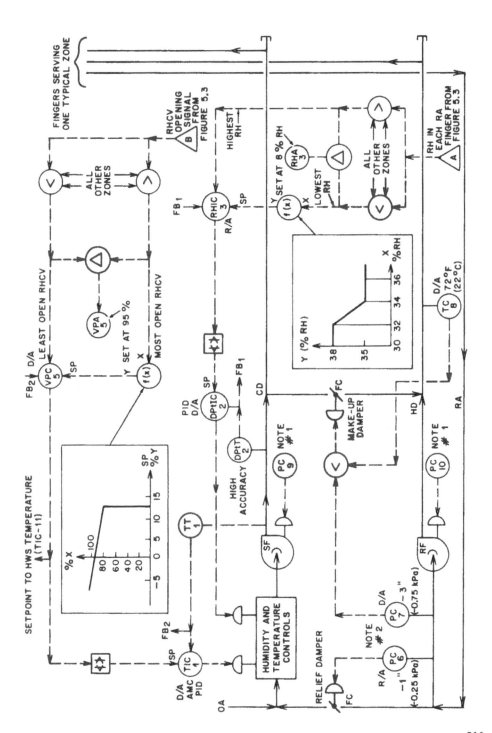

proaching (that is, when the difference between the opening of the most open and the least open valves reaches 95 percent), a valve position alarm (VPA-5) is actuated. This allows the operator to check the causes of such excessively dissimilar loads between zones and to take corrective action by revising processes, relocating tools, modifying air supply ducts, and/or adding or removing mixing boxes.

By keeping all RHCV valves from being nearly closed, this control system simultaneously accomplishes the following goals:

- Eliminates unstable (cycling) valve operation by not allowing valves to operate nearly closed
- Minimizes pumping costs by minimizing pressure losses through throttling valves
- Minimizes heat pump operating costs by minimizing the required hot water temperature
- Provides a means of detecting and thereby smoothly following the variations in the plant-wide load

Humidity Controls

At the zone level, the RA humidity is controlled by RHIC-3 in figure 5.3. In addition, the dew point of the main CD air supply must be modulated. The main CD supply temperature is already being modulated to follow the load. This is accomplished by measuring the relative humidity in all RA fingers (RHT-3 in figure 5.3) and selecting the one finger that is farthest away from the control target of 35-percent RH.

The control system is similar to the TIC set point optimization loop discussed earlier. The first elements in the loop are the selectors shown in figure 5.5. They pick out the return air fingers with the highest and the lowest relative humidities. The purpose of the loop is to "herd" all RH transmitter readings in such a direction that *both* the highest and the lowest readings will fall within the acceptable control gap limits of 35% ± 3% RH. This is accomplished by sending the highest reading to RHIC-3 in figure 5.5 as its measurement and also sending the lowest reading to this RHIC-3 as its modified set point.

A humidity change in either direction can be recognized and corrected through this herding technique. The set point of RHIC-3 is produced by the function generator, f(x). Its dual purposes are to prevent the most humid RA finger from exceeding 38-percent RH and to keep the driest RA finger humidity from dropping below 32-percent RH, as long as this can be accomplished without violating the first goal.

These purposes are accomplished by keeping the set point of RHC-3 at 35 percent as long as the driest finger reads 34-percent RH or more. If it drops below that value, the set point is raised to the limit of 38-percent RH in order to overcome this low humidity condition, without allowing excessive humidity in the return air of some other zone.

This control strategy and the RHIC-3 controller in figure 5.3 will automatically respond to seasonal changes and will give good RH control as long as the loads in the various zones are similar. If the humidity loads are substantially dissimilar, this control system is also subject to mechanical equipment limitations. In other words, the me-

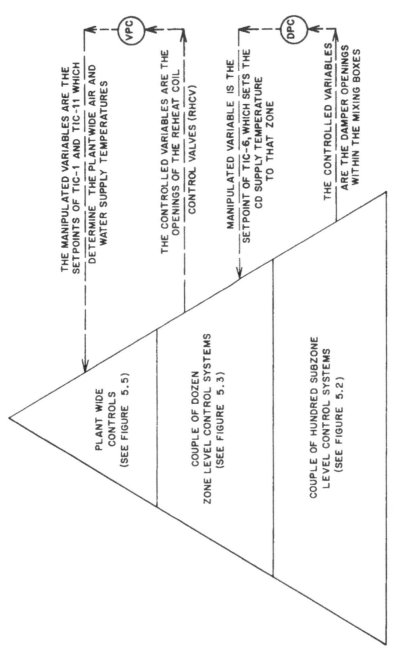

THE MANIPULATED VARIABLES ARE THE SETPOINTS OF TIC-1 AND TIC-11 WHICH DETERMINE THE PLANTWIDE AIR AND WATER SUPPLY TEMPERATURES

THE CONTROLLED VARIABLES ARE THE OPENINGS OF THE REHEAT COIL CONTROL VALVES (RHCV)

MANIPULATED VARIABLE IS THE SETPOINT OF TIC-6, WHICH SETS THE CD SUPPLY TEMPERATURE TO THAT ZONE

THE CONTROLLED VARIABLES ARE THE DAMPER OPENINGS WITHIN THE MIXING BOXES

VPC

DPC

PLANT WIDE CONTROLS (SEE FIGURE 5.5)

COUPLE OF DOZEN ZONE LEVEL CONTROL SYSTEMS (SEE FIGURE 5.3)

COUPLE OF HUNDRED SUBZONE LEVEL CONTROL SYSTEMS (SEE FIGURE 5.2)

Fig. 5.6 The overall control architecture in a semiconductor manufacturing plant is hierarchical in structure. (From Lipták [6])

chanical equipment is so configured that addition or removal of moisture is possible only at the main air supply to the CD. Consequently, if some zones are moisture-generating and others require humidification, this control system can respond to only one of these needs. Therefore, if the lowest RH reading is below 32 percent while the highest is already at 38 percent, the low humidity condition will be left uncontrolled, and the high humidity zone will be controlled to prevent it from exceeding the 38 percent RH upper limit. Allowing the minimum limit to be temporarily violated while maintaining control on the upper limit is a logical and safe response to humidity load dissimilarities between zones. It is safe because the intermixing of the return airs will allow self-control of the building by transferring moisture from zones with excess humidity to ones with humidity deficiency.

Whenever the difference between minimum and maximum humidity readings reaches 8-percent RH, the RHA-3 alarm is actuated. This will alert the operator that substantial moisture load dissimilarities exist and will cause him to find and eliminate the causes.

Exhaust Air Controls

The exhaust air controls at the subzone level are shown in figure 5.2. At this level, the control element is a two-position damper, EAD-5. When the work station is functioning, its operable sash is open. This condition is detected by the door position limit switch, ZSC-5, which in that case fully opens the two-position damper. When the work station is out of service, its operable sash is closed, and therefore the door switch closes EAD-5 to its minimum position. This minimum damper position still provides sufficient air exhaust flow from the work station to guarantee the face velocity needed for operator safety. In other words, when the sash is closed, the air will enter the work station over a smaller area; therefore, less exhaust air flow is required to provide the air inflow velocity that is needed to keep the chemical fumes from leaking out. Through this technique, the safety of operation is unaffected, and operating costs are lowered because less outside air needs to be conditioned if the exhaust air flow is lowered.

In order for EAD-5 to maintain the required exhaust air flows accurately, it is necessary to keep the vacuum in the exhaust air collection ductwork at a constant value. This pressure control loop is depicted in figure 5.7. In each zone exhaust finger, the PC-1 shown in figure 5.7 keeps the vacuum constant by throttling the EAD-1 damper as required. In order for these dampers to stay in control while the exhaust air flow varies, it is important to keep them from fully opening. This is accomplished first by identifying the most open finger damper (low selector) and then by comparing its opening with the set point of the damper position controller, DPC-2. The task of this controller is to limit the opening of the most open EAD-1 to 80 percent and to increase the vacuum in the main EA header if this opening would otherwise exceed 80 percent. Therefore, if the measurement of DPC-2 exceeds 80 percent (drops below 9 PSIG, or 62.5 kPa), the vacuum set point of PC-3 is increased (pressure setting lowered). This in turn will increase the operating level of the exhaust fan station (EAF).

The limits on the set point of DPC-2 will prevent mechanical damage, such as

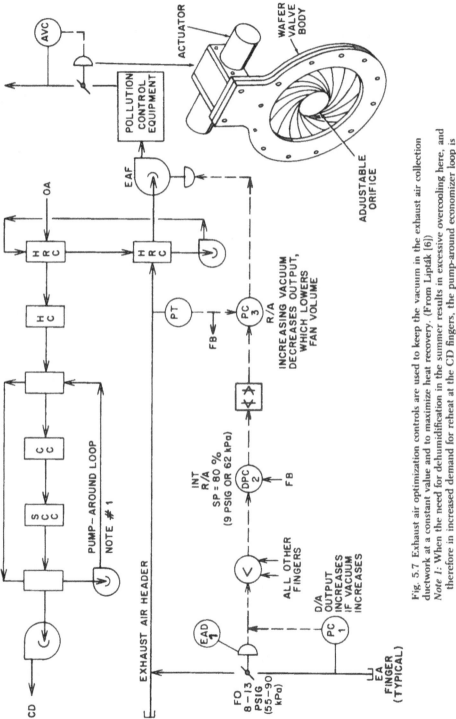

Fig. 5.7 Exhaust air optimization controls are used to keep the vacuum in the exhaust air collection ductwork at a constant value and to maximize heat recovery. (From Lipták [6])

Note 1: When the need for dehumidification in the summer results in excessive overcooling here, and therefore in increased demand for reheat at the CD fingers, the pump-around economizer loop is started to lower operating cost and increase efficiency by reducing the degree of overcooling and thereby lowering the need for reheat.

215

collapsing of the ducts as a result of excessive vacuum or manual misoperation. The reasons for integral action and external feedback are the same as in the case of all other damper and valve position controllers discussed earlier.

Figure 5.7 also shows a glycol-circulating heat recovery loop that can be used as shown to preheat the entering outside air or that can be used as a heat source to a heat pump in the winter (not shown). In either case, the plant operating cost is lowered by recovering the heat content of the air before it is exhausted in the winter.

The discharging of chemical vapors into the atmosphere is regulated by pollution considerations. The usual approach is to remove most of the chemicals by adsorption and scrubbing prior to exhaustion of the air. An added measure of safety is provided by exhausting the air at high velocity, so as to obtain good dispersion in the atmosphere. Because the volume of air being exhausted varies, an air velocity controller (AVC in figure 5.7) is used to maintain the velocity of discharge constant. This is accomplished by modulation of a variable orifice iris damper, also illustrated in figure 5.7.

Conclusions

The productivity of semiconductor manufacturing plants can be greatly increased and the operating costs can be lowered through the instrumentation and control methods described previously and by such added control system features as:

- Low leakage dampers (0.5 CFM/ft^2 at 4″ H$_2$O ΔP, or 2.5 l/s/m^2 at 1 kPa ΔP)
- Accurate air flow metering for material balance and pressurization loss control
- Pump-around economizers (see figure 5.7)

The initial cost of the previously described control system is not higher than the cost of conventional systems, because the added expense of more accurate sensors is balanced by the much-reduced installation cost of distributed shared controls. Therefore, the benefit of increased productivity is a result not of higher initial investment but of better control system design.

It should be emphasized that the described load-following optimization envelope control strategy is far from being typical of today's practices in the semiconductor manufacturing industry. In fact, it has never been fully implemented, and many plants are still operated under conventional HVAC controls [6]. Therefore, the conclusion that yields will increase and operating costs will drop when such improved control systems are installed is an assumption. This appears to be supported by the results of partial system implementations and by the experiences in other industries but will remain an assumption until the first semiconductor manufacturing plant with such modern controls is operating.

REFERENCES

1. Daryanani, S., "Design Engineer's Guide to Variable Air-Volume Systems," *Actual Specifying Engineer*, July 1974.

2. DHO/Atlanta Corporation, "Conserving Energy," Powered Induction Unit, data sheet #1, May 1, 1974.

3. Kusuda, T. "Intermittent Ventilation for Energy Conservation," ASHRAE symposium #20, paper #4.

4. Lipták, B. G., ed., "Carbon Dioxide Sensors," in *Environmental Engineer's Handbook*, vol. 2

(Radnor, PA: Chilton Book Co., 1975), section 4.6.

5. Lipták, B. G., "Envelope Control for Clean Room Improves Product Quality," *I&CS*, September 1982.

6. Lipták, B. G., *Instrument Engineer's Handbook, Process Control* (Radnor, PA: Chilton Book Company, 1985), section 8.14.

7. Lipták, B. G., *Instrument Engineer's Handbook, Process Measurement* (Radnor, PA: Chilton Book Company, 1982), section 4.12.

8. Lipták, B. G., "Reducing Operating Costs of Buildings by the Use of Computers," *ASHRAE Transactions*, vol. 83, part 1, 1977.

9. Lipták, B. G., "Save Energy by Optimizing Your Boilers, Chillers, and Pumps," *Instrumentation Technology*, March 1981.

10. Lipták, B. G., "Savings Through CO_2 Based Ventilation," *ASHRAE Journal*, July 1979.

11. Nordeen, H., "Control of Ventilation Air in Energy Efficient Systems," ASHRAE symposium #20, paper #3.

12. Shih, J. Y., "Energy Conservation and Building Automation," ASHRAE Paper #2354.

13. Spielvogel, L. G., "Exploding Some Myths about Building Energy Use," *Architectural Record*, February 1976.

14. Stillman, R. B., *Systems Simulation Engineering Report*, prepared for IBM-RECD, December 10, 1971.

15. Woods, J. E., "Impact of ASHRAE Ventilation Standards 62–73 on Energy Use," ASHRAE symposium #20, paper #1.

Chapter 6

OPTIMIZATION
OF
COMPRESSORS AND FANS

THE TRANSPORTATION of vapors and gases is a major element in the operating cost of processing plants. The optimization of these unit operations (fans, blowers, compressors) can substantially lower the total operating cost of the plant. The goal of this chapter is to describe the optimization strategies available to increase the safety and energy efficiency in these systems. Prior to the discussion of state-of-the-art advanced controls, the basic equipment and the conventional control strategies will also be briefly described.

The gas transportation equipment will be discussed in four major categories: fans, rotary blowers, reciprocating compressors, and centrifugal compressors. The emphasis will be on the centrifugal units, whose load, surge, and override controls will be described in some detail; multiple units will also be considered.

Fans

Fans are classified into axial flow and radial (or centrifugal) flow types according to the nature of the flow through the blade passages. They move large volumes of air at relatively low pressures ($"H_2O$), with pressure-flow characteristics as shown in figure 6.1.

If the operating point is to the left of maximum pressure on the fan curve, and particularly if the fan pressure exceeds $10"H_2O$, pulsations and unstable operation can occur. This maximum point on the curve is referred to as the *surge point* or *pumping limit*. All fans must always operate to the right of the surge point. Under low load conditions the fan can be kept out of surge either by artificially increasing the load through venting or by modifying the fan curve by the use of pitch, speed, or vane control.

Figure 6.2 illustrates how the characteristic curve of the fan (A) is modified as the inlet vane is throttled down (from A to E), shifting the surge point to the left. This then allows stable operation at lower flows.

Tubeaxial and vaneaxial fans are made with adjustable pitch blades in order to permit the balancing of the fan against the system or to make infrequent adjustments. Vaneaxial fans are also produced with controllable pitch blades (that is, pitch that can be varied while the fan is in operation) for use when frequent or continuous adjustment is needed. Varying pitch angle retains high efficiencies over a wide range of conditions.

Figure 6.3 illustrates both the variable blade pitch design and its performance curves. The efficiency is near maximum at zero pitch angle, and it drops off as the pitch angle is increased. From the standpoint of power consumption, the most desirable method of control is to vary the fan speed to produce the desired performance. If the

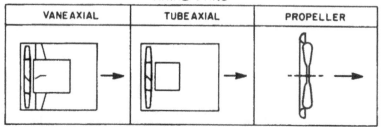

AXIAL FANS

VANE AXIAL	TUBE AXIAL	PROPELLER

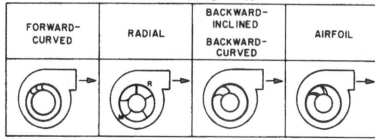

CENTRIFUGAL FANS

FORWARD-CURVED	RADIAL	BACKWARD-INCLINED / BACKWARD-CURVED	AIRFOIL

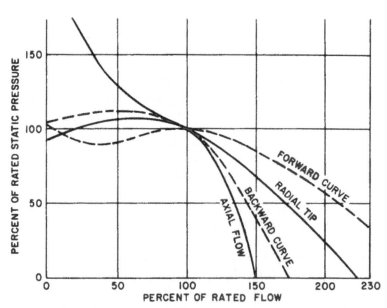

Fig. 6.1 Different fan types exhibit different pressure-flow characteristics. (From ASHRAE [2])

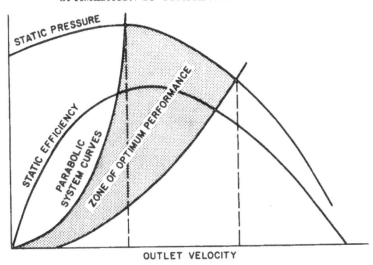

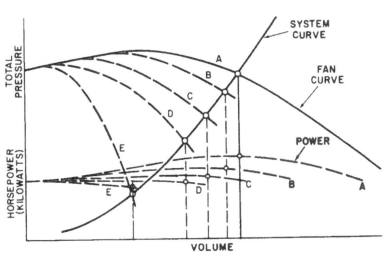

Fig. 6.2 Upper curve shows optimum operating zone for fans; lower curves describe the effect of suction vane throttling on the fan curve. (From ASHRAE [2])

frequency of change is sufficiently low, belt-driven units may be adjusted by changing the pulley on the drive motor of the fan. Variable speed motors or variable speed drives, whether electrical or hydraulic, may be used when frequent or essentially continuous speed modulation is desired.

Figure 6.4 illustrates both the fan curves and the power consumption of fans at partial loads. From the standpoint of noise, variable speed is somewhat better than the

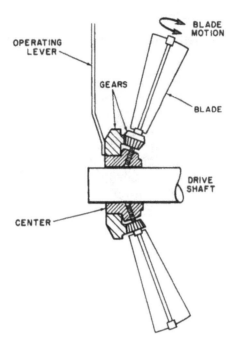

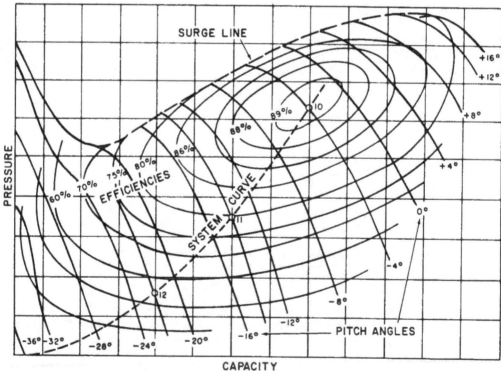

Fig. 6.3 The design and characteristics of axial-flow fans with variable pitch control are described here. (Upper part from Lipták [9]; lower part from Mark's Handbook [14])

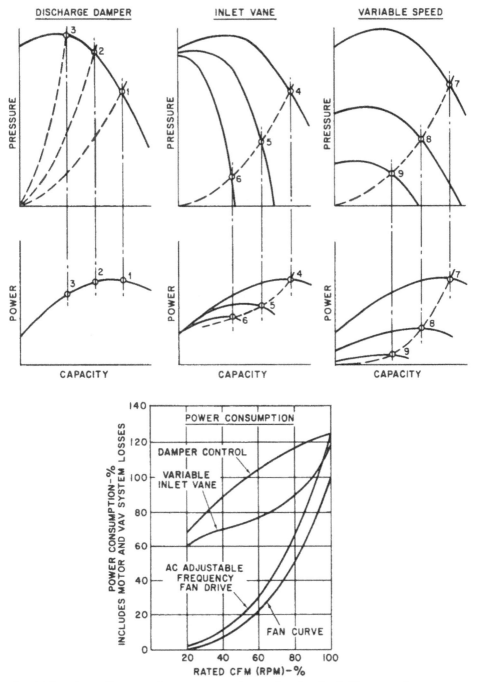

Fig. 6.4 Turndown efficiency and energy conservation are directly related. (Upper part from Mark's Handbook [14])

variable blade pitch design. On the other hand, both the variable speed and the variable pitch designs are much quieter and more efficient than the discharge damper or suction vane throttling systems.

Safety Interlocks

All fans should be provided with safety interlocks, such as the ones illustrated in figure 6.5. Interlock #1 will stop the fan if either excessively high pressures develop on its discharge side (PSH-02) or excessively high vacuums are detected on its suction side (PSL-03). Both of these pressure switches protect the ductwork from bursting or collapsing due to extreme pressure conditions, which can be caused by accidental blockage of the air flow.

Once the fan is shut down, it cannot be restarted until the reset button (R-04) is pressed. This gives the operator an opportunity to eliminate the cause of the abnormal pressure condition before restarting the fan. Whenever the fan is stopped, its discharge damper (XCD-06) is automatically closed to protect from flow reversal when several fans are connected in parallel. This discharge damper is designed for fast opening and slow closure, to make sure that it is always open when the fan is running.

The time delay (TD-05) guarantees that once the fan is started, it will run for a preset period, unless the safety interlocks turn it off. This protects the motor from overheating as a result of excessive cycling. The fan cycling interlocks (#2) are described in connection with figure 6.6.

Cycling of Multiple Fans

The fan station is optimized when it is meeting the process demand for air or other gases at the lowest possible cost. The minimum cost of operation is achieved by first minimizing the number of fans in operation and then minimizing the power consumption of the operating fans. Figure 6.6 illustrates a control system that fulfills both of these goals. The lower part describes the control loops, and the fan curves are shown in the upper left and the interlock table in the upper right.

The second fan is started and stopped in response to load variations according to the logic in this interlock table. As the process demand increases, the speed of the operating fan (Fan #1) is gradually increased. When its maximum speed is reached (point A on the fan curve) PSH-07 activates interlock #2 to start the second fan. If the speed control signal was unchanged when the second fan was started an upset would occur, because the operating point would instantaneously jump from point A to point C. In order to eliminate this temporary surge of pressure, which otherwise could shut down the station, PY-07 is introduced. This is a signal generator, which upon actuation by interlock #2 drops its output to x. This is the required speed for the two fan operation at point A. The low signal selector PY-09 immediately selects this signal x for control; thus, the upset is avoided. After actuation, the output signal of PY-07 slowly rises to full scale. As soon as it rises above the output of PIC-10, PY-07's signal is disregarded, and control is returned to PIC-10.

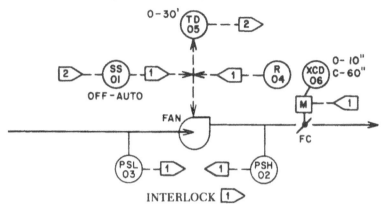

	Conditions				Actions	
SS-01	PSH-02	PSL-03	R-04 (Reset)	Fan	XCD-06 (Damper)	TD-06 (Note #1)
Off	—	—	—	Stop	Close	Time
Auto-On	High	—	—	Stop	Close	Time
	Low	—	Stop	Close	Time
	Not high	Not low	Not reset	Stop	Close	Time
			Reset	Run	Open	Reset to start

— = Any state or condition (don't care)
Note #1: Shut-down delay used in interlock ⌷2▷

Fig. 6.5 Safety interlocks should be provided on all fans.

Once both fans are smoothly operating, the next control task is to stop the second fan when the load drops to the point at which it can be met by a single fan. This is controlled by the low flow switch FSL-08, which is set at 90 percent of the capacity of one fan (point B on the system curve). When the flow drops below the setting of FSL-08, interlock #2 is actuated to stop the second fan. The stopping of the second fan is delayed until time delay TD-05 times out. This shut-down delay protects the fan from overheating due to excessively frequent cycling. Therefore, if TD-05 is set for, say, 20 minutes, the fan cannot be started more than three times an hour.

The fan cycling control described here can be used to cycle any number of parallel fans. For each additional fan, another FSL-08 and PY-07 needs to be added. If n is the number of fans in operation, then FSL-08 is to be set for 90 percent of the capacity of $(n-1)$ fans and PY-07 is to be set for x corresponding to the required speed of $(n+1)$ fans at point A.

Optimized Discharge Pressure

The optimum discharge pressure is the minimum pressure that is still sufficient to satisfy all the users. This minimum pressure is found by observing the opening of the most open user damper. If even the most open damper is not fully open, the supply

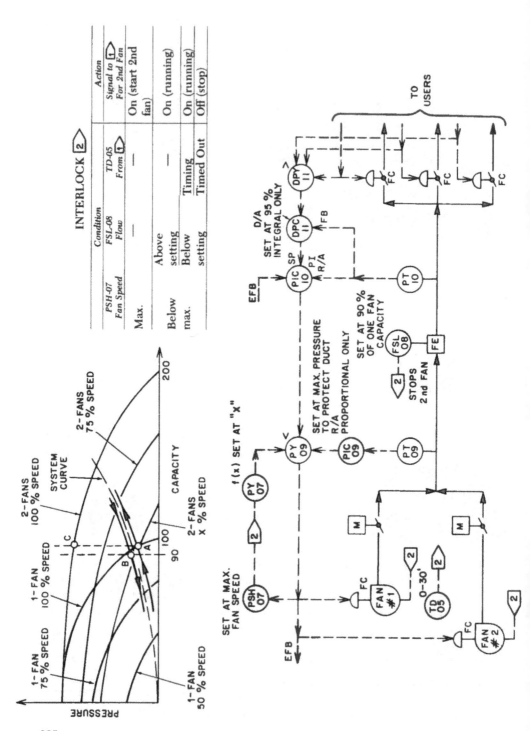

228

pressure can be safely lowered, whereas if the most open damper is fully open, the supply pressure must be raised. This supply-demand matching strategy not only minimizes the use of fan power but also protects the users from being undersupplied.

The optimization loop functions as follows: DPY-11 selects the opening of the most open damper and sends it as the measurement signal to the damper position controller DPC-11. This controller opens up all the dampers by lowering the supply pressure set point, until one of the dampers reaches 95 percent. Opening up all the dampers lowers the resistance to flow and saves fan power.

DPC-11 is an integral-only controller with an integral time (minutes/repeat) of ten times that of PIC-10. This guarantees smooth, stable control even if the dampers are unstable or cycling. The feedback signal (FB) protects the damper optimizer controller from reset wind-up when PIC-10 is switched to manual set point.

PIC-10 is the load-following controller. It compares the optimized set point with the actual header pressure and adjusts the speed or the blade pitch angle of the fan. When its output signal reaches maximum, PSH-07 acts to start another fan. The role of PIC-09 is to provide over-pressure protection at the fan discharge. Under normal conditions the pressure is much below the set point of PIC-09, and its output is saturated at its maximum value. Therefore, under normal conditions PY-09 will select the output signal from PIC-10 for control.

As can be seen from figure 6.4, if the average load is 60 percent of full capacity, the above-described optimization strategy can reduce the yearly operating cost to less than 50 percent of current costs.

Rotary Compressors, Blowers

The rotary compressor is essentially a constant-displacement, variable-discharge pressure machine. Common designs include the helical-screw, the lobe, the sliding vane, and the liquid ring types. The characteristic curves for a lobe type unit are shown in figure 6.7. As shown by curves I and II, the inlet flow varies linearly with the speed of this positive displacement machine. The small decrease in capacity at constant speed with an increase in pressure is a result of gas slippage through impeller clearances. It is necessary to compensate for this by small speed adjustments as the discharge pressure varies. For example, when the compressor is operating at point 1, it delivers the design volume of 60 ACFM (0.028 m³/s) and 3½ PSIG (24 kPa). In order for the same flow to be maintained when the discharge pressure is 7 PSIG (48 kPa), the speed must be increased from 1420 rpm to 1550 rpm at point 2 by the flow controller in the discharge line.

In addition to speed variation, the discharge flow can also be varied by throttling the suction, the bypass, or the vent line from the rotary blower. In figure 6.8 excess gas is vented to the atmosphere as the temperature control valve closes. The temperature of the outlet gas is controlled to prevent product degradation and to provide the proper product dryness. In systems in which the gas is not vented, it may be returned to the suction of the blower on pressure control.

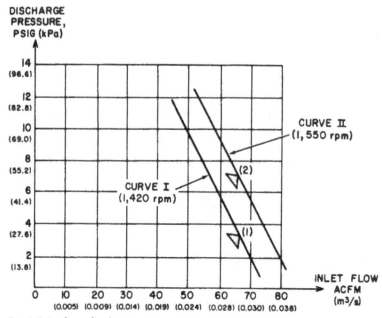

Fig. 6.7 As shown by the rotary compressor curves, the inlet flow varies linearly with the speed of this positive displacement machine. (From Lipták [9])

An important application of the liquid ring rotary compressor is in vacuum service. The suction pressure is often the independent variable and is controlled by bleeding gas into the suction on pressure control. This is shown in figure 6.9, where suction pressure control is used on a rotary filter maintaining the proper drainage of liquor from the cake on the drum.

The optimization of rotary compressors is discussed together with that of reciprocating compressors.

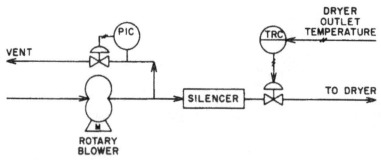

Fig. 6.8 A rotary blower with bypass capacity control vents excess gas as the temperature control valve closes. (From Lipták [9])

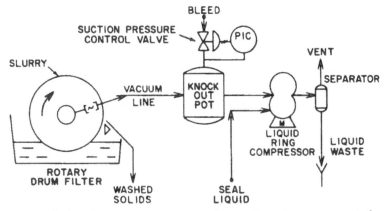

Fig. 6.9 The liquid ring rotary compressor is shown with suction pressure control. (From Lipták [9])

Reciprocating Compressors

The reciprocating compressor is a constant-volume, variable-discharge pressure machine. A typical compressor curve is shown in figure 6.10 for constant-speed operation. The curve shows no variation in volumetric efficiency in the design pressure range, which may vary by 8 PSIG (53.6 kPa) from unloaded to fully loaded.

The volumetric inefficiency is a result of the clearance between piston end and cylinder end on the discharge stroke. The gas that is not discharged re-expands on the suction stroke, thus reducing the intake volume.

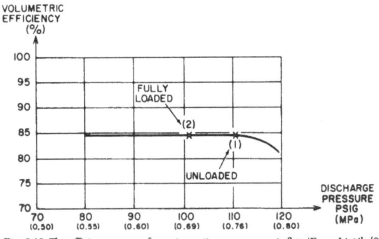

Fig. 6.10 The efficiency curve of a reciprocating compressor is flat. (From Lipták [9])

The relationship of speed to capacity is a direct ratio, because the compressor is a displacement-type machine. The typical normal turndown with gasoline or diesel engine drives is 50 percent of maximum speed, in order to maintain the torque within acceptable limits.

First the conventional and then the optimized load-following controls are described in sections below. The abbreviations used in these descriptions are as follows:

Abbreviations and Terminology

APSL: Low air pressure switch with contact that closes as pressure drops below setting

ATA: Air temperature alarm, actuated on high temperature

ATSH: High air temperature switch with two contacts

ATSH-1: First contact of ATSH is normally closed; it opens on high temperature

ATSH-2: Second contact of ATSH is normally open; it closes on high temperature

H: Fused, 120-volt, hot power supply of the stand-alone compressor control circuit

Ho: Fused, 120-volt, hot power supply of the integrated remote controls

M: Motor

N: Neutral wire of the stand-alone compressor control circuit

No: Neutral wire of the integrated remote control system

OFF DELAY: This type of time delay introduces a delaying action when the relay is deenergized. A two-minute off-delay means that for the first two-minute period after deenergization, the relay contacts will remain as if the relay was energized. When the delay expired, the contacts will switch to their deenergized state.

ON DELAY: This type of time delay introduces a delaying action when the relay is energized. A three-second on-delay means that for the first three seconds after energization, the relay contacts will remain as if the relay was still deenergized. When the delay expires, the contacts will switch to their energized state.

TR: Time delay relay. The delay time setting is marked next to it.

TTC: Time to close action. This contact is open, then the TR relay is deenergized. This contact stays open when the TR relay is energized until the delay time setting expires. Then it closes.

TTO: Time to open action. This contact is closed when the TR relay is deenergized. This contact stays closed when the TR relay is energized until the delay time setting expires. Then it opens.

USV: Unloading solenoid. The compressor is loaded when this solenoid is energized.

WFA: Water flow alarm. This alarm is energized if the flow drops below the minimum allowable.

WFSL: Low water flow switch, with two contacts

WFSL-1: This contact opens on low flow.

WFSL-2: This contact closes on low flow.

WSV: Cooling water solenoid valve; opens when energized

WTA: Water temperature alarm, energized on high temperature

WTSH: High water temperature switch, with two contacts

WTSH-1: This contact opens on high temperature.

WTSH-2: This contact closes on high temperature.

The Stand-Alone Air Compressor

A typical air compressor, together with the controls normally provided by its manufacturer, is shown in figure 6.11. Such a stand-alone compressor usually operates as follows:

When the operator turns the control switch to "On," this will energize the time delay (TR) and will open the water solenoid valve (WSV) if neither the air temperature (ATSH) nor the cooling water temperature (WTSH) is high.

After seven seconds, TR-1 opens. Assuming that this time delay was sufficient for the oil pressure to build up, the opening of TR-1 will have no effect, because OPSL will have closed in the meantime.

If the opening of WSV resulted in a cooling water flow greater than the minimum setting of the low water flow switch, then WFSL closes and the compressor motor (M) is started. Whenever the motor is on, the associated "Run" pilot light is energized. This signals to the operator that oil (OPSL), water (WTSH, WFSL), and air (ATSH) conditions are all acceptable and therefore the unit can be loaded.

When the operator turns the other control switch to "Load," the TR-2 contact is already closed because the seven seconds have passed. Therefore, the APSL contact will determine the status of the machine. If the demand for air at the users is high, the pressure at APSL will drop, which in turn will cause the APSL contact to close and the compressor to load. As a result of loading the compressor, the pressure will rise until it exceeds the control gap of APSL, causing its contact to reopen and the machine to unload.

The compressor continues to load and unload automatically as a function of plant demand. As the demand rises, the loaded portion of the cycle will also rise. Once the load reaches the full capacity of the compressor, the unit stops cycling and stays in the loaded state continuously. If the demand rises beyond the capacity of this compressor, it is necessary to start another one. The following paragraphs describe how this is done automatically, requiring no operator participation.

If at any time during operation the cooling water flow (WFSL) drops too low, the motor will stop. If either water or air temperatures rise to a high value, this condition not only will stop the motor but will also close the water solenoid valve (WSV). The above three conditions (WFSL, WTSH, ATSH) will also initiate remote alarms, as shown in figure 6.11, to advise the operators of the possible need for maintenance. If the oil pressure drops below the setting of OPSL, it will also cause the stoppage of the motor and the closure of WSV, but after such an occurrence, the compressor will not be allowed to restart automatically when the oil pressure returns to normal. In order to restart the machine, the operator will have to go out to the unit and reset the system by turning the control switch "Off" and then "On" again to repeat the complete start-up procedure.

Local/Remote Switch

The first step in integrating several compressors into a single system is the addition of a local/remote switch at each machine. As shown in figure 6.12, when this switch is

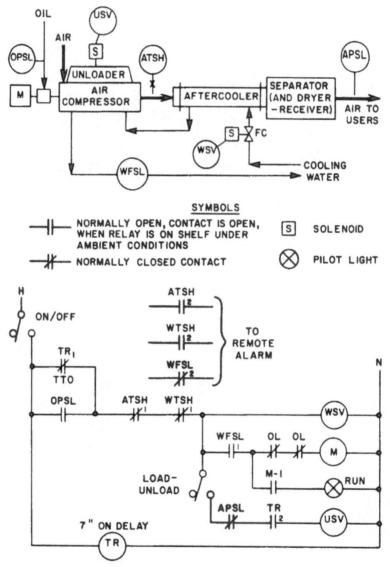

Fig. 6.11 A typical stand-alone air compressor is provided with this type of equipment layout and control logic. (From Lipták [11])

turned to "local," the compressor operates in the stand-alone mode, as was described in connection with figure 6.11. Turning this switch to "Remote" the compressor becomes a part of the integrated plant-wide system, consisting of several compressors.

As shown in figure 6.12, this two-position switch has 9 sets of contacts and is mounted near the compressor. It can be installed in a few hours, and once installed, the compressor can again be operated in the "local" mode.

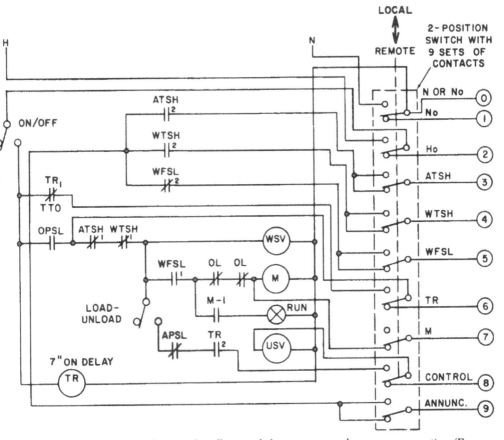

Fig. 6.12 The local/remote switch is wired to allow stand-alone or integrated compressor operation. (From Lipták [11])

Only ten wires need to be run from each compressor to the remote controls. These ten wires serve the following functions:

#0 The working neutral, N in local, No in the remote mode of operation
#1 The common neutral (No) of the integrated controls
#2 The common hot (Ho) of the integrated controls
#3 The high air temperature (ATSH) alarm
#4 The high water temperature (WTSH) alarm
#5 The low water flow (WFSL) alarm
#6 The 7 sec. time delay (TR) of the integrated controls
#7 The motor (M) status indication
#8 The load/unload control signal from the remote system
#9 The common hot for the remote annunciator (Annunc.)

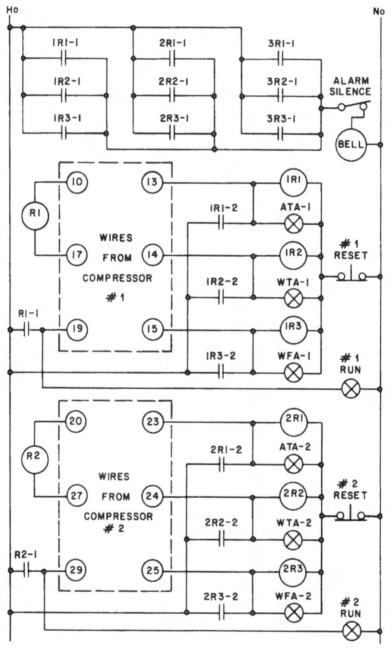

Fig. 6.13 This annunciator is wired to accommodate two compressors. (From Lip-ták [11])

When integrating several compressors into a single system, it is advisable to number these ten wires in a consistent manner, such as:

Compressor #1	Wire #10 to #19
Compressor #2	Wire #20 to #29
Compressor #3	Wire #30 to #39

etc.

In this system, the first digit of the wire number indicates the compressor and the second digit describes the function of the wire. Immediately knowing, for example, that wire number 45 comes from compressor #4 and serves to signal a low water flow condition on that machine simplifies check-out and start-up.

Annunciator

Of the ten wires from each of the compressors, six are used for remote alarming. Figure 6.13 shows a remote annunciator for two compressors. This alarm system can be expanded to serve any number of compressors.

The position of the local/remote switch in figure 6.12 does *not* affect the operation of the annunciator. It provides the following remote indications all the time, for each compressor:

- "Run" light
- High air temperature light
- High water temperature light
- Low water flow light
- Audible alarm bell with silencer
- Alarm reset buttons

The only time these circuits are deactivated is when the associated compressor is off.

Lead-Lag Selector

As plants grow, their compressed air requirements also tend to increase. As a result of such evolutionary growth, many existing plants are served by several uncoordinated compressor stations. When, because of space limitations, the new compressors are installed in different locations, the manual operation of such systems becomes not only inefficient but also unsafe.

The steps involved in integrating such stand-alone compressor stations into an automatically operating, load-following single system are described here. In such integrated systems, the identity of the "lead" and "lag" compressors, or the ones requiring maintenance ("Off"), can all be quickly and conveniently altered, while the system continues to efficiently meet the total demand for air. Thus, air supply shortage or interruption is eliminated, together with the need for continuous operator's attention.

When it is desirable to combine two compressors into an integrated lead-lag station, all that needs to be added are two pressure switches as shown in figure 6.14. APSL-A is the lead and APSL-B is the lag control pressure switch. To maintain a 75 PSIG air supply at the individual users, these normally closed pressure switches can be set as follows:

Pressure Control Switch	Pressure Below Which Switch Will Close (PSIG)	Differential Gap (PSI)
APSL-A	90	10
APSL-B	85	10

With the above settings, the system performance will be as shown in table 6.1.

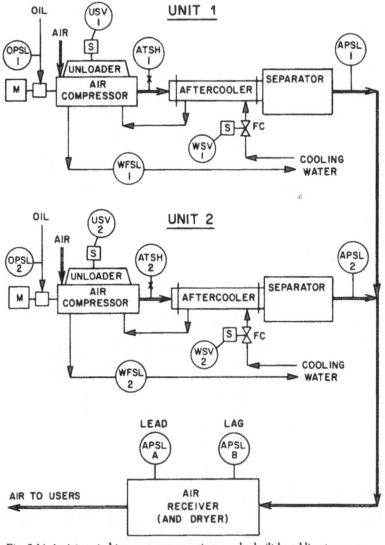

Fig. 6.14 An integrated two-compressor system can be built by adding two pressure switches. (From Lipták [11])

Table 6.1

SYSTEM PERFORMANCE

WITH APSL-A SET TO CLOSE AT 90 PSIG

APSL-B SET TO CLOSE AT 85 PSIG

AND DIFFERENTIAL GAP SET AT 10 PSI

Pressure in Air Receiver (PSIG)	APSL-A Contact	APSL-B Contact	Lead Compressor	Lag Compressor
Over 100	Open	Open	Off	Off
100	Open	Open	Off	Off
95	Open	Open	Off	Off
90	Closed	Open	On	Off
85	Closed	Closed	On	On
80	Closed	Closed	On	On
85	Closed	Closed	On	On
90	Closed	Closed	On	On
95	Closed	Open	On	Off
100	Open	Open	Off	Off

When two stand-alone compressors are integrated into such a lead-lag system, only five wires need to be brought from each compressor to the lead-lag switch, shown in figure 6.15. Turning this single switch automatically reverses the lead-lag relationship between the two compressors. In addition to the two-position lead-lag switch, running lights are also provided; they show whether either or both compressors are unloaded (standby) or loaded.

As the demand for air increases, the lead machine will load and unload to meet that demand. If the lead machine is continuously loaded and the demand is still rising, the lag compressor will be started automatically.

The interlocks provided are as follows:

The on-time delays 2TR and 3TR are provided for stabilizing purposes only. They guarantee that a system configuration (or reconfiguration) will be recognized only if it is maintained for at least three seconds. Responses to quick changes are thus eliminated.

The on-time delays 1TR and 4TR are provided to give time for the oil pressure to build up in the system. If after seven seconds the oil pressure is not yet established, and therefore OPSL in figure 6.12 is still open, the contacts 1TR-1 and 4TR-1 will open and the corresponding compressor will be stopped.

The off-time delay 5TR guarantees that the lag machine will not be cycled on and off too frequently. Once started, the lag compressor will not be turned off (but will be kept on standby) until the off delay of two minutes has passed.

Off Selector

Three stand-alone compressors can be combined into an integrated load-following single system, by the addition of a three-position "off" selector switch, illustrated in figure 6.16. With the addition of the off-selector, any of the compressors can be selected for load, lag or off duties, just by turning these conveniently located switches.

Depending on the position of the off selector, the identities of Compressor A and B in figure 6.15 will be as follows:

Selected Off	Compressor A	Compressor B
1	3	2
2	1	3
3	1	2

For an integrated three-compressor system, only thirty wires need to be run if the remote annunciator is included. Without remote alarms, only fifteen wires are needed (five per compressor). The face of a remote control cabinet for a three-compressor lead-lag system is illustrated in figure 6.17.

Large Systems

From the building blocks discussed in the previous paragraphs, an integrated remote controller can be configured for every compressor combination. Figure 6.18 illustrates how six stand-alone compressors might be integrated into a single load-following system. As the system pressure drops below 90 PSIG, APSL-A closes, starting the first compressor. As the load rises, causing the system pressure to drop further, more switches will close, until at 80 PSIG all six switches will be closed and all six compressors will be running.

The "priority selector" shown in figure 6.18 is provided for the convenient recon-figuration of compressor sequencing. When the selector is in position #1, as the load increases, the compressors will be started in the following order: A, B, C, D, E, F. When switched to position #2, the sequence becomes C, D, E, F, A, B; in position #3, the sequence is E, F, A, B, C, D.

If a system consists of nine compressors, and full flexibility in integrated remote control is desired, the control cabinet might look like figure 6.19. With this system, the operator can set the lead, lag or off status for all nine compressors and can also select the order in which they are to come on.

Flexibilities and Costs

From the previous discussion it can be seen that the building-block approach to compressor system design is very flexible. Any number of stand-alone compressors can be integrated into an automatic load-following control system, with complete flexibility for priority, lead-lag or off selection. The logic blocks described can be implemented either in hardware or in software. All wires and terminals are prenumbered, which minimizes installation errors.

The cost elements of implementation on a per-compressor basis can be estimated as follows:

Prepare design drawings $1,000
Make local tie-ins and install local/remote switch $1,000
Run ten wires from compressor to control cabinet and add lead-lag
   switches required $1,000
Apportioned cost of control cabinet $1,000
    Total cost per compressor $4,000

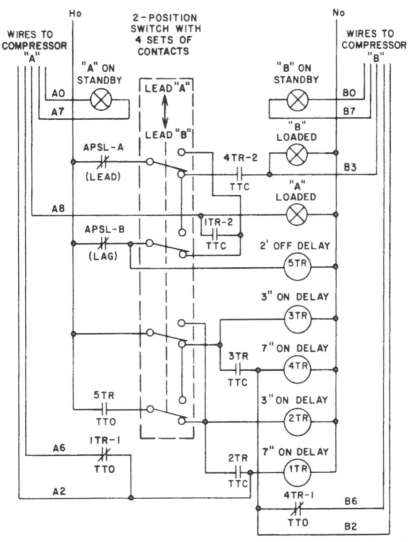

Fig. 6.15 The lead-lag selector switch automatically reverses the lead-lag relationship between the two compressors. (From Lipták [11])

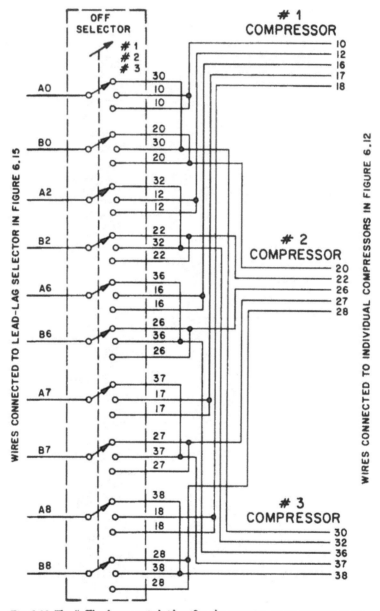

Fig. 6.16 The "off" selector switch identifies the common spare compressor, or the unit which is undergoing maintenance. (From Lipták [11])

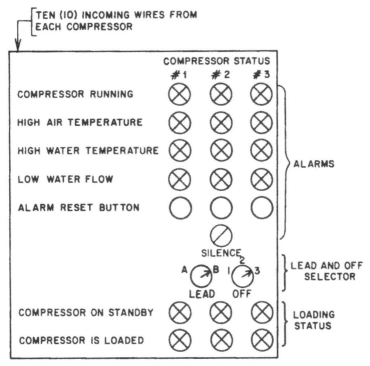

Fig. 6.17 A remote control panel is used to operate an integrated three-compressor system. (From Lipták [11])

This estimate is relatively unaffected by the distance between compressors or by the variations between features desired by the user.

Assuming that the cost of $4,000/compressor is correct, this investment should be evaluated against the potential benefits to be derived from system integration. The potential benefits are listed below:

- Unattended, automatic operation relieves operators for other tasks.
- Automatic load-following eliminates the possibility of accidents caused by the loss of the air supply.
- Because the supply and the demand are continuously and automatically matched, the energy cost of operation is minimized. (The energy cost of each horsepower-year is about $500.)

If the cost-benefit balance is favorable, the next decision that must be made is whether hardware or software implementation is the proper choice. The availability of spare computer inputs and outputs on existing equipment can be an argument in favor of digital implementation. If priority or lead-lag switching is under computer control, this would also make the digital implementation more desirable. On the other hand,

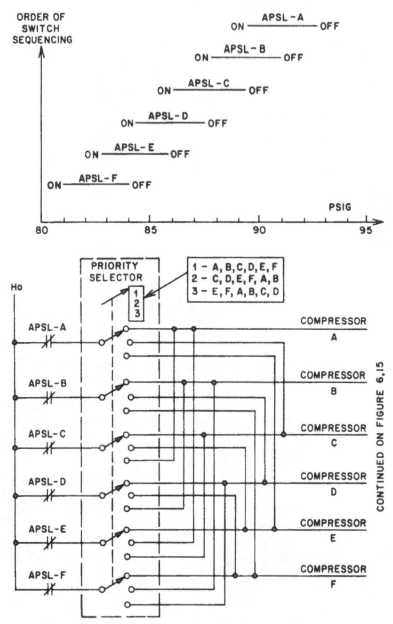

Fig. 6.18 A priority selector can be used to change the order of priorities in large compressor systems. (From Lipták [11])

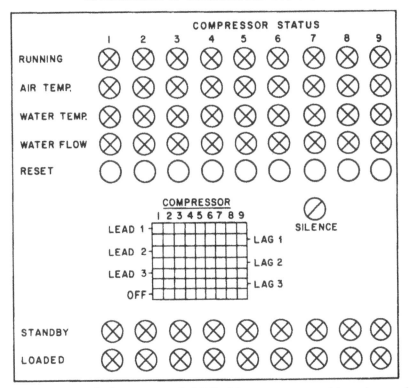

Fig. 6.19 A control cabinet can provide full flexibility in remote control for an integrated nine-compressor system. The mechanically interlocked pushbutton station is so designed that only one button can be pressed in each column or row, except in the bottom one. In the bottom row any or all buttons can be simultaneously pressed. (From Lipták [11])

if a dedicated, hard-wired system is preferred—a system that is familiar to the average operator—hardware implementation is the proper choice.

Axial and Centrifugal Compressors

Positive displacement compressors pressurize gases through confinement. Dynamic compressors pressurize them by acceleration. The axial compressor drives the gas parallel to the shaft; in the centrifugal compressor, the gas receives a radial thrust toward the wall of the casing where it is discharged. The axial compressor is better suited for constant flow applications, whereas the centrifugal design is more applicable for constant pressure applications. This is because the characteristic curve of the axial design is steep, and that of the centrifugal design is flat (figure 6.20). The characteristic curve of a compressor plots its discharge pressure as a function of flow, and the load curve relates the system pressure to the system flow. The operating points (L1 or L2 in figure 6.20) are the intersections of these curves. Axial compressors are more efficient; centrifugal ones are better suited for dirty or corrosive services.

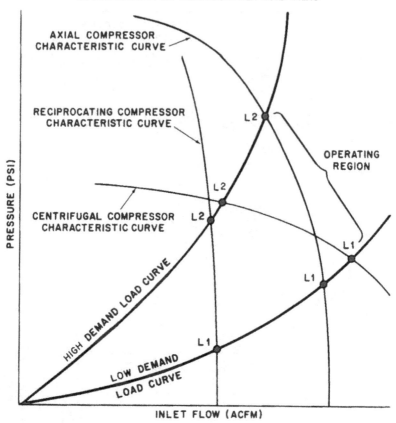

Fig. 6.20 The characteristic curve of the axial compressor is steep; that of the centrifugal compressor is flat. (Adapted from McMillan [15])

The normal operating region falls between the low and the high demand load curves in figure 6.20. The operating ranges of various compressor and fan designs are shown in figure 6.21.

Compressor Throttling

Compressor loading can be reduced by throttling a discharge or a suction valve by modulating a prerotation vane or reducing the speed. As was shown in figure 6.4, discharge throttling is the least energy efficient and speed modulation is the most energy efficient method of turndown. Suction throttling is a little more efficient and gives a little better turndown than discharge throttling, but it is still a means of wasting that transportation energy, which should not have been introduced in the first place.

Guide vane positioning, which provides prerotation or counter rotation to the gas, is not as efficient as speed modulation, but it does provide the greatest turndown. As was shown in figure 6.4, speed control is the most efficient, as small speed reductions

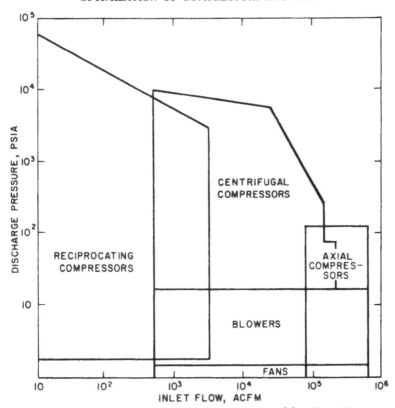

Fig. 6.21 Operating ranges vary for different compressors and fans (From Mc-Millan [15])

result in large power savings because of the cubic relationship between speed and power. If the discharge pressure is constant, flow tends to vary linearly with speed. If the discharge head is allowed to vary, it will change with the square of flow and, therefore, with the square of speed as well. This square relationship between speed and pressure tends to limit the speed range of compressors to the upper 30 percent of their range [19].

The operating ranges of the most popular compressor drives are shown in figure 6.22. Also given in that figure is the classification of two-speed governors. Regulation error is related to the difference in speed at zero power output and at the rated power output. Variation error is related to the effect of whether the speed change occurred above or below set point.

Constant-speed steam turbine governors can be converted into variable-speed governors by revising the quick-opening characteristics of the steam valve into a linear one (figure 6.23). The efficiency of variable speed drives varies substantially with their design (figure 6.24). Electronic governors tend to eliminate the dead bands that are

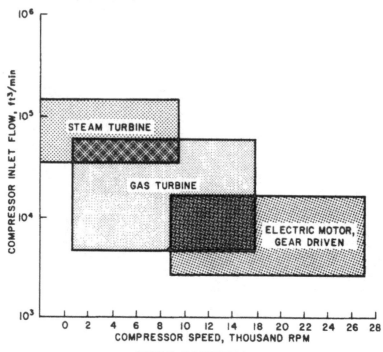

SPEED GOVERNOR
CLASSIFICATION

Governor class	Regulation	Variation
A	10%	¾%
B	6%	½%
C	4%	¼%
D	½%	¼%

Fig. 6.22 Each drive has its own throughput and speed range. (From McMillan [15])

present in mechanical designs. They also require less maintenance because of the elimination of mechanical parts. Electronic governors also give better turndowns and are quicker and simpler to interface with surge or computer controls.

The error in following the load increases as the speed of process disturbances increases or as the speed control loop speed is reduced. Therefore, it is desirable to make the speed loop response as fast as possible. On turbines, this goal is served by the use of hydraulic actuators, and the motor response is usually increased by the use of tachometer feedback.

For the purposes of the control systems shown in this chapter, it will be assumed

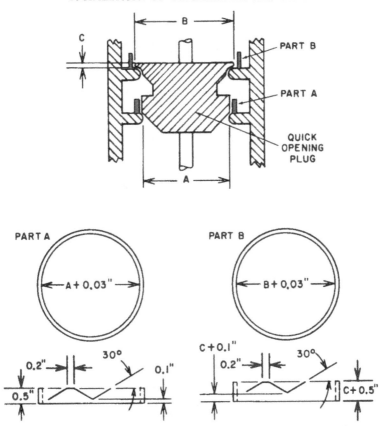

Fig. 6.23 A double-seated steam governor valve can be rebuilt for optimized variable-speed service. The notched rings provide the necessary rangeability. (From Lipták [12])

that the compressor throughput is controlled through speed modulation with tachometer feedback.

The Phenomenon of Surge

In axial or centrifugal compressors, the phenomenon of momentary flow reversal is called *surge*. During surging, the compressor discharge pressure drops off and then is reestablished on a fast cycle. This cycling, or surging, can vary in intensity from an audible rattle to a violent shock. Intense surges are capable of causing complete destruction of compressor parts, such as blades and seals.

The characteristic curves of compressors are such that at each speed they reach a maximum discharge pressure as the flow drops (figure 6.25). A line connecting these points (A to F) is the surge line. If flow is further reduced, the pressure generated by

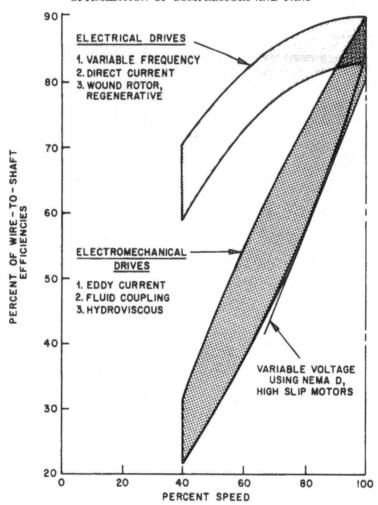

Fig. 6.24 Wire-to-shaft efficiencies differ for various types of variable-speed drives. (From Lipták [11])

the compressor drops below that which is already existing in the pipe, and momentary flow reversals occur. The frequency of these oscillations is between 0.5 and 10 hertz. The surge frequency of most compressor installations in the processing industries is slightly less than 1 hertz [14]. Surge is usually preceded by a stall condition, which is caused by localized flow oscillations around the rotor at frequencies of 50 to 100 hertz.

At the beginning of surge, the total flow drops off within 0.05 seconds, and then it starts cycling rapidly at a period of less than 2 seconds [15]. This period is usually shorter than that of the flow control loop that controls the capacity of the compressor.

If the flow cycles occur faster than the control loop can respond to it, this cycling will pass through undetected as uncontrollable noise. Therefore, fast sensors and instruments are essential for this loop.

As is shown in figure 6.25, the surge line is a parabolic curve on a plot of pressure rise (discharge pressure minus suction pressure) versus flow. This function shows an increasing nonlinearity as the compression ratio increases. If the surge line is plotted as Pd-Pi versus the square of flow (orifice differential = h), it becomes a straight line (figure 6.26) if the compression ratio is low (less than 4:1).

On a plot of ΔP versus Q (figure 6.25), the following changes will increase the probability of surge by reducing the distance between the operating point and the surge line:

- A decrease in suction pressure
- An increase in suction temperature
- A decrease in molecular weight
- A decrease in specific heat ratio

On a plot of ΔP versus h (figure 6.26), these effects are more favorable. A decrease in suction pressure moves the surge line in the safe direction, temperature has no effect, and the effect of the other variables is also less pronounced. Therefore, although the ΔP versus h plot is accurate only at low compression ratios, it does have the advantage of being independent from the effects of composition and temperature changes. However, suction pressure should be included in the model in order to be exact; most ΔP versus h plots disregard it.

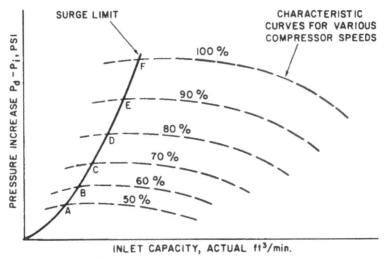

Fig. 6.25 The surge line of a speed controlled centrifugal compressor is parabolic. (From Magliozzi [13])

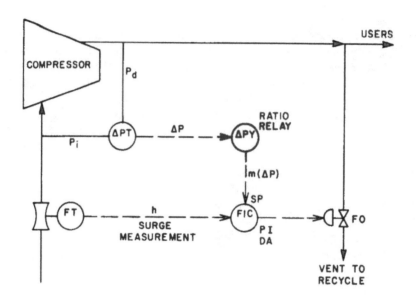

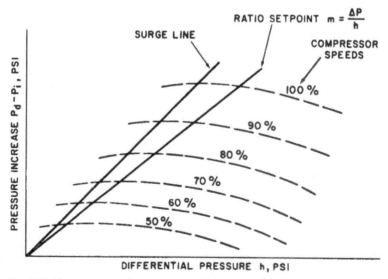

Fig. 6.26 The surge curve becomes a straight line on a plot of ΔP versus h. (Adapted from Magliozzi [13])

Variations in the Surge Curve

Figure 6.25 shows the surge and speed characteristic curves of a centrifugal compressor; figure 6.27 shows these curves for an axial compressor. The characteristic curves of the axial compressor are steeper, which makes it better for constant flow services. The centrifugal design is better for constant pressure control.

The effect of guide vane throttling is also shown for both centrifugal and axial compressors (figure 6.28). As can be seen, the shape of the surge curve varies with the type of equipment used. The surge curve of speed controlled centrifugal compressors bends up (figure 6.25), whereas for axial (figure 6.27) and vane controlled machines (figure 6.28), the surge line bends over. It is this negative slope of the axial compressor's surge curve that makes it sensitive to speed variations, because an increase in speed at constant flow can quickly bring the unit into surge.

As can be seen from the above information the shape of the surge curve varies with compression ratio and with equipment design. It should also be noted that in case of multistage compressors, the surge line is discontinuous. If the compressor characteristics are as shown in figure 6.29, the addition of a compressor stage causes a break

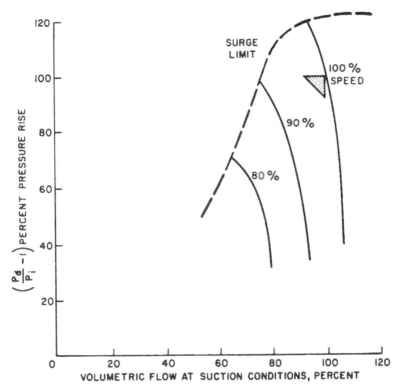

Fig. 6.27 The surge line for an axial compressor goes through a slope reversal at higher flows. (From Shinskey [19])

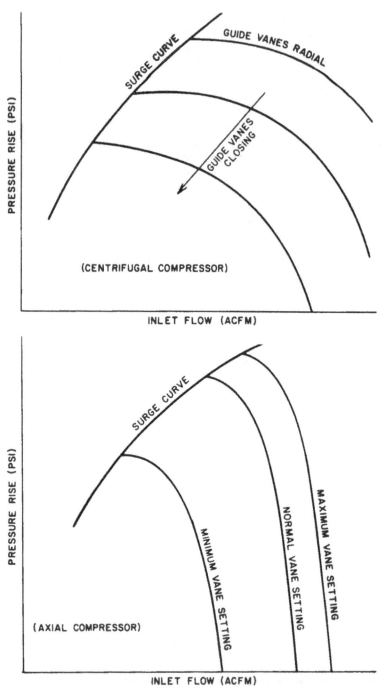

Fig. 6.28 The surge curve bends over when guide vanes are used for throttling.
(From McMillan [15])

254

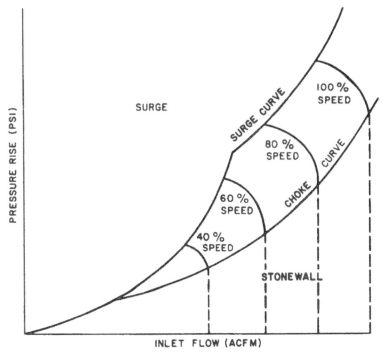

Fig. 6.29 With multi-stage compressors, operation must be confined between the surge and choke curves. (From McMillan [15])

point in the surge curve. With more stages, more break points would also be added, and the resulting net effect is a surge curve that bends over instead of bending up, as does the surge curve for the single-stage compressor in figure 6.25.

Figure 6.29 also shows the choke curve. This curve connects the points at which the compressor characteristic lines become vertical. Below this curve, flow will stay constant even if pressure varies, as long as the compressor is at a constant speed [15]. As speed is reduced, the surge and choke lines intersect. Below this intersection, the traditional methods of surge protection (venting, recycling) are ineffective; only a quick raising of the compressor speed can bring the machine out of surge.

This is similar to the situation when the compressor is being started up. As the operating point is moving on the load curve (figure 6.30), it must pass through the unstable region on the left of the surge curve as fast as possible in order to avoid damage from vibration.

Flow Measurement

The two critical components of a surge control loop are the flow sensor and the surge valve. The capabilities that both must possess are speed and reliability.

Flow oscillations under surge conditions occur on a cycle of little more than a

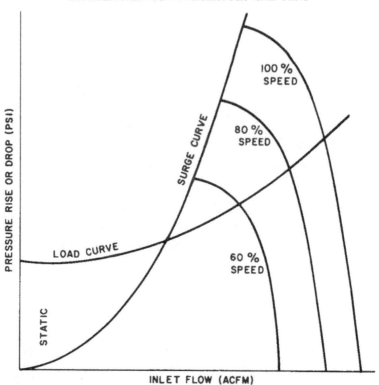

Fig. 6.30 The speed and surge curves are characteristics of the compressor, while
the load curve is a function of the load. (From McMillan [15])

second. The flow transmitter should be fast enough to detect these. The time constants
of various transmitter designs are as follows [15]:

Pneumatic with damping	up to 16 sec.
Electronic d/p	0.2 to 1.7 sec.
Diffused silicone d/p	down to 0.005 sec.

Only the diffused silicone design is fast enough to follow the precipitous flow drop
at the starting of surge or the oscillations during surge.

Measurement noise is another serious concern, because it necessitates a greater
margin between the surge and the control lines. Noise can be minimized by the use
of twenty pipe diameters of upstream and five diameters of downstream straight runs
around the streamlined flow tube type sensor. Noise will also be reduced if the low
pressure tap of the d/p cell is connected to a piezometric ring in the flow tube. The
addition of straightening vanes will also contribute to the reduction of noise.

Antisurge control usually requires a flow sensor on the suction side of the compressor. If good, noise-free flow measurement cannot be obtained on that side, a cor-

rected discharge side differential pressure reading (*hd*) can be substituted. The *hs* reading can be obtained from a reading of *hd* with the following correction for pressure and temperature differences:

$$hs = hd(Pd/Ps)(Ts/Td)$$

Surge Control Valves

Surge valves are usually fail-open, linear valves that are tuned for fast and precise throttling. Because the valve is the weak link in the total surge protection system, and because testing and maintenance cannot be done on-line if only one valve is used, total redundancy is recommended. Each of the two parallel surge valves should be sized for the full flow of the compressor but for only 70 percent of the discharge pressure. This pressure reduction is caused by the flow reversals during surge.

Globe valves are preferred to rotary designs because of hysteresis and breakaway torque considerations. Boosters (figure 6.31) can cut the sum of prestroke dead time and full-scale stroking to under one second. Actually, valves as large as 18″ have been made to throttle to any position in under a half second [15].

Digital valves are even faster: they can stroke in 0.1 second without any overshoot. Their limitations are in the plugging of the smaller ports and in the difficulty of inspecting them.

Positioners are frequently required to reduce hysteresis (packing friction), dead band (boosters), disc flutter (butterfly valves) and overshoot; they are also often needed just to provide the higher air pressures required by piston actuators.

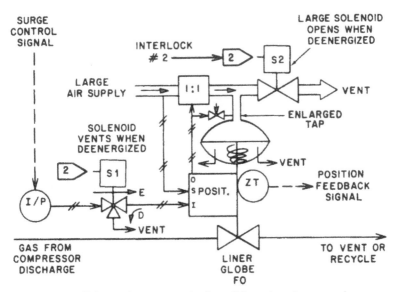

Fig. 6.31 A well-designed surge control valve will have these features and accessories. (Adapted from McMillan [15])

Figure 6.31 shows some of the main components of a surge control valve [15]. The electronic control signal is converted to a pneumatic one by the I/P converter, and this pneumatic signal is sent to the positioner through a three-way solenoid (S1). The output signal from the positioner is sent simultaneously into a booster relay (1:1) and an adjustable bypass. The output from the booster relay is connected by large-diameter, short tubes to the valve actuator with an enlarged pressure tap and to the large venting solenoid (S2). Some of the operational features of the system are as described below.

When interlock #2 is deenergized, requiring instantaneous full opening of the surge valve (because of discharge line blockage or other reasons), S1 vents the positioner inlet and S2 vents the actuator topwork. Because S2 is large, air removal is fast and the valve opens quickly. On return to normal, both solenoids are energized. This closes S2 and opens S1 to the control signal. Depending on the size or restriction in the line from S1, air might enter the system more slowly and the valve might close more slowly than it opened. Such fast-opening, slow-closing designs can respond to the first precipitous drop in flow and thus can prevent the second surge cycle from developing. It is important not to slow the speed of valve opening too much, because quick throttling is still required.

The valve must also respond quickly to the control signal and throttle quickly in either direction. In order to speed up the air movement into and out of the actuator, the vent and signal ports on the actuator are both drilled out, an air booster is installed, and a large-capacity path is established between the booster and the actuator. In order to reduce the dead band of the 1:1 booster, it is advisable to use a signal range of 6-30 PSIG instead of the usual 3-15 PSIG.

The bypass needle valve around the booster in figure 6.31 is required because without it the volume on the outlet of the positioner would be much smaller than on the outlet of the booster. This would allow the positioner to change the input to the booster faster than the booster could change its output, resulting in a limit cycle. This limit cycle is eliminated by the addition of the adjustable bypass.

All surge valves should be throttle-tested before shipment. It is also desirable to monitor the surge valve opening through the use of a position transmitter.

Surge Control Curves

As was shown in figure 6.26, a parabolic surge curve with a positive slope will appear as a straight line on a ΔP versus h plot. The purpose of surge control is to establish a surge control line to the right of the actual surge line, so that corrective action can be taken *before* the machine goes into surge. Such a control system is shown in figure 6.32.

The biased surge control line is implemented through a biased ratio relay (ΔPY), which generates the set point for FIC as follows:

$$SP = m(\Delta P) + b$$

where

SP = desired value of h in $''H_2O$
m = slope of the surge line at the operating point

ΔP = compressor pressure rise in PSI
b = bias of the surge set point in $''H_2O$

The offset between the surge and the control lines should be as small as possible for maximum efficiency, but it must be large enough to give time to correct upsets without violating the surge line [22]. The slower the upsets and the faster the control loop, the less offset is required for safe operation. In a good design, the bias b is about 10 percent, but in bad ones, as disturbances get faster or instruments slower, it can grow to 20 percent.

A second line of defense is the back-up interlock FSL. It is normally inactive, as the value of h is normally above the set point SP. When the surge controller FIC is not fast enough to correct a disturbance, h will drop below SP. FSL is set to actuate when h has fallen 5 percent below SP, and at that point it fully opens the surge valve through interlock #1.

The shape of the control and backup lines shown in figure 6.32 is applicable only when the surge curve is parabolic. As was shown in figures 6.27, 6.28, and 6.29, surge curves can also be linear or have negative slopes and discontinuities. If the surge line is not parabolic, the following techniques can be considered for generating the surge control lines.

If the surge curve is linear, the signal h should be replaced by a signal representing flow. This can be obtained by adding a square root extractor to the FT in figure 6.32.

If the surge curve bends down, the use of two square root extractors in series [15] can be used to approximate its shape.

On multistage machines with high compression ratios, it might be necessary to substitute a signal characterizer for ΔPY in figure 6.32.

On variable vane designs, it has been proposed [19] that the flow meter differential (h) could be corrected to be a function of actual vane position to better approximate the surge curve (see figure 6.33). If a flow meter differential cannot be obtained from either the suction or the discharge side of the compressor, the surge curve can also be approximated on the basis of speed, power, or vane position measurements [19].

The Surge Flow Controller

The surge controller (FIC) in figure 6.32 is a direct-acting controller with proportional and integral action and with anti-reset wind-up features. It is a flow controller with a remote set point and a narrower proportional band than would be used for flow control alone. In a well-tuned surge controller, both the proportional band and the integral settings must be minimized (PB~50%, 1–3 sec./repeat). With this level of responsiveness, the FIC set point can be less than 10 percent from the surge curve, as shown in figure 6.32.

Under normal conditions, the actual flow (h) is much higher than the surge control limit (SP), and therefore the integral mode of the FIC has a tendency to wind up on its positive error toward a saturated maximum output. If this was allowed to happen, the controller could not respond to a surge condition (a quick drop in h), because it would first need to develop a negative error equal to that which caused the saturation.

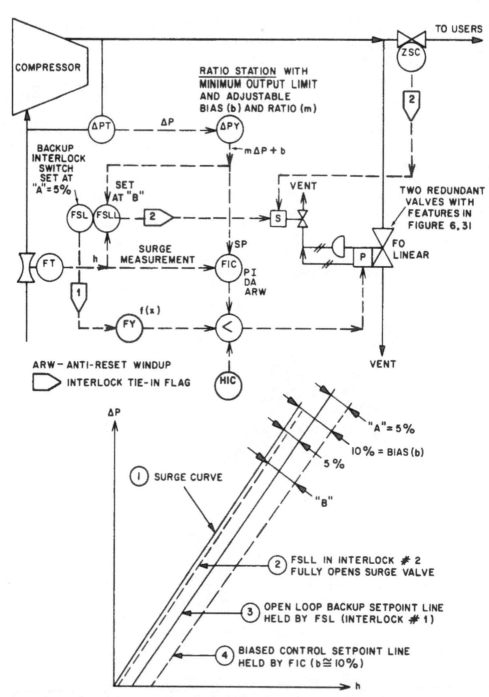

Fig. 6.32 The illustrated surge control loops and lines are used to protect centrifugal machines at low compression ratios. (Adapted from McMillan [15])

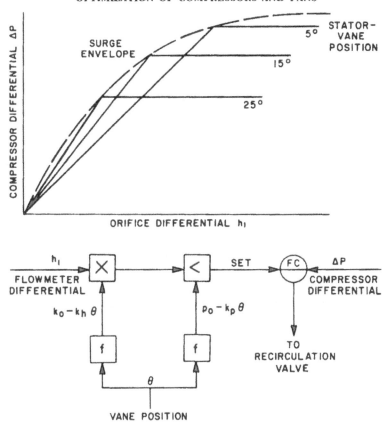

Fig. 6.33 When the surge curve bends over, the control model should include vane position. (Adapted from Shinskey [19])

The aim of an anti-reset wind-up (ARW) feature in this controller is to hold the FIC output under normal conditions at around 105 percent, so that the valve is closed but the signal is just lingering above the 100-percent mark without saturation. As soon as h starts dropping toward the set point, the valve starts to open, and it reaches full opening *before* the set point is reached. As soon as h starts dropping, the proportional contribution to the FIC output decreases. If the approach is slow, the increase in reset contribution will be greater than the decrease caused by the proportional contribution. Therefore, if h does not drop to SP, the valve slowly returns to the closed position. The effect of the internal feedback in the FIC is to activate the integral action only when the valve is *not* closed, while keeping the proportional action operational all the time.

The main goal of the feedback surge controller is to protect the machine from going into surge. Once the compressor is in surge, the FIC is not likely to be able to bring it out of it, because the surge oscillations are too fast for the controller to keep up with them. This is why back-up systems are needed.

The surge controller is usually electronic, and the more recent installations tend to be microprocessor-based. Such digital units can memorize complex, nonlinear surge curves and can also provide adaptive gain, which is a means of increasing controller gain as the operating point approaches the surge curve. In digital controllers, it is desirable to set the sample time at about one quarter of the period of surge oscillation, or at 0.3 seconds, whichever is shorter. If a flow derivative back-up control (figure 6.34) is used, the sample time must be 0.05" or less, because if it is slower, the back-up system would miss the precipitous drop in flow as surge is beginning.

Tuning the Controller

The *closed loop ultimate oscillation* method cannot be used for tuning the surge loop, because there is a danger of this oscillation triggering the start of surge. The *open loop reaction curve* technique of tuning is used with both the surge and the throughput controller in manual. The output of each controller is changed by 5 or 10 percent, and the resulting measurement response is analyzed on a high-speed recorder, yielding the

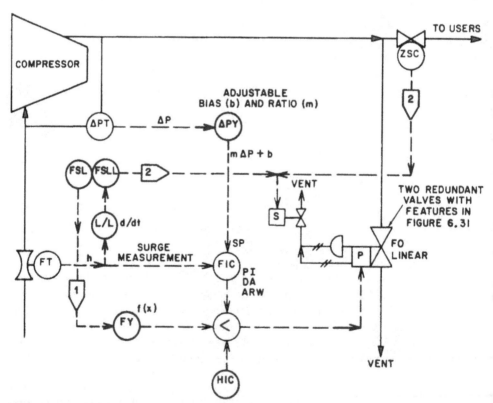

Fig. 6.34 In the flow derivative method of back-up, the sample time must be 0.05" or less. (Adapted from McMillan [13])

time constant and the dead time of the process. The tuning settings of the surge controller can be tested by first increasing the bias (b in figure 6.32) until the surge valve opens and then analyzing the loop response. The FIC must be so tuned that its overshoot is *less* than the distance between the FSL and FIC set points (A in figure 6.32). The throughput controller should be kept in automatic while the surge controller is tested so that interaction problems will be noted. Once the surge controller is tuned, it should not be switched to manual but should remain always in automatic.

The anti-reset wind-up (ARW) feature is not effective on slow disturbances, because the integral contribution that increases the FIC output outweighs the proportional contribution that lowers it; thus, the valve stays closed until the surge line is crossed. In order for the ARW to be effective, the time for the controller error to drop from its initial value to zero (T) must fall within two limits. It must be slower than the stroking time of the surge valve (Tv), but it must be faster than two integral times of the controller [15]:

$$Tv<T<2Ti$$

If the above requirement cannot be satisfied—because, for example, one of the potential disturbances is a slow-closing discharge valve—an "anticipator" control loop should be added (figure 6.35).

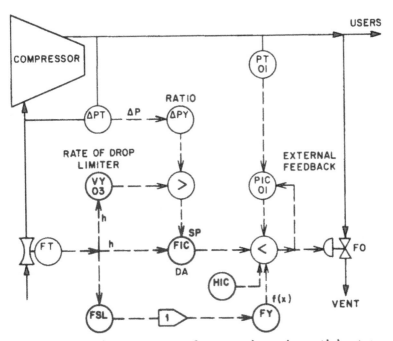

Fig. 6.35 Velocity limiter is one type of protection that can be provided against sudden load changes. (From Baker [3])

Surge Protection Back-Up

The purpose of the FIC in figure 6.32 is to protect the compressor from going into surge. The feedback surge controller is usually not fast enough to bring the machine out of surge once it has developed, because the surge oscillations are too fast for this controller. Yet even with the best feedback surge controller design, the machine will go into surge. This can be intentionally initiated to verify the surge curve, it can be caused by errors or shifts in the surge curve as the unit ages, and it can occur because of misoperation or failure of either the controller or downstream block and check valves. For these reasons, a second line of defense—a back-up system—is required in addition to the surge FIC.

The back-up system usually consists of two subsystems, identified as interlocks #1 and #2 in figure 6.32. In interlock #2, FSLL detects conditions that will cause the complete stopping of all forward flow from the compressor; when such a condition is detected, FSLL instantaneously and fully opens the surge valve. One such condition in figure 6.32 is signaled by a closed position limit switch (ZSC) on the block valve.

Interlock #1 in figure 6.32 provides open loop back-up as follows: the FSL detects an approach to surge that is closer than the FIC set point; when it detects such a condition, it takes corrective action. Under normal conditions, FSL senses an h value that is higher than SP, meaning that the operating point is to the right of curve ④. As surge approaches, the operating point crosses the control line (curve ④) as h drops below SP. The FSL is set to actuate interlock #1 when h has crossed curve ③. This occurs when the flow has dropped below the FIC set point by the amount "A" (usually 5%).

When interlock #1 is actuated, it triggers the signal generator FY to drop its output signal to zero and then, when the operating point has returned to the right of curve ③, gradually increase it back to 100 percent. When the FY output drops to zero, the low signal selector (in the output of the FIC) will select it and send it to the valve. This causes the surge valve to open fully in less than a second, followed by a slow closure according to the program in the signal generator FY. As the FY output rises, it will reach the output of FIC, at which point control is returned from the back-up system to the feedback controller.

This type of back-up is called the *operating point method*. Its advantage is that it takes corrective action *before* the surge curve (curve ①) is reached. Its disadvantages are that it is not usable on multistage machines or on systems with recycle and that the protection it provides is lost if ΔPY in figure 6.32 fails by dropping its output to zero. For this reason, it is desirable to provide a minimum limit on the output of ΔPY.

The manual loader input (HIC) to the low signal selector is used during start-up. Because its output passes through the low selector, it can increase but cannot decrease the valve opening.

Flow Derivative Back-Up

The difference between the operating point technique of back-up (figure 6.32) and the flow derivative method (figure 6.34) is that the first method acts before the surge curve is reached, whereas the second is activated by the beginning of surge. Because flow drops off very quickly at the beginning of surge, the instruments that measure the rate of this drop must be very fast. If implemented digitally, the sample time of the flow derivative loop must be 0.05" or less in order not to miss the initial drop in flow. The lead-lag station (L/L) that detects the rate of flow reduction is adjusted so that the lag setting will filter out noise and the lead setting will give a large output when the flow drops.

The function of interlock #1 is the same as was described in connection with figure 6.32. When FSL detects the drop in flow, it causes the signal generator (FY) output to drop to zero, fully opening the valve within a second. If this action succeeds in arresting the surge, the signal generator output slowly rises and control is returned to the feedback FIC.

If interlock #1 does not succeed in bringing the machine out of surge, then after a preset number of oscillations FSLL is actuated. This triggers interlock #2, which keeps the surge valve fully open until surge oscillations stop.

The flow derivation method of back-up has the advantage of providing protection even if the compressor pressure-rise instruments (ΔPT, ΔPY) fail, but it also has the disadvantage of not being usable when the flow signal is noisy.

Optimized Adaptations of Surge Curve

The surge curve shifts with wear and with operating conditions. Figure 6.36 describes a surge control loop that recognizes such shifts and automatically adapts to the new curve. The adaptation subroutine consists of two segments, the set point adaptation section (blocks ① to ⑥) and the output back-up section (blocks ⑦ to ⑪).

The purpose of the output back-up section is to recognize the approach of a surge condition that the feedback controller (FIC) was unable to arrest and to correct such a condition when it occurs. As long as the surge measurement (h) is not below the set point (SP), no corrective action is needed, and therefore blocks ⑦ and ⑨ will set the FIC output modifier signal (A) to zero. Once h is below SP, the operating point is to the left of curve ④ in figure 6.32, and the output of block ⑦ in figure 6.36 is switched to "Yes." Next block ⑧ checks if the adjustable time delay D1—having typical values of 0.3 to 0.8 seconds [15]—has passed. If it has not, signal A remains at its last value. If D1 has passed, A is increased by an increment X. Typical values of the X increment range from 15 percent to 30 percent [15]. As A is subtracted from the FIC output signal, this back-up loop will open up the surge valve at a speed of 15 percent to 30 percent per 0.3 to 0.8 second. When h is restored to above set point, the signal A is slowly returned to zero, allowing the surge valve to reclose.

The purpose of the set point adaptation loop (blocks ① to ⑥) is to recognize shifts in the surge curve and, as a response, to move the control line ④ in figure 6.32 to the right by increasing to total bias b of the ratio relay ΔPY. The logic of blocks ② to ⑥ is similar to that described for blocks ⑦ to ⑪, except that the resulting variable bias signal (b_1) is added to the fixed bias (b_0) to arrive at the adapted new bias (b). The speed of set point adaptation does not need to be as fast as the opening of the surge valve. Therefore, the time delay D2 tends to be longer than D1, and the increment Y is smaller than X. The purpose of the reset button in block 1 is to provide a means for the operator to reset the b_1 signal back to zero.

Override Controls

Figures 6.32 and 6.34 show back-up and overrides implemented by a low signal selector on the outlet of the feedback FIC. Figure 6.37 shows some additional overrides, which might be desirable in some installations. Under normal conditions all override signals are above 100 percent, and they leave the feedback FIC in control. Sending the override signals through the low signal selector guarantees that they can only increase the opening of the surge valve, not decrease it.

The high pressure override controller (PIC-01) protects downstream equipment from overpressure damage. Normally its measurement is far below its set point and therefore its output signal is high. In order to prevent reset wind-up, external feedback is used.

The other override loop (FIC-03) guarantees that the shut-down of a user or a sudden change in user demand will not upset the mass balance (balance between inlet and outlet flow) around the compressor. This is an important feature, because without it the resulting domino effects could shut down other user reactors or the whole plant.

In figure 6.37, when a user is shut down, FT-02 detects the drop in total flow within milliseconds, if a fast flow transmitter is used. This drop in flow reduces the measurement signal (h) to FIC-03, while its set point is still unaltered because of the lag (FY-03). Therefore, the surge valve is opened to maintain the mass balance. As FY-03 slowly lowers the set point to the new user flow rate (plus bias), FIC-03 recloses the valve. The bias b is used to keep the set point of FIC-03 above its measurement, thus keeping the surge valve closed when conditions are normal.

Figure 6.35 above illustrates another method of maintaining mass balance as users are suddenly turned off. The velocity limiter (VY-03) limits the rate at which the surge signal (h) can be lowered. If the surge signal drops off suddenly, the VY-03 output remains high and, through the high signal selector, keeps the surge controller set point up. This opens the surge valve to maintain the mass balance until it is reclosed by the limiter. This method is called an *anticipator algorithm*, because it opens the surge valve sooner than the time when the operating point (h) would otherwise drop to the set point, computed by the ratio station (ΔPY).

Optimized Load Following

When a compressor is supplying gas to several parallel users, the goal of optimization is to satisfy all users with the minimum investment of energy. The minimum required header pressure is found by the valve position controller (VPC) in figure 6.38. It compares the highest user valve opening (the needs of the most demanding user) with a set point of 90 percent, and if even the most open valve is not yet 90-percent open, it lowers the header pressure set point. This supply-demand matching strategy not only minimizes the use of compressor power but also protects the users from being undersupplied, because it protects any and all supply valves from the need to open

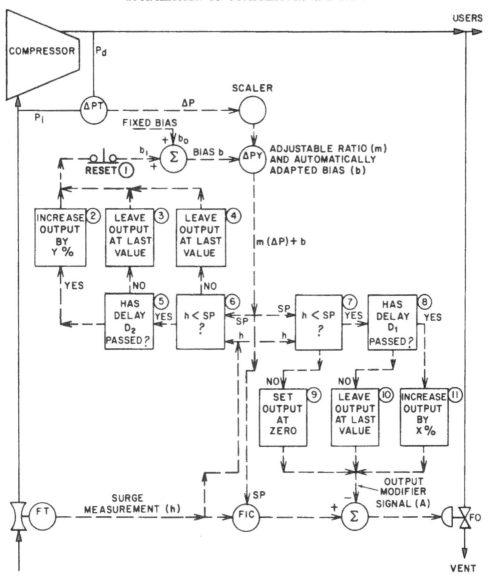

Fig. 6.36 This control system recognizes changes in the surge curve and adapts the surge controller to the new curve. (Adapted from McMillan [15], describing U.S. Patent 856.302, owned by Naum Staroselsky)

fully. On the other hand, as the VPC causes the user valves to open farther, these valves not only reduce their pressure drops but also become more stable.

The VPC is an integral-only controller, with its integral time set for about ten times that of the PIC. This guarantees that the VPC will act more slowly than all the user controllers, thus giving stable control even if the user valves are unstable. The external

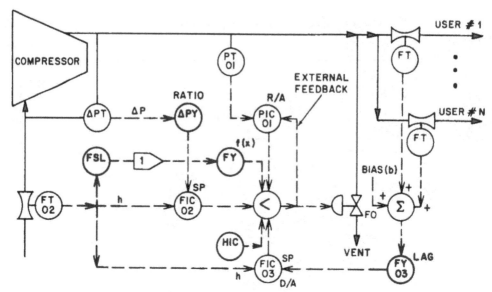

Fig. 6.37 This control system includes high pressure and user shut-down overrides. (Adapted from Mc-Millan [15])

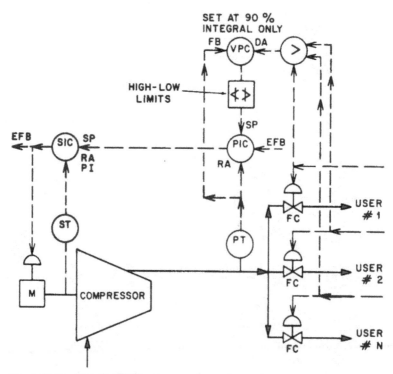

Fig. 6.38 Optimized load-following controls are designed to satisfy all users with a minimum energy investment.

feedback (FB) protects the VPC from reset wind-up when its output is limited or when the PIC has been switched to manual.

In addition to following the load, it is also necessary to protect the equipment. In figure 6.39, one protective override prevents the development of excessively low suction pressures (PIC-02), which could result in drawing oil into the compressor. The other override (KWIC-03) protects from overloading the drive motor and thereby tripping the circuit breaker.

In order to prevent reset wind-up when the controller output is blocked from affecting the SIC set point, external feedback (EFB) is provided for all three controllers. This arrangement is typical for all selective or selective-cascade control systems.

Interaction

The load and surge control loops together, as shown in figure 6.40, both affect the same variable: compressor throughput. Under normal conditions, there is no problem of interaction; because the surge loop is inactive, its valve is closed. Yet, when point "A" in figure 6.41 is reached, FIC-02 quickly opens the surge valve, which causes the discharge pressure to drop off as the flow increases. PIC-01 responds by increasing the vane setting or speed of the machine. The faster the PIC-01 loop instrumentation and the higher its gain (the narrower the proportional band), the larger and faster will be the increase in the speed of the compressor. If the load curve is steeper than the surge

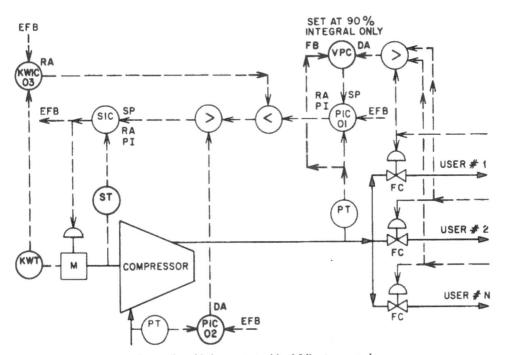

Fig. 6.39 Protective overrides can be added to optimized load following controls.

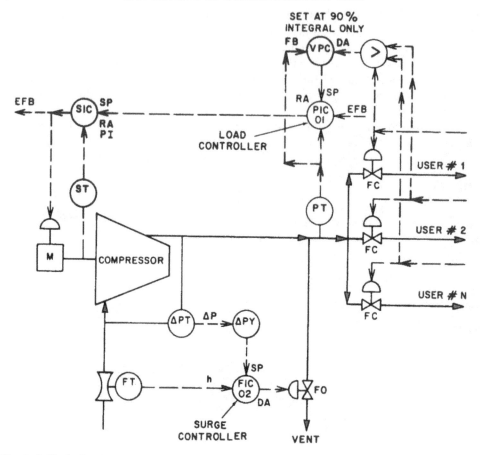

Fig. 6.40 The load and surge controllers are both effecting compressor throughput.

curve (as in figure 6.41), the action of PIC-01 will bring the operating point closer to the surge line. In this case, a conflict exists between the two loops, because as FIC-02 acts to correct an approaching surge situation, PIC-01 responds by worsening it. The better tuned (narrower proportional band) and faster (electronic or hydraulic speed governors) the PIC-01 loop is, the more dangerous is its effect of worsening the approach of surge.

The throughput controller (PIC-01 in figure 6.40) moves the operating point on the path of the load curve. The surge controller (FIC-02 in figure 6.40) moves the operating point on the path of the characteristic curve at constant speed. If the characteristic curve is steep, the action of the surge controller will cause a substantial upset in the operation of the pressure controller (figure 6.40). If the load curve is flat, the effect of the pressure controller on the surge controller will be the greatest.

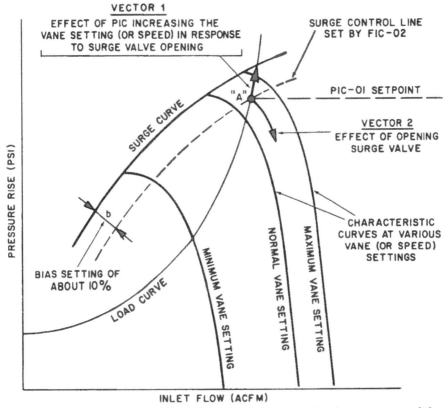

INLET FLOW (ACFM)

Fig. 6.41 Load and surge loops conflict if the load curve is steeper than the surge curve and if the controlled variable is pressure. (Adapted from McMillan [15])

If the throughput is under flow control, the opposite effects will be observed. The effect of the surge controller on the flow controller will be the greatest when the characteristic curves of the compressor are flat, and the effect of the flow controller on the surge controller will be greatest when the load curve is steep.

In determining whether the loops will assist or conflict with each other, both the relative slopes of the load and surge curves and the variable selected for process control need to be considered:

Controlled Variable	Load Curve Steeper than Surge Curve	Nature of Interaction
Pressure	Yes	Conflict
	No	Assist
Flow	Yes	Assist
	No	Conflict

As both loops try to position the operating point on the compressor map, the resulting interaction can cause the type of inverse response that was described in figure 6.41, or it can cause oscillation and noise. The oscillating interaction is worst if the proportional bands and time constants or periods of oscillation are similar for the two loops. Therefore, if tight control is of no serious importance, one method of reducing interactions is to reduce the response (widen the proportional band) of the load controller. Similarly, the use of slower actuators (such as pneumatic ones) to control compressor throughput will also reduce interaction, but at the cost of less responsive overall load control.

Relative Gain

In evaluating the degree of conflict and interaction between the load and surge loops, it is desirable to calculate the relative gain between the two loops. The relative gain is the ratio between the open loop gain when the other loop is in manual, divided by the open loop gain when the other loop is in automatic. The open loop gain of PIC-01 (figure 6.40) is the ratio of the change in its output to a change in its input, which caused it. Therefore, if a 1-percent increase in pressure results in a 0.5-percent decrease in compressor speed, the open loop gain is said to be -0.5. Assuming that the open loop gain of PIC-01 is -0.5 when FIC-02 is in manual and that it drops to -0.25 when FIC-02 is in automatic, the relative gain is $0.5/0.25 = 2$. The relative gain (RG) values can be interpreted as follows:

RG Value	Effect of Other Loop
0 to 1.0	Assists the primary loop
Above 1.0	Conflicts with the primary loop
Below 0	Conflict that also reverses the action of primary loop

If the calculated relative gain values are put in a 2-by-2 matrix, the best pairing of controlled and manipulated variables is selected by choosing those that will give the least amount of conflict. These are the RG values between 0 and 1.0, preferably close to 1.0.

Decoupling

Decoupling is provided to reduce the interaction between the surge and the load control loops. If the goal of decoupling is to maintain good pressure control even during a surge episode, the system shown in figure 6.42 can be used. In this system, as the antisurge controller opens the vent valve, a feedforward signal (X) proportionately increases the speed of the compressor. The negative sign at the summing device is necessary because the surge valve fails open. If the two vectors shown in figure 6.41 are correctly weighed in the summer, the end result of their summation will be a horizontal vector to the right, and PIC-01 will stay on set point.

On the other hand, if the main goal of decoupling is to temporarily reduce the size

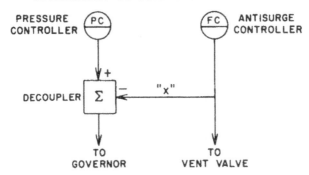

Fig. 6.42 The opening of the vent valve is automatically converted into a proportional increase in speed to avoid upsetting pressure. (From Shinskey [19])

of vector #1 in figure 6.41 so that it will not contribute to the worsening of the surge condition, then the half decoupling shown in figure 6.43 can be considered. Here the decoupling summer also receives a feedforward signal (X) from the surge valve, but it acts to slow down (not speed up) the machine. As the output of FIC-02 drops, the compressor speed is temporarily lowered, because the added value of X is reduced. This brings the operating point further away from the surge line, as shown in figure 6.41. The lagged signal Y later on eliminates this bias, because when its time constant has been reached, the values of X and Y will be equal and will cancel each other. Therefore, this decoupler will serve to temporarily desensitize a fast and tightly tuned PIC-01 loop, which otherwise might worsen the situation by overreacting to a drop in pressure when the vent valve opens.

Multiple Compressor Systems

Compressors are connected in series to increase their discharge pressure (compression ratio) or they are connected in parallel to increase their flow capacity. Series compressors on the same shaft can usually be protected by a single antisurge control system. When driven by different shafts, they require separate antisurge systems, although an overall surge bypass valve can be common to all surge controllers through the use of a low signal selector [19]. This eliminates the interaction that otherwise would occur between surge valves, as the opening of a bypass around one stage not only would increase the flow through the higher stages but would also decrease it through the lower ones. In some installations, this interaction has been found to be of less serious consequence than the time delay caused by the use of a single overall bypass surge valve, which cannot quickly increase flow in the upper stages [15]. In such installations it is best to duplicate the complete surge controls around each compressor in series.

If streams are extracted or injected between the compressors in series and their flows are not equal, a control loop needs to be added to keep both compressors away from surge by automatically distributing the total required compression ratio between

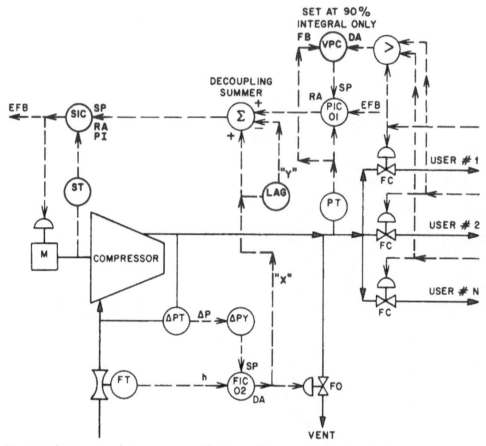

Fig. 6.43 The interaction between surge and load control loops can also be decoupled by temporarily reducing speed when the surge valve opens. (Adapted from McMillan [15])

them. Such a control system is shown in figure 6.44. In this system, PIC-01 controls the total discharge pressure by adjusting the speed of compressor #1. The speed of compressor #2 is set by FFIC-02; both compressors are thus kept at equal distances from their respective surge lines. This is accomplished by maintaining the following equality:

$$(K_1(P_1)(\Delta P_1)/(T_1)) + B_1 = (K_2(P_2)(\Delta P_2)/(T_2)) + B_2$$

In this equation, K is the slope and B is the bias of the surge set points of the respective compressors.

Control of Parallel Compressors

Figure 6.45 illustrates how two compressors can be proportionally loaded and unloaded, while keeping their operating points at equal distance from the surge curve.

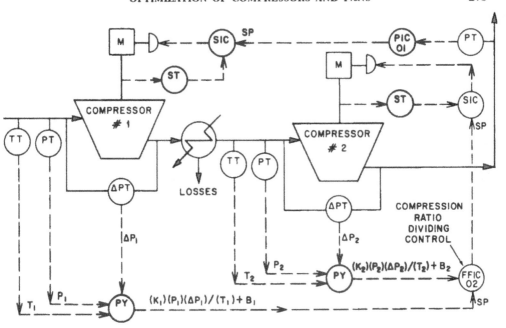

Fig. 6.44 The compression ratio distribution is controlled between two compressors when they operate in series. (Adapted from McMillan [15])

The lead compressor (#1 in figure 6.45) is selected either as the larger unit or as the one that is closer to the surge curve when the load rises or is further from it when the load drops. In figure 6.45 it is assumed that the compressors were so chosen that their ratio of bias to slope (b/K) of the surge set point is equal. In that case,

$$h_2 = h_1 \, (K_2/K_1)$$

where h is the flow meter differential and K is the surge set point slope of the respective compressors.

Because of age, wear, or design differences, no two compressors are identical. A change in load will not affect them equally and each should therefore be provided with its own antisurge system. Another reason for individual surge protection is that check valves are used to prevent backflow into idle compressors. Therefore, the only way to start up an idle unit is to let it build up its discharge head while its surge valve is partially open. If this is not done and the unit is started against the head of the operating compressors, it will surge immediately. The reason why the surge valve is usually not opened fully during start-up is to protect the motor from overloading.

Improper distribution of the load is prevented by measuring the total load (summer #9 in figure 6.46) and assigning an adjustable percentage of it to each compressor by adjusting the set points of FFIC-01 and FFIC-02.

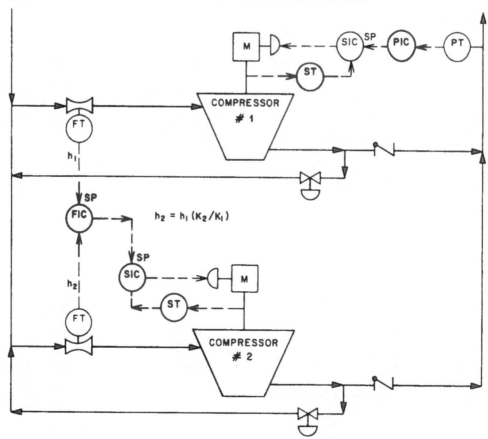

Fig. 6.45 Two parallel compressors can be proportionally loaded. (Adapted from McMillan [15])

Optimization Maximizes Efficiency

The load distribution can be computer-optimized by calculating compressor efficiencies (in units of flow per unit power) and loading the units in the order of their efficiencies. The same goal can be achieved if the operator manually adjusts the ratio settings of FFIC-01 and 02.

In the control system of figure 6.46, the pressure controller (PIC-01) directly sets the set points of SIC-01 and 02 while the balancing controllers (FFIC-01 & 02) slowly bias that setting. This is a more stable and responsive configuration than a pressure-flow cascade, because the time constants of the two loops are similar.

The output of PIC-01 must be corrected as compressors are started or stopped. One method of handling this was shown in figure 6.6; another technique is illustrated in figure 6.46. Here a high-speed integrator (item #5) is used on the summed speed signals to assure a correspondence between the PIC output signal and the number of

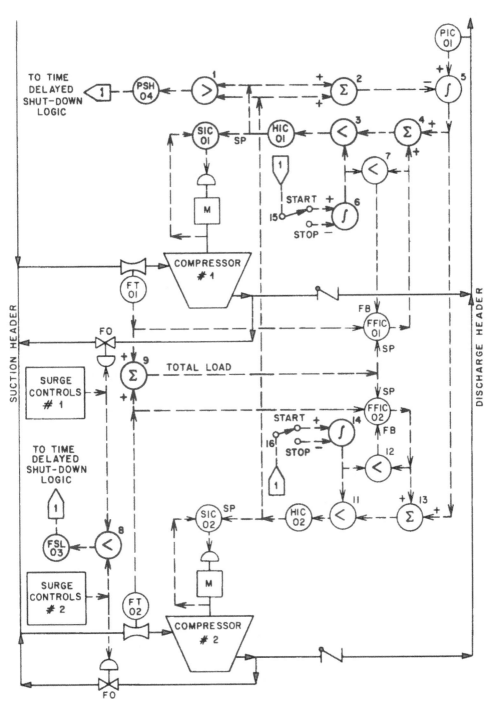

Fig. 6.46 A flow-balancing bias can be superimposed on direct pressure control. (Adapted from Shinskey [19])

compressors (and their loading) used. The integrator responds in a fraction of a second and therefore does not degrade the speed of response of the PIC loop.

Figure 6.46 also illustrates the automatic starting and stopping of individual compressors as the load varies. When the total flow can be handled by one fewer compressor or when any of the surge valves open, FSL-03 triggers the shut-down logic interlock circuit #1 after a time delay. Operation with an open surge valve would be highly inefficient, because the recirculated gas is redistributed among all the operating units. When a compressor is to be stopped, item #15 (or #16) is switched to the stop position, causing the integrator #6 (or #14) to drive down until it overrides the control signal in low selector #3 (or #11) and reduces the speed until the unit is stopped.

Automatic starting of an additional compressor is also initiated by interlock #1 when PSH-04 signals that one of the compressors has reached full speed. When a compressor is to be started, interlock #1 switches item #15 (or #16) to the start position, causing the integrator #6 (or #14) to drive up (by applying supply voltage to the integrator) until PIC-01 takes over control through the low selector #3 (or #11). The integrator output will continue to rise and then will stay at maximum, so as not to interfere with the operation of the control loop. The ratio flow controller (FFIC-01 and 02) are protected from reset wind-up by receiving an external feedback signal through the low selector #7 (or #12), which selects the lower of the FFIC output and the ramp signal.

Interlock #1 is also provided with "rotating sequencer" logic, which serves to equalize run times between machines and protects the same machine from being started and stopped frequently. A simple approximation of these goals is achieved if the machine that operated the longest is stopped and the one that was idle the longest is started.

If only one of the compressors is variable-speed, the PIC-01 output signal can be used in a split range manner. For example, if there were five compressors of equal capacity, switches would be set to start an additional constant-speed unit as the output signal rises above 20 percent, 40 percent, 60 percent and 80 percent of its full range. The speed setting of the one variable-speed compressor is obtained by subtracting from the PIC output the sum of the flows developed by the constant-speed machines and multiplying the remainder by five. The gain of five is the result of the capacity ratio between that of the individual compressor and the total [12].

Conclusions

As was shown in figures 4.26 and 6.4, the energy cost of operating a single compressor at 60 percent average loading can be cut in half by optimized variable-speed design. In the case of multiple compressor systems, similar savings can be obtained by optimized load-following and supply-demand matching.

The full automation of compressor stations—including automatic start-up and shut-down—not only will reduce operating cost but will also increase operating safety as human errors are eliminated. Figure 6.47 illustrates the use of microprocessor-based compressor unit controllers. Such units can be provided with all the algorithms that were described in this chapter; therefore, they provide flexibility by allowing changes in validity checks, alteration of limit stops, or reconfiguration of control loops.

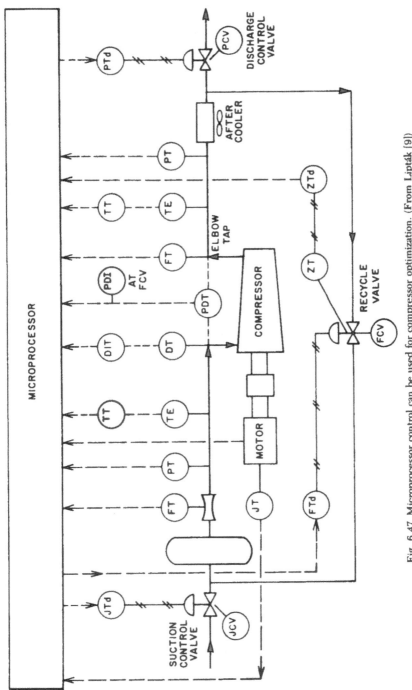

Fig. 6.47 Microprocessor control can be used for compressor optimization. (From Lipták [9])

References

1. Adamski, R. S., "Improving Reliability of Rotating Machinery," *InTech*, February 1982.
2. American Society of Heating, Refrigerating, and Air-Conditioning Engineers, *ASHRAE Handbook, 1983 Equipment Volume* (Atlanta, GA: ASHRAE, 1983).
3. Baker, D. F., "Surge Control for Multistage Centrifugal Compressors," *Chemical Engineering*, May 31, 1982, pp. 117–22.
4. Bentley, D. E., *Machinery Protection Systems for Various Types of Rotating Equipment* (Minden, NV: Bentley Nevada Corp., 1980).
5. Bogel, G. D. and R. L. Rhodes, "Digital Control Saves Energy in Gas Pipeline Compressors," *InTech*, July 1982, pp. 43–45.
6. Gaston, J. R., "Improved Flow/Delta-P Anti-Surge Control System," *Compressor-Engine Workshop* (published by Pacific Energy Association), March 1981, pp. 1–13.
7. Hansen, K. E., et al., "Experimental and Theoretical Study of Surge in a Small Centrifugal Compressor," *Journal of Fluids Engineering*, September 1981, pp. 391–95.
8. Langill, A. W., Jr., "Microprocessor-Based Control of Large Constant-Speed Centrifugal Compressors," in *Advances in Test Measurement*, vol. 19 (Research Triangle Park, NC: ISA, 1982).
9. Lipták, B. G., *Instrument Engineer's Handbook, Process Control* (Radnor, PA: Chilton Book Co., 1985), section 8.5.
10. Lipták, B. G., *Instrument Engineer's Handbook, Process Measurement* (Radnor, PA: Chilton Book Co., 1982).
11. Lipták, B. G. "Integrating Compressors into One System," *Chemical Engineering*, Jan. 19, 1987.
12. Lipták, B. G., "Optimizing Plant Chiller Systems," *Instrumentation Technology*, September 1977.
13. Magliozzi, T. L., "Control System Prevents Surging," *Chemical Engineering*, May 8, 1967.
14. *Mark's Standard Handbook for Mechanical Engineers*, eighth edition, Theodore Baumeister, ed. (New York, NY: McGraw-Hill 1978), pp. 14–53.
15. McMillan, G. K., *Centrifugal and Axial Compressor Control* (Instrument Society of America, 1983).
16. Rammler, R., "Energy Savings Through Advanced Centrifugal Compressor Performance Control," in *Advances in Instrumentation*, vol. 37 (Research Triangle Park, NC: ISA, 1982).
17. Roberts, W. B. and W. Rogers, "Turbine Engine Fuel Conservation by Fan and Compressor Profile Control," Symposium on Commercial Aviation Energy Conservation Strategies, National Technical Information Service Document PC A 16/MF A01, April 2, 1981, pp. 231–56.
18. Shinskey, F. G., *Controlling Multivariable Processes* (ISA, 1981), p. 159.
19. Shinskey, F. G., *Energy Conservation Through Control* (Academic Press, 1978).
20. Staroselsky, N., "Better Efficiency and Reliability for Dynamic Compressors Operating in Parallel or in Series," paper 80-PET-42, ASME, New York, 1980.
21. Van Omer, H. P., Jr., "Air Compressor Capacity Controls: A Necessary Evil, Volume I, Reciprocating Compressors," *Hydraulics and Pneumatics*, June 1980, pp. 67–70.
22. Waggoner, R. C., "Process Control for Compressors," paper given at Houston Conference of ISA, October 11–14, 1976.

Chapter 7

OPTIMIZATION OF COOLING TOWERS

COOLING TOWERS are water-to-air heat exchangers that are used to discharge waste heat into the atmosphere. Their cost of operation is a function of the water and air transportation costs. The goal of cooling tower optimization is to maximize the amount of heat discharged into the atmosphere per unit of operating cost invested.

In this chapter the types of cooling tower design are described and the various techniques of load-following capacity controls are discussed. The optimization strategies discussed are not limited to standard cooling towers but also include the evaporative types and the free cooling configurations. Cooling tower winterizing and blowdown controls are also covered.

Definitions

Before the control aspects of cooling towers can be discussed, some related terms must be defined [6]:

Approach: The difference between the wet bulb temperature of the ambient air and the water temperature leaving the tower. The approach is a function of cooling tower capability; a larger cooling tower will produce a closer approach (colder leaving water) for a given heat load, flow rate, and entering air condition (figure 7.1). (Units: °F or °C.)

Bay: The area between two bents of lines of framing members; usually longitudinal.

Bent: A line of structural framework composed of columns, girts, or ties; a bent may incorporate diagonal bracing members; usually transverse.

Blowdown: Water discharged to control concentration of impurities in circulated water. (Units: percentage of circulation rate.)

Brake horsepower (BHP): The actual power output of an engine or motor.

Cold water basin: A device underlying the tower to receive the cold water from the tower and direct its flow to the suction line or sump.

Counterflow water cooling tower: A cooling tower in which air flow is in opposite direction from the fall of water through the water cooling-tower.

Cross-flow water cooling tower: A cooling tower in which air flow through the fill is perpendicular to the plane of the falling water.

Dew point temperature (Tdp): The temperature at which condensation begins if air is cooled under constant pressure (see figure 7.2).

Double-flow water cooling tower: A cross-flow tower with two fill sections and one plenum chamber that is common to both fill sections.

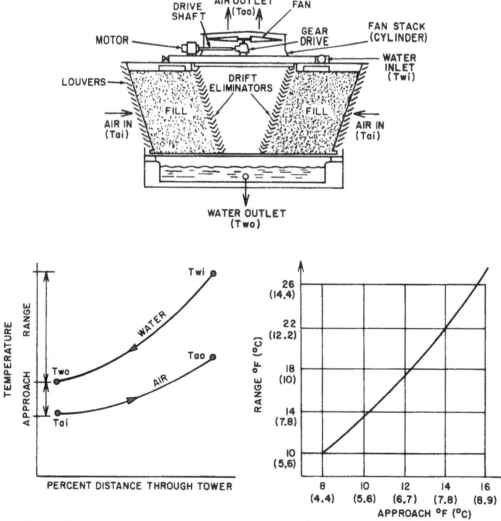

Fig. 7.1 The mechanical cooling tower is a water-to-air heat exchanger. A direct relationship exists between water and air temperature in a counterflow cooling tower. An increase in range will cause an increase in approach, if all other conditions remain unaltered. (From Lipták [11])

Drift: Water loss due to liquid droplets entrained in exhaust air. Usually under 0.2 percent of circulated water flow rate.

Drift eliminator: An assembly constructed of wood or honeycomb materials which serves to remove entrained moisture from the discharged air.

Dry-bulb temperature (Tdb): The temperature of air measured by a normal thermometer (see figure 7.2).

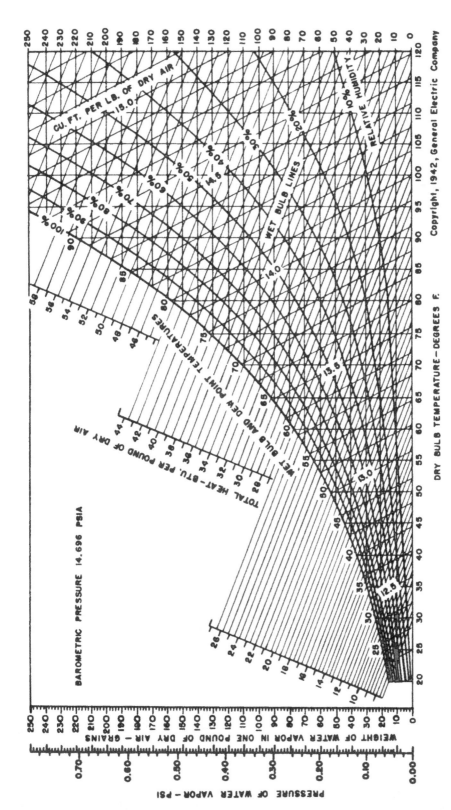

Fig. 7.2 The psychrometric chart describes the properties of air. (From Lipták [12])

Evaporation loss: Water evaporated from the circulating water into the atmosphere in the cooling process. (Unit: percentage of total GMP.)

Fan-drive output: Actual power output (BHP) of driver to shaft.

$$BHP = \frac{(\text{Motor Efficiency}) (\text{Amps}) (\text{Volts}) (\text{Power Factor}) \, 1.73}{746}$$

Fan pitch: The angle a fan blade makes with the plane of rotation. (Unit: degrees from horizontal.)

Fan stack (cylinder): Cylindrical or modified cylindrical structure in which the fan operates. Fan cylinders are used on both induced draft and forced draft axial-flow propeller type fans.

Filling: That part of an evaporative tower consisting of splash bars, vertical sheets of various configurations, or honeycomb assemblies that are placed within the tower to effect heat and mass transfer between the circulating water and the air flowing through the tower.

Forced draft cooling tower: A type of mechanical draft water cooling tower in which one or more fans are located at the air inlet to force air into the tower.

Heat load: Heat removed from the circulating water within the tower. Heat load may be calculated from the range and the circulating water flow. (Unit: BTU per hour = GPM × 500 × [Tctwr-Tctws].)

Liquid gas ratio (L/G): Mass ratio of water to air flow rates through the tower. (Units: lbs/lbs or kg/kg.)

Louvers: Assemblies installed on the air inlet faces of a tower to eliminate water splash-out.

Make-up: Water added to replace loss by evaporation, drift, blowdown, and leakage. (Unit: percentage of circulation rate.)

Mechanical draft water cooling tower: A tower through which air movement is effected by one or more fans. There are two general types of such towers: those that use forced draft with fans located at the air inlet and those that use induced draft with fans located at the air exhaust.

Natural draft water cooling tower (air movements): A cooling tower in which air movement is essentially dependent upon the difference in density between the entering air and internal air. As the heat of the water is transferred to the air passing through the tower, the warmed air tends to rise and draw in fresh air at the base of the tower.

Nominal tonnage: One nominal ton corresponds to the transfer of 15,000 BTU/hr (1.25 kW) when water is cooled from 95°F to 85°F (35°C to 29.4°C) by ambient air having a wet-bulb temperature of 78°F (25.6°C) and when the water circulation rate is 3 GPM (0.054 L/s) per ton.

Performance: The measure of a cooling tower's ability to cool water. Usually expressed in gallons per minute cooled from a specified hot water temperature to a specified cold water temperature with a specific wet-bulb temperature. Typical performance curves for a cooling tower are shown in figure 7.3. This

is a "7500 GPM at 105-85-78" tower, meaning that it will cool 7500 GPM
of water from 105°F to 85°F when the ambient wet bulb temperature is 78°F.

Plenum: The enclosed space between the eliminators and the fan stack in induced
draft towers, or the enclosed space between the fan and the filling in forced
draft towers.

Power factor: The ratio of true power (watts) to the apparent power (amps × volts).

Psychrometer: An instrument used primarily to measure the wet-bulb temperature.

Range (cooling range): The difference between the temperatures of water inlet
and outlet, as shown in figure 7.1. For a system operating in a steady state,

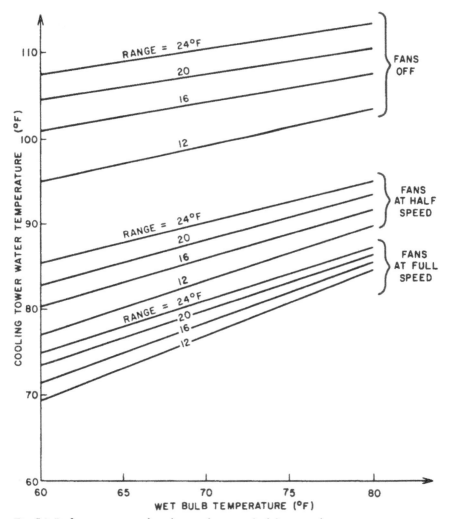

Fig. 7.3 Performance curves describe a cooling tower's ability to cool water.

the range is the same as the water temperature rise through the load heat exchanger. Accordingly, the range is determined by the heat load and water flow rate, not by the size or capability of the cooling tower. On the other hand, the range does affect the approach, and as the range increases, a corresponding increase in the approach will occur if all other factors remain unaltered (see figure 7.1).

Riser: Piping that connects the circulating water supply line from the level of the base of the tower or the supply header to the tower inlet connection.

Speed reducer: A device for changing the speed of the driver in order to arrive at the desired fan speed.

Standard air: Dry air having a density of 0.075 lbs./cu. ft. at 70°F and 29.92 in. Hg.

Story: The vertical dimension or area between two lines of horizontal frame work ties, girts, joists, or beams. In wood frame structures, the story will vary from 5' to 7' in height.

Sump (basin): Lowest portion of the basin to which cold circulating water flows: usually the point of suction connection.

Tonnage: See Nominal Tonnage.

Total pumping head: The total head of water, measured above the basin curb, required to deliver the ciculating water through the distribution system.

Tower pumping head: Same as the total pumping head minus the friction loss in the riser. It can be expressed as the total pressure at the centerline of the inlet pipe plus the vertical distance between the inlet centerline and the basin curb.

Water loading: Water flow divided by effective horizontal wetted area of the tower. (Unit: GPM/ft² or m³/[hr] [m²].)

Wet-bulb temperature (Twb): If a thermometer bulb is covered by a wet, water absorbing substance and is exposed to air, evaporation will cool the bulb to the wet-bulb temperature of the surrounding air. This is the temperature read by a psychrometer. If the air is saturated with water, the wet bulb, dry bulb, and dew point temperatures will all be the same. Otherwise, the wet bulb temperature is higher than the dew point temperature but lower than the dry bulb temperature (see figure 7.2).

Cooling Tower Operation

In a cooling tower, heat and mass transfer processes combine to cool the water. The mass transfer due to evaporation does consume water, but the amount of loss is only about 5 percent of the water requirements for equivalent once-through cooling by river water.

Cooling towers are capable of cooling within 5°F to 10°F (2.8°C to 5.6°C) of the ambient wet bulb temperature. The larger the cooling tower for a given set of water

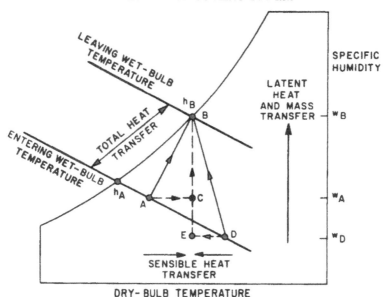

Fig. 7.4 Heat and mass transfer processes combine to cool water. (From ASH-RAE [2])

and air flow rates, the smaller this approach will be. The mass transfer contribution to the total cooling is illustrated in the psychrometric chart in figure 7.4. Ambient air might enter in various conditions, such as illustrated by points A or D. After it transfers heat and absorbs mass from the evaporating water, the air leaves in a saturated condition at point B. The total heat transferred in terms of the difference between entering and leaving enthalpies ($h_B - h_A$) is the same for both the AB and the DB process.

The AC component of the AB vector represents the heat transfer component (sensible air heating), and the CB vector represents the evaporative mass transfer component (latent air heating). If the air enters at condition D, the two component vectors do not complement each other, because only one of them heats the air (EB). Therefore, the total vector DB results from the latent heating (EB) and sensible cooling (DE) of the air. In terms of the water side of the DB process, the water is being sensibly heated by the air and cooled by evaporation. Therefore, as long as the entering air is not saturated, it is possible to cool the water with air that is warmer than the water.

The Cooling Process

The basic cooling tower designs are described in figure 7.5. For operational purposes, tower characteristic curves for various wet bulb temperatures, cooling ranges,

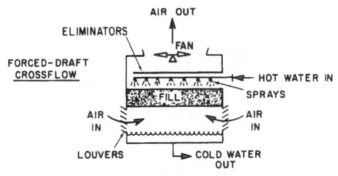

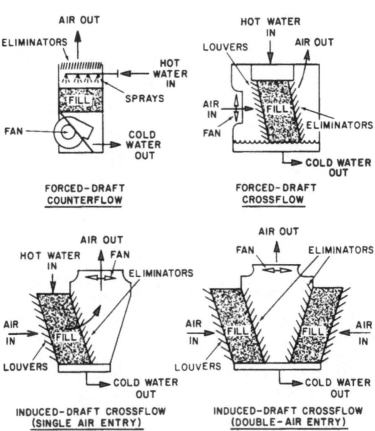

Fig. 7.5 Five basic cooling tower designs. (From ASHRAE [2])

and approaches can be plotted against the water flow to air flow ratio as shown in the following equation:

$$\frac{KaV}{L} \sim \frac{(L)^n}{(G)}$$

where

K = overall unit of conductance, mass transfer between saturated air at mass water temperature, and main air stream: lb per hour (ft²) (lb/lb)

a = area of water interface per unit volume of tower: ft² per ft³

V = active tower volume per unit area: ft³ per ft²

L = mass water flow rate: lb per hour

G = air flow rate: lb dry air per hour

n = experimental coefficient: varies from -0.35 to -1.1 and has an average value of -0.55 to -0.65

From these curves, it is possible to determine tower capability, effect of wet bulb temperature, cooling range, water circulation, and air delivery, as shown in figure 7.6. The ratio KaV/L is determined from test data on hot water and cold water, wet bulb temperature, and the ratio of the water flow to the air flow.

The thermal capability of cooling towers for air conditioning applications is usually stated in terms of nominal tonnage based on heat dissipation of 1.24 kW (15,000 Btu/h) per condenser kW (ton) and a water circulation rate of 0.054 l/s per kW (3 GPM per ton) cooled from 95°F to 85°F (35°C to 29.4°C) at 78°F (25.6°C) wet bulb temperature. It may be noted that the subject tower would be capable of handling a greater heat load (flow rate) when operating in a lower ambient wet bulb region. For operation at other flow rates, tower manufacturers usually provide performance curves covering a range of at least 2 to 5 GPM per nominal ton (0.036 to 0.09 lps/kW).

The initial investment required for a cooling tower is essentially a function of the required water flow rate, but this cost is also influenced by the design criteria for approach, range, and wet bulb. As shown in figure 7.7, cost tends to increase with range and to decrease with a rise in approach or wet bulb. The initial investment is about $20 per GPM of tower capacity, whereas the energy cost of operation is approximately 0.01 BHP/GPM. This means that the cost of operation reaches the cost of initial investment in five to ten years, depending on energy costs.

Capacity Controls

The cooling tower is an air-to-water heat exchanger. The controlled variables are the supply and return water temperatures, and the manipulated variables are the air and water flow rates. Manipulation of these flow rates can be continuous, through the use of variable speed fans and pumps, or incremental, through the cycling of single- or multiple-speed units.

If the temperature of the water supplied by the cooling tower is controlled by modulating the fans, a change in cooling load will result in a change in the operating level of the fans. If the fans are single-speed devices, the only control available is to

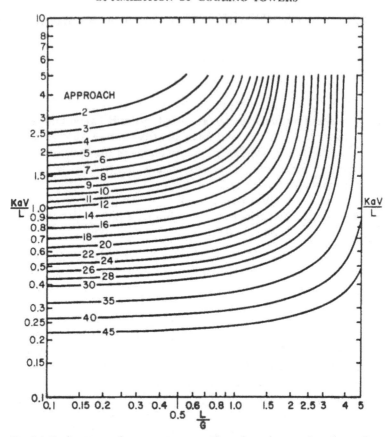

Fig. 7.6 Cooling tower characteristic curves (those shown here are based on 64°F [17.8°C] air wet-bulb temperature and 18°F [10°C] cooling range). (From Lipták [11])

cycle the fan units on or off as a function of load, as illustrated in figure 7.8. This can be done by simple sequencers or by more sophisticated digital systems. In most locations, the need to operate all fans at full speed applies for only a few thousand hours every year. Fans running at half speed consume approximately one-seventh of design air horsepower but produce over 50 percent of the design air rate (cooling effect). Because 98 percent of the time, the entering air temperature is less than the design, two-speed motors are a wise investment for minimizing operating costs.

Interlocks

Figure 7.9 describes a simple fan starter. It is controlled by the three-position "hand/off/automatic" switch. When the switch is placed in "automatic" (contacts 3 and 4 connected), the status of the interlock contact I determines whether or not the circuit

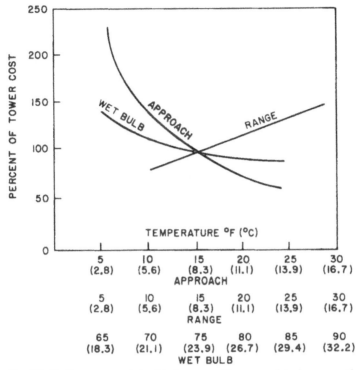

Fig. 7.7 Cooling-tower costs tend to increase with range and to decrease with a rise in approach or wet bulb. (Courtesy of Foster Wheeler Corp.)

is energized and the fan is on. The sequencing of this contact 1 was described in figure 7.8. The purpose of the auxiliary motor contact M is to energize the running light R whenever the fan motor is on. The parallel hot lead to contact 6 allows the operator to check quickly to see whether the light has burned out.

The amount of interlocking is usually greater than that shown in figure 7.9. Some of the additional interlock options are described in the paragraphs below.

Most fan controls include some safety overrides in addition to the overload (OL) contacts shown in figure 7.9. These overrides might stop the fan if fire, smoke, or excessive vibration is detected. They usually also provide a contact for remote alarming.

Most fan controls include a reset button that must be pressed after a safety shutdown condition is cleared before the fan can be restarted.

When a group of starters is supplied from a common feeder, it may be necessary to make sure that the starters are not overloaded by the inrush currents of simultaneously started units. If this feature is desired, a 25-second time delay is usually provided to prevent other fans from starting until that time has passed.

As was illustrated in figure 6.6, larger fans should be protected from overheating

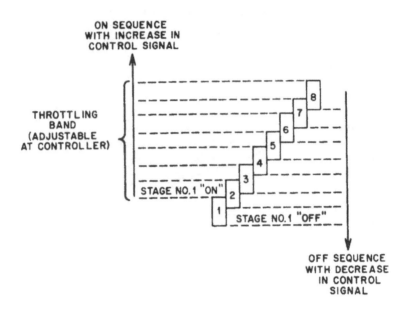

ON SEQUENCE
WITH INCREASE IN
CONTROL SIGNAL

THROTTLING
BAND
(ADJUSTABLE
AT CONTROLLER)

STAGE NO.1 "ON"

STAGE NO.1 "OFF"

OFF SEQUENCE
WITH DECREASE
IN CONTROL
SIGNAL

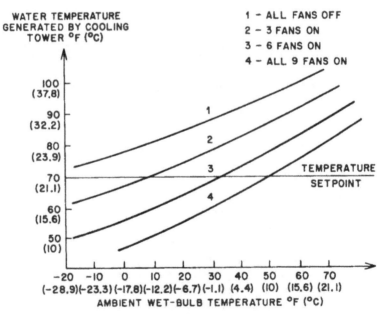

WATER TEMPERATURE
GENERATED BY COOLING
TOWER °F (°C)

1 - ALL FANS OFF
2 - 3 FANS ON
3 - 6 FANS ON
4 - ALL 9 FANS ON

TEMPERATURE
SETPOINT

AMBIENT WET-BULB TEMPERATURE °F (°C)

Fig. 7.8 Fans can be cycled to provide load-following cooling tower control.
Above: The output of the approach controller (TDIC-1 in Figure 7.11) can be sent
to a sequencer, which turns fan stages on and off. Below: Fan cycling has an ap-
preciable effect on a 9-cell tower designed for 160,000 GPM (608,000 l/m) with an
84°F (29°C) inlet and 70°F (21°C) outlet temperature, operating at 50% load.
(From Wistrom [18])

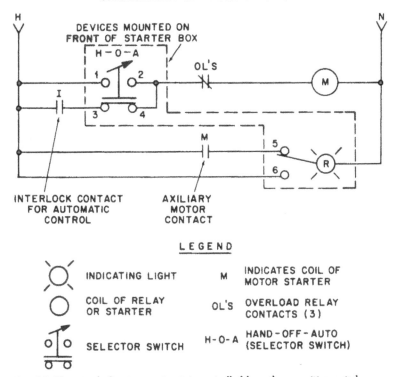

Fig. 7.9 This simple fan starter circuit is controlled by a three-position switch.

caused by too frequent starting and stopping. This protection can be provided by a 0-to-30-minute time delay guaranteeing that the fan will not be cycled at a frequency faster than the delay time setting.

When two-speed fans are used, added interlocks are frequently provided. One interlock might guarantee that even if the operator starts the fan in high, it will operate for 0 to 30 seconds in low before advancing automatically to high. This makes the transition from off to high speed more gradual. Another interlock might guarantee that when the fan is switched from high to low, it will be off the high speed for 0 to 30 seconds before the low-speed drive is engaged. This will give time for the fan to slow before the low gears are engaged.

When the fan can also be operated in reverse, further interlocks are needed. One interlock should guarantee that the fan is off for 0 to 2 minutes before a change in the direction of rotation can take place. This gives time for the fan to come to rest before making a change in direction. Another interlock will guarantee that the fan can operate only at slow speed in the reverse mode and that reverse operation cannot last more than 0 to 30 minutes. This limitation is desirable because in reverse ice tends to build up on the blades in winter and the gears wear excessively.

If the same fans can be controlled from several locations, interlocks are needed to resolve the potential problems with conflicting requests. One method is to provide an interlock that determines the location that is in control. This can be a simple local/ remote switch or a more complicated system. When many locations are involved, conflicts are resolved by interlocks that either establish priorities between control locations or select the lowest, highest, safest, or other specified status from the control requests. In such installations with multiple control centers, it is essential that feedback be provided so that the operator is always aware not only of the actual status of all fans but also of any conflicting requests coming from other operators or computers.

Evaporative Condensers

A special cooling tower type is the evaporative condenser illustrated in figure 7.10. Induced air and sprayed water flow concurrently downward, and the process vapors travel upward in the multipass condenser tubes. The PIC modulates the amount of cooling to match the load. The continuous water spray on the tube surfaces improves heat transfer and maintains tube temperatures. The approach of these cooling towers is around 20°F (11°C), and the wet-bulb temperature of the cooling air can be varied by the recirculation of humid (or saturated) air. The resulting condensing temperatures are lower than those in dry air cooled condensers. The higher heat capacity and lower resistance to heat transfer make these units more efficient than the dry air coolers.

The turndown capability of these units is also very high, because the fresh-air flow can be reduced all the way to zero. Both the exhaust and the fresh air dampers close while the recirculation damper simultaneously opens to maintain a constant internal air flow rate. As more humid air is recirculated, the internal wet-bulb temperature rises and the heat flow drops off. In figure 7.10 the heat flow is effectively linear with fresh-air flow [11].

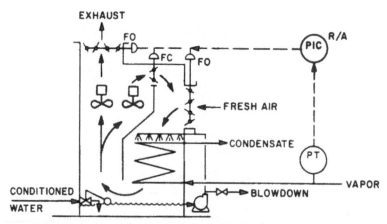

Fig. 7.10 Evaporative condensers can be controlled by the recirculation of humid air. (Adapted from Lipták [11])

Cooling Tower Optimization

The optimization of cooling towers as parts of chiller systems has already been discussed in chapter 4. The discussion here will concentrate on cooling towers that directly cool the process. The goal of optimization is to minimize the fan and pump operating costs and thereby minimize the unit cost of cooling.

The cost of fan operation can be reduced by allowing the cooling tower temperature (Tctws in figure 7.11) to rise, thereby increasing the approach (Tctws-Twb). As was shown in figure 7.3, the approach can be increased to a point at which the fan cost is zero (the fans are off). Under most load conditions, however, this would not produce a low enough temperature for the process. The cooling tower water temperature must always be lower than the desired process temperature (Tp) on the other side of the process heat exchangers.

As shown in figure 7.11, as the approach, and therefore Tctws, rises, the exchanger Δ T (Tp-Tctws) is reduced; thus, the process temperature controller (TIC-4) must open its coolant valve, CV-4. As the exchanger Δ T is reduced, more and more water must be pumped. Consequently, as the approach increases, the pumping costs rise.

If the actual total operating cost is plotted against approach (as in the right side of figure 7.11), the fan costs tend to drop and the pumping costs tend to rise with an increase in approach. If the operating cost of some of the users is also affected by cooling water temperature (this was the case with chiller condensers in figures 4.6 and 4.10), then the total cost model should also consider that effect. Once a total operating cost curve is found, the optimum approach is that which corresponds to the minimum point on that curve.

The data for the cost curves is empirically collected and is continually updated through the actual measurement of fan and pump operating costs. Consequently, for any combination of load and ambient conditions there is a reliable prediction of optimum approach setting. Once the initial setting is made, it can be refined by adjusting it in 0.5°F increments in the direction that lowers the total operating cost.

As is also shown in the right side of figure 7.11, there is an empirical relationship between the values of approach and range, if all other conditions remain unaltered. Therefore, if the optimum value of approach (SP1) has been determined, the corresponding range value (SP2) under the prevailing load conditions can be read from this curve. This is how the optimized set points (SP1 and SP2) of the approach and range controllers are determined and updated.

Supply Temperature Optimization

As shown in figure 7.11, an optimization control loop is required in order to maintain the cooling tower water supply continuously at an economical minimum temperature. This minimum temperature is a function of the wet bulb temperature of the atmospheric air. The cooling tower cannot generate a water temperature that is as low as the ambient wet bulb, but it can approach it.

Figure 7.11 illustrates the fact that as the approach increases, the cost of operating

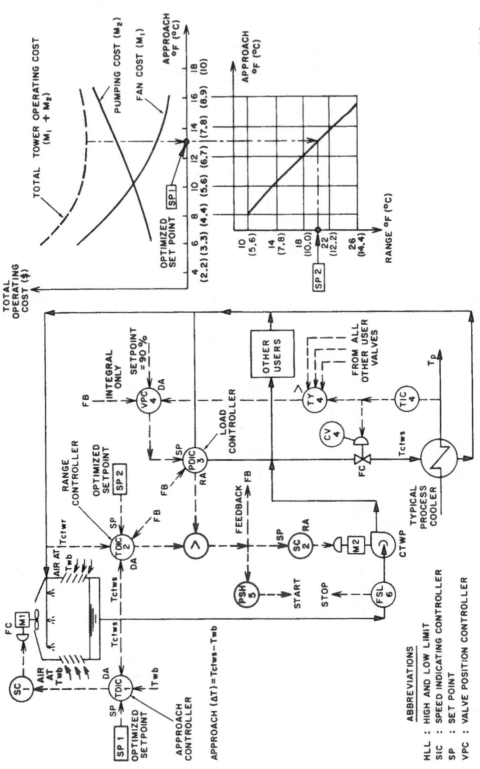

Fig. 7.11 Cooling tower optimization can reduce the fan and pump operating costs, thereby minimizing the unit cost of cooling. (Adapted from Lipták [11])

ABBREVIATIONS

HLL : HIGH AND LOW LIMIT
SIC : SPEED INDICATING CONTROLLER
SP : SET POINT
VPC : VALVE POSITION CONTROLLER

APPROACH (ΔT) = Tctws − Twb

the cooling tower fans drops and the cost of pumping increases. The optimum approach is the one that will allow operation at an overall minimum cost. This Δ T automatically becomes the set point of TDIC-1. The optimum approach increases if the load on the cooling tower increases or if the ambient wet bulb decreases.

If the cooling tower fans are centrifugal units or if the blade pitch is variable, the optimum approach is maintained by continuous throttling. If the tower fans are two-speed or single-speed units, the output of TDIC-1 will incrementally start and stop the fan units in order to maintain the optimum approach as was described in figure 7.8.

In cases in which a large number of cooling tower cells constitute the total system, it is also desirable to balance the water flows to the various cells automatically as a function of the operation of the associated fans. In other words, the water flows to all cells whose fans are at high speed should be controlled at equal high rates; cells with fans operating at low speeds should receive water at equal low flow rates; and cells with their fans off should be supplied with water at equal minimum flow rates.

Water Flow Balancing

The normal water flow rate ranges from 2 GPM to 5 GPM per ton when the fan is at full speed. Figure 7.12 illustrates the relationship between water flow and water temperature (Tctws) or approach. Under the illustrated conditions, a 20-percent change in water flow rate will affect the approach by about 2°F.

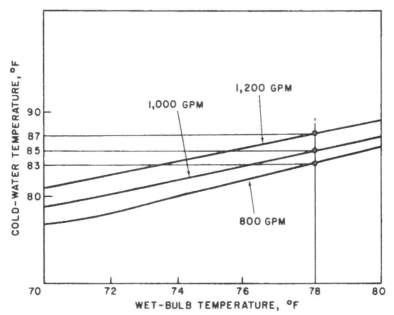

Fig. 7.12 Performance of a cooling tower is described under the following design conditions: 85°F cold water, 95°F hot water, 78°F wet bulb, 10°F range, and 1,000 gal/min cell. (Adapted from *Marks' Handbook* [5])

Water balancing is usually done manually, but it can also be done automatically, as shown in figure 7.13. Here the total flow is used as the set point of the ratio flow controllers (FFIC's). If the ratio settings are the same, the total flow is equally distributed. The ratio settings can be changed manually or automatically to reflect changes in fan speeds. Naturally, the total of the ratio settings must always be 1; therefore, if one ratio setting is changed, all the others should also be modified. This, too, can be done automatically.

The purpose of the control system in figure 7.13 is not only to distribute the returning water correctly between cells but also to make sure that this is done at minimum cost. The cost in this case is pumping cost, and it will be minimum when the pressure drop through the control valves is minimum. This is the function of the

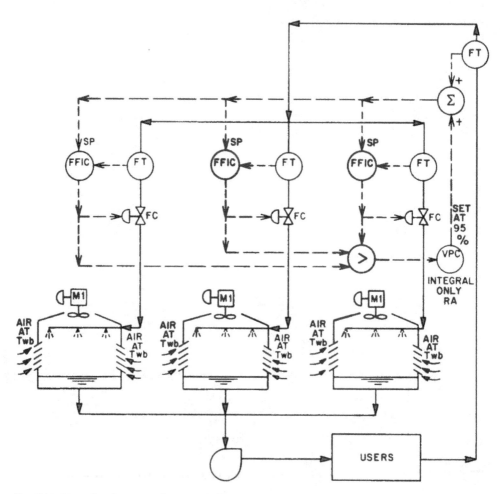

Fig. 7.13 Water distribution can be controlled automatically.

valve position controller (VPC) in figure 7.13. As long as even the most open valve is not nearly fully open, the VPC adds a positive bias to all the setpoint signals of all the FFIC's. As a result, all valves open and keep opening until the most open valve reaches the desired 95-percent opening. This technique enables correct flow distribution to be achieved at a minimum cost in pumping energy.

When cells are manually balanced, it is not unusual to find all balancing valves throttled and to see the same water flow when the fan is on or off. Both of these conditions will increase operating costs, and the savings from automatic balancing can more than justify the instruments required.

Return Temperature Optimization

Figure 7.11 also shows the controls for return water temperature optimization. TDIC-2 is the range controller having the optimized set point of SP2, which is the range value corresponding to the optimum approach. TDIC-2 throttles the water circulation rate in order to maintain the range at its optimum value. The output signals of TDIC-2 and PDIC-3 are sent through a high signal selector, which guarantees that the needs of the users take priority over range control. The user demand is established by PDIC-3, and if its output is higher than that of the range controller, it will be selected for setting the pump speed. The load-following optimization loop guarantees that all cooling water users in the plant will always be satisfied while the range is being optimized. This is done by selecting (with TY-4) the most-open cooling water valve in the plant and comparing that signal with the 90-percent set point of the valve position controller, VPC-4. If even the most-open valve is less than 90-percent open, the set point of PDIC-3 is decreased; however, if the valve opening exceeds 90 percent, the PDIC-3 set point is increased. In this way, a condition is maintained that allows all users to obtain more cooling (by further opening their supply valves) if needed while the differential pressure of the water is continuously optimized. The VPC-4 set point of 90 percent is adjustable. Lowering it gives a wider safety margin, which might be required if some of the cooling processes served are very critical. Increasing the set point maximizes energy conservation at the expense of the safety margin.

An additional benefit of this load-following optimization strategy is that because all cooling water valves in the plant are opened up as the water ΔP across the users is minimized, valve cycling is reduced, and pumping costs are lowered. The reduction in pumping costs is a direct result of the opening of all cooling water valves, which require less pressure drop in a more open condition. Valve cycling is eliminated when the valve opening is moved away from the unstable region near the closed position [13].

In order for the control system in figure 7.11 to be stable, it is necessary to use an integral-only controller for VPC-4, with an integral time that is tenfold that of the integral setting of PDIC-3. This mode selection is needed to allow the optimization loop to be stable when the valve opening signal selected by TY-4 is either cycling or noisy. A high limit setting on the output of VPC-4 guarantees that it will not drive the cooling water pressure to unsafe or undesirable levels. Because this limit can block the VPC-4 output from affecting the pump speed, it is necessary to protect against reset wind-up in VPC-4. This is done through the external feedback signal (FB) shown in figure 7.11, which protects TDIC-2, PDIC-3, and VPC-4 from reset wind-up.

A side benefit of this optimization system is that it brings attention to design errors. For example, if PDIC-3 is in control most of the time, it shows that the control valves and water pipes are undersized. If TY-4 consistently selects the same user, it shows that the water supply to that user is undersized. When such design errors are corrected, either by adding local booster pumps or by replacing undersized valves and pipes, control will automatically return to TDIC-2. Therefore, in a well-designed water distribution network, the range optimizer TDIC-2 will be in control most of the time; PDIC-3 will act only as a safety override that becomes active under emergency conditions, guaranteeing that no user will ever run out of cooling water.

When the cooling tower water pump station consists of several pumps, only one of which is variable-speed, additional pump increments are started when PSH-5 signals that the pump speed controller set point is at its maximum. When the load is dropping, the excess pump increments can be stopped on the basis of flow, detected by FSL-6. In order to eliminate cycling, the excess pump increment is turned off only when the actual total flow corresponds to less than 90 percent of the capacity of the remaining pumps.

This load-following optimization loop will float the total cooling tower water flow to achieve maximum overall economy.

Operating Mode Selection

When a refrigerant compressor or heat pump is operated, the process is said to be cooled by *mechanical refrigeration* (figure 7.14). When the process is cooled directly by the cooling tower water, without the use of chillers or refrigeration machines, the process is said to be cooled by *free cooling* (figure 7.13). Most modern cooling systems are provided with the capability for operating in several different modes as a function of load, ambient conditions, and utility costs.

The decision to switch from one operating mode to another is usually based on economic considerations. The control system will select the mode of operation that will allow the meeting of the load at the lowest total cost. Figure 4.13 illustrates the process of selecting the heat recovery or the mechanical refrigeration mode on the basis of cost-effectiveness. The same kind of logic is applied when the choice is free cooling or mechanical refrigeration.

Once the decision is made to switch modes, the actual reconfiguration of the associated piping and valving is done automatically. In larger, more complex cooling systems, there can be more than fifty different modes of operation (such a system was designed by the author for IBM Corporation Headquarters at 590 Madison Avenue in New York). In this chapter, only four modes will be discussed.

The cooling system illustrated in figure 7.14 is configured as a mechanical refrigeration system. The controls of this configuration have been discussed in figures 4.8 to 4.11. The heavily drawn pipelines show the active flow paths in that mode of operation. When the load drops off or when the ambient temperature decreases, it becomes

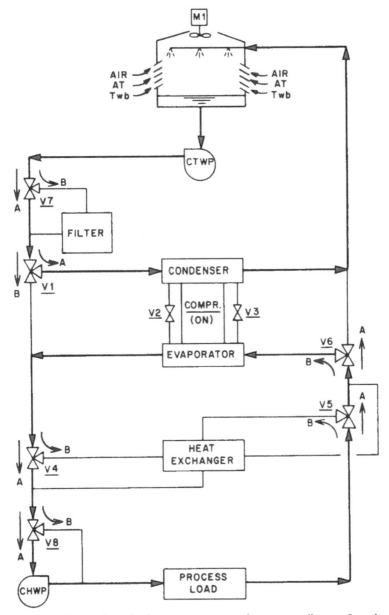

Fig. 7.14 This mechanical refrigeration system can be automatically reconfigured to operate in any of three free-cooling configurations.

possible to switch to one of the free cooling modes after the compressor has been stopped. The mode reconfiguration logic is summarized in table 7.1, in which the valve and equipment tags are those used in figure 7.14.

Indirect Free Cooling

Free cooling can be direct or indirect. Indirect free cooling maintains the separation between the cooling tower water and the chilled water circuits. Mode 2 of table 7.1 describes one of the indirect free cooling configurations, in which the compressor is off and the heat is transferred from the evaporator to the condenser through the natural migration of refrigerant vapors (figure 7.15). Opening the refrigerant migration valves equalizes the pressures in the evaporator and condenser. Because the condenser is at a lower temperature, the freon that is vaporized in the evaporator is recondensed there and returns to the evaporator by gravity flow. The cooling capacity of a chiller is about 10 percent of full load in this mode of operation.

The cooling system is operated in this mode when the load is low and the cooling tower water temperature is about 10°F (5.6°C) below the required chilled water temperature. This usually means 40°F–45°F (4.4°C–7.2°C) cooling tower and 50°F–60°F (10°C–15.6°C) chilled water temperatures. Such high chilled water temperatures are not unrealistic in the winter, because the air tends to be dry and dehumidification is not required.

Mode 2 is frequently implemented without controls. This means that both sets of pumps are operated and the temperatures are allowed to float as a function of load and ambient conditions. To optimize this operation, the control strategy shown in figure 7.15 can be implemented. Here TDIC-01 maintains the optimum approach and TDIC-02 maintains the optimum range, unless these controls are overridden by VPC-04 when the refrigerant migration valves approach their full opening, signifying that increased heat transfer is needed at the condenser in order to meet the load. The migration valves V1 and V2 are throttled by TIC-05. The set point of TIC-05 is maximized by VPC-06

Table 7.1
MODE RECONFIGURATION LOGIC

Equipment	Mode 1 (Mechanical refrigeration)	Mode 2 (Vapor migration based free cooling)	Mode 3 (Indirect free cooling by the use of heat exchanger)	Mode 4 (Direct free cooling with full flow filtering)
Compressor	On	Off	Off	Off
Cooling tower pumps	On	On	On	On
Chilled water pumps	On	On	On	Off
Valve V1	A	A	B	B
Valves V2 and 3	Closed	Open	Closed	Closed
Valve V4	A	A	B	A
Valve V5	A	A	B	A
Valve V6	A	B	A	A
Valve V7	A	A	A	B
Valve V8	A	A	A	B

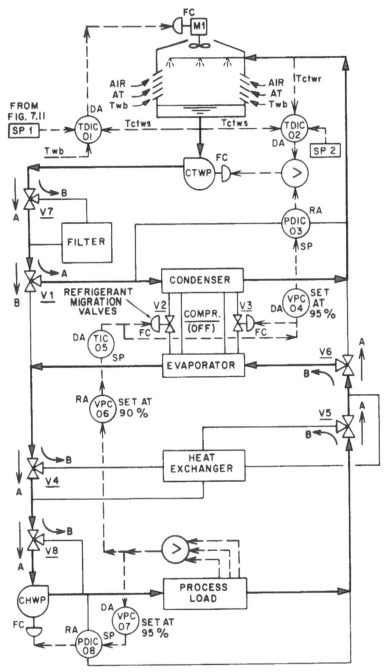

Fig. 7.15 Free cooling can be achieved by allowing the refrigerant vapor to migrate. (mode 2).

by allowing all user valves to open until the most-open valve reaches 90-percent opening. Chilled water pumping rate is minimized by VPC-07, which provides the set point of PDIC-08. As VPC-06 normally keeps the most open user valve at 90-percent opening, the set point of VPC-07 is not reached under normal conditions. Therefore, the PDIC-08 set point is kept at its allowable minimum. This in turn results in minimizing the pumping cost as long as all the users are satisfied. If any of the user valves opens to over 95 percent, VPC-07 will quickly increase the pumping rate to guarantee sufficient coolant to that valve.

At this point the system is using up the cooling capacity stored in the circulated chilled water. Once this is exhausted, control will be lost. To respond to this condition, it is advisable to switch to a different mode of operation: one that can provide more cooling.

Plate Type Heat Exchanger Control

Another method of indirect free cooling is to bring the cooling tower water and the chilled water into heat exchange through a plate and frame type heat transfer unit (mode 3). This allows heat to be transferred from the cooling tower to the chilled water system, completely bypassing the refrigeration machine. There is no heat transfer capacity limitation on this system, because any cooling load can be handled as long as the plate type heat exchanger is large enough. Therefore, the main advantage of mode 3 over mode 2 is that it is not restricted to loads of 10 percent or less. Its main disadvantage is the need for an additional major piece of equipment; the heat exchanger.

Indirect free cooling using the plate-type heat exchanger can be operated manually, without any controls. In that case, the chilled water temperature floats with load and ambient temperature while the pump stations operate at full capacity.

If it is desired to optimize the operation under mode 3, the control system described in figure 7.16 can be implemented. Here TDIC-01 maintains the optimum approach and TDIC-02 maintains the optimum range, unless overridden by PDIC-03 or by TIC-05. The pumping rate of cooling tower water circulation will be set by the highest of the three controller outputs. The purpose of PDIC-03 is to guarantee the minimum pressure differential required for the exchanger; TIC-05 can override TDIC-02 and PDIC-03 if more cooling is required.

The set point of TIC-05 is adjusted by VPC-06 to prevent the most-open user valve from exceeding a 90-percent opening. Under normal conditions, the pumping rate of chilled water is kept at a minimum by PDIC-08, which maintains the minimum ΔP across the load. When VPC-06 is unable to keep the most-open user valve at a 90-percent opening and it rises to 95 percent, VPC-07 will start raising the set point of PDIC-08, thereby increasing the pumping rate. This is only a temporary cure, because the added cooling capacity is available only at the expense of heating up the stored chilled water in the pipe distribution system. Therefore, it is advisable to detect this condition; when it occurs for more than a minute or so, the system should be automatically switched to a cooling mode that can handle the increased load. This can be mechanical refrigeration (mode 1) or free cooling through interconnection (mode 4).

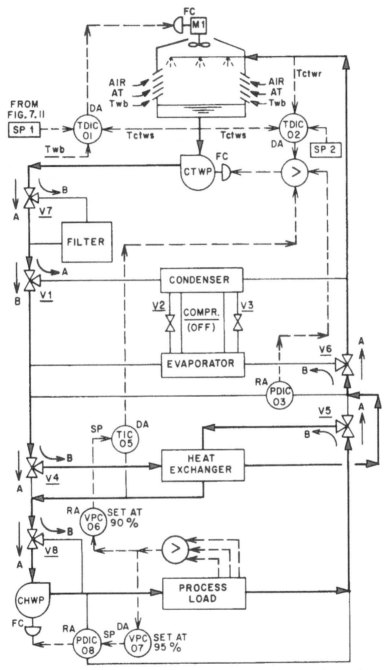

Fig. 7.16 Indirect free cooling can be achieved by using a plate-type heat exchanger (mode 3).

Direct Free Cooling

In direct free cooling, the cooling tower water is piped directly to the process load, as shown in figure 7.17. This method (mode 4) is the most cost-effective mode of cooling, because both the compressor and the chilled water pump station are off. This mode of operation can handle high process loads, as it is limited only by the size of the towers and their pumps.

The main disadvantage of direct free cooling is that it brings potentially dirty cooling tower water to the process users, causing plugging and build-up on the heat transfer surfaces. This problem is solved either by full flow filtering (also called strainer cycle), as shown in figure 7.17, or by the use of closed circuit, evaporative cooling towers, as illustrated in figure 7.10. In such "non-contact" or "closed-loop" cooling towers, the water has no opportunity to pick up contaminants from the air.

This configuration is also frequently operated without automatic controls. In that case, the cooling water temperature floats as the load and ambient conditions vary, and the fan and pumping rates are not optimized. If optimization is desired, the controls shown in figure 7.17 can be implemented. In this figure, TDIC-01 serves to keep the approach at an optimum value and TDIC-02 optimizes the range. The range controller can be overridden by PDIC-03 or by TIC-05 when either of these controllers require a higher pumping rate than does the range controller. The set point of TIC-05 is optimized to keep the most-open user valve from exceeding a 90-percent opening.

Mode Reconfiguration

In cooling systems that also include optional storage and heat recovery systems plus alternate types of motor drives, the number of possible modes of operation can be rather high. Switching from one mode to another is not as simple as it might appear at first glance, because as a result of reconfiguration, equipment needs to be started or stopped, control loops must be reconfigured, and pump discharge heads are also modified. For example, in the case of mode 4 in figure 7.17, the water circulation loop served by the cooling tower pumps becomes much longer when the system is switched to direct free cooling. This in turn shifts the operating point of the pump and can lower its efficiency if the system is not carefully designed and evaluated for each mode.

The control loop configuration requirements for modes 1 to 4 are tabulated in table 7.2.

When the control loops are automatically reconfigured, it is important to revise their tuning constants as their outputs are directed to different manipulated variables, because the time constants of the loop are also changed. This should preferably be done automatically to minimize the potential for human error.

The switching from one mode to another is done in a stagnant state, while the equipment is turned off, if an interruption of a few minutes can be tolerated. This usually is acceptable for all motors except for the pumps serving the process load. These pumps can be left running, utilizing the coolant storage capacity of the water distribution piping. When transfer to the next mode is initiated, dynamic transfer can be accom-

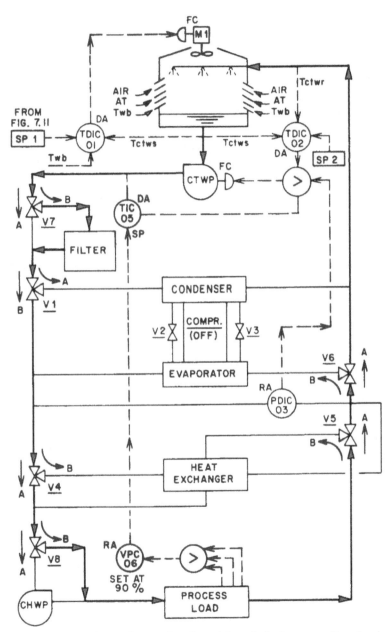

Fig. 7.17 Direct free cooling can be achieved through the interconnection of cooling circuits (mode 4).

Table 7.2
CONTROL LOOP CONFIGURATION REQUIREMENTS

| Controller | Associated Manipulated Variables | | | |
	Mode 1	Mode 2	Mode 3	Mode 4
TDIC-01	Fan	Fan	Fan	Fan
TDIC-02	CTWP	CTWP	CTWP	CTWP
PDIC-03	CTWP	CTWP	CTWP	CTWP
VPC-04	—	PDIC-03 Set point	—	—
TIC-05	Compr.	V2/3	CTWP	CTWP
VPC-06	TIC-05 Set point	TIC-05 Set point	TIC-05 Set point	TIC-05 Set point
VPC-07	PDIC-08 Set point	PDIC-08 Set point	PDIC-08 Set point	—
PDIC-08	CHWP	CHWP	CHWP	—

plished, because the slowly diverting three-way valves will never block the pump discharge but will only gradually change the destination of the water.

If the mode changes are frequent or if the coolant capacity of the piping headers is very small (insufficient to meet the process load for even a few minutes), all systems must be switched while running. Dynamic switching requires more planning and a higher level of automation, because the automatic starting of certain pieces of equipment (such as a chiller driven by a steam turbine) require a more comprehensive set of safety interlocks than do some other devices.

Automatic operating mode reconfiguration is one of the most powerful tools of optimization, because it makes a previously rigid system flexible. This technique is not limited to cooling systems but is effective in any unit operation in which the system must adapt to changing conditions.

Winter Operation

In case of a power failure during subfreezing weather, electrically heated tower basins should be provided with an emergency draining system. Another method of freeze protection is the bypass circulation method illustrated in figure 7.18. In this system, a thermostat detects the outlet temperature of the cooling tower water, and when it drops below 40°F (4.4°C), the bypass circulation (located in a protected indoor area) is started. This circulation is terminated when the water temperature rises to approximately 45°F (7.2°C).

The thermostat is wired to open the solenoids S1, S2, and S3 first and then to start the circulating pump (P2), which usually is a small, ¼ or ⅓-HP (186.5 or 248.7 W) unit. The bypass line containing solenoid S1 is sized to handle the flow capacity of P2, with a pressure drop (h) that is less than the height of the riser pipe. This makes sure that when P1 is off, no water will reach the top of the column.

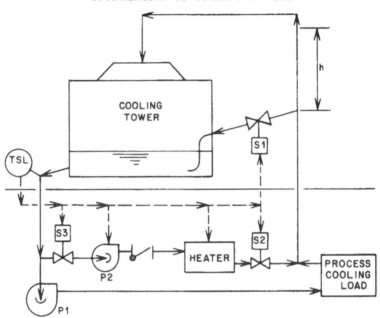

Fig. 7.18 The bypass circulation method can be used to protect against freezing in cooling towers. (From Lipták [11])

In the northern regions, in addition to the S1 bypass shown in figure 7.18, there is a full-sized bypass (not shown in figure 7.18). When the main (P1) pump is started in the winter, this full bypass is opened; it does not allow the water to be sent to the top of the tower until its temperature is approximately 70°F (21°C). Once the water reaches that temperature, the bypass is closed because the water is warm enough to be sent to the top of the tower without danger of freezing. As the water temperature rises further, fans are turned on to provide added cooling.

When open cooling towers are operated at freezing temperatures, the induced draft fans must be reversed periodically in order to de-ice the air intakes. The need for de-icing can be determined through visual observation, through remote closed-circuit TV inspection, or by comparing the load and ambient conditions to past operating history. The safety requirements of reverse operation have already been discussed under interlocks.

All exposed pipes that will contain water when the tower is down should be protected by electric tape or cable tracing and by insulation. The sump either should be drainable to an indoor auxiliary sump or should be provided with auxiliary steam or electric heating.

Closed cooling towers should either operate on antifreeze solutions or be provided with supplemental heat and tracing. If the second method is used, the system must be drained if power failure occurs.

Blowdown Controls

The average water loss by blowdown is 0.5 to 3.0 percent of the circulating water rate. The loss is a function of the initial quality of water and the amount and concentration of the dissolved natural solids and chemicals added for protection against corrosion or build-up of scale on the heat transfer surfaces. Because the cooling tower is a highly effective air scrubber, it continuously accumulates the solid content of the ambient air on its wet surfaces, which are then washed off by the circulated water.

The "normal" condition of the circulated water can be defined arbitrarily as:

pH	6 to 8
Chloride as NaCl	under 750 ppm
Total dissolved solids	under 1500 ppm

The blowdown requirement can be determined as follows:

$$B = \frac{([E + D]/N) - D}{1 - (1/N)}$$

where

B = Blowdown in percentages (typically, for N = 2, blowdown is 0.9 percent; for N = 3, 0.4 percent; for N = 4, 0.24 percent; etc.)

E = Evaporation in percentages (typically 1 percent for each 12.5°F (7°C) of cooling range)

D = Drift in percentages (typically 0.1 percent)

N = Number of concentrations relative to initial water quality

In cooling towers, the major causes for concern are delignification (loss of the binding agent for the cellulose) caused by the use of oxidizing biocides, such as chlorine; excessive bicarbonate alkalinity; biological growth, which can clog the nozzles and foul the heat exchange equipment; corrosion of the metal components (this corrosion should be less than 3 mils per year without pitting); general fouling by a combination of silt, clay, oil, metal oxides, calcium and magnesium salts, organic compounds, and other chemical products that can cause reduced heat transfer and enhanced corrosion; and scaling by crystallization and precipitation of salts or oxides (mainly calcium carbonate and magnesium silicate) on surfaces. The treatment techniques to prevent these conditions from occurring are listed in table 7.3. The blowdown must meet the water quality standards for the accepting stream and must not be unreasonably expensive. It is also possible to make the water system a closed-cycle system (no blowdown) by ion-exchange treatment.

Miscellaneous Controls

The minimum basin level is maintained by the use of float-type make-up level control valves. The maximum basin level is guaranteed by overflow nozzles sized to handle the total system flow. In critical systems, additional safety is provided by the use of high/low level alarm switches [12, ch. 3].

The operational safety of the fan and the fan drive can be safeguarded by torque

Table 7.3
COOLING TOWER PROBLEMS AND SOLUTIONS [11]

Problem	Factors	Causative Agents	Corrective Treatments
Wood deterioration	Microbiological	Cellulolytic fungi	Fungicides
	Chemical	Chlorine	Acid
Biological growths	Temperature	Bacteria	Chlorine
	Nutrients	Fungi	Chlorine donors
	pH	Algae	Organic sulfurs
	Inocula		Quatenary ammonia
General fouling	Suspended solids	Silt	Polyelectrolytes
	Water	Oil	Polyacrylates
	Velocity		Lignosulfonates
	Temperature		Polyphosphates
	Contaminants		
	Metal oxides		
Corrosion	Aeration	Oxygen	Chromate
	pH	Carbon dioxide	Zinc
	Temperature	Chloride	Polyphosphate
	Dissolved solids		Tannins
	Galvanic couples		Lignins
			Synthetic organic compounds
Scaling	Calcium	Calcium carbonate	Phosphonates
	Alkalinity	Calcium sulfate	Polyphosphates
	Temperature	Magnesium silicate	Acid
	pH	Ferric hydroxide	Polyelectrolytes

and vibration detectors [12, ch. 9]. Low-temperature alarm switches [12, ch. 4] can also be furnished as warning devices signaling the failure of freeze protection controls.

The need for flow balancing among multiple cells is another reason for using added valving and controls. The warm water is usually returned to the multi-cell system through a single riser, and a manifold is provided at the top of the tower for water distribution to the individual cells. Balancing valves should be installed to guarantee equal flow to each operating cell. If two-speed fans are used, further economy can be gained by lowering the water flow when the fan is switched to low speed (figure 7.13). It is also advisable to install equalizer lines between tower sumps to eliminate imbalances caused by variations in flow rates or pipe layouts.

Another reason for considering the use of two-speed fans is to have the ability to lower the associated noise level at night or during periods of low load. Switching to low speed will usually lower the noise level by approximately 15 dB.

Conclusions

One of the least understood and most neglected unit operations in the processing industries is the cooling tower. Its pumps are frequently constant speed with three-way or bypass valves used to circulate the excess water, which should not have been pumped in the first place because the process does not require it. Meeting a variable load with a constant supply by wasting the excess is also frequently practiced in fan operation. In some installations, fan speeds cannot be changed at all; in others, they are changed only manually, which can mean seasonal adjustments in some extreme cases.

As the rating of the tower fans and pumps usually adds up to several hundred horsepowers, their yearly operating cost is in the hundreds of thousands of dollars. As optimization can cut this in half, the added cost for controls is justified.

References

1. American Society of Heating, Refrigerating, and Air-Conditioning Engineers, "Chilled and Dual-temperature Water Systems," in *ASHRAE Handbook, 1983 Equipment Volume* (Atlanta, GA: ASHRAE, 1983), chapter 18.
2. American Society of Heating, Refrigerating, and Air-Conditioning Engineers, "Cooling Towers," in *ASHRAE Handbook, 1983 Equipment Volume* (Atlanta, GA: ASHRAE, 1983), chapter 21.
3. Bacchler, R. H., Blew, J. O., and Duncan, C. G., "Cause and Prevention of Decay of Wood in Cooling Towers," ASME Petroleum Division Conference, New York, 1961.
4. Baker, D. R., "Durability of Wood in Cooling Towers," Technical Bulletin R-62-W-1 (Kansas City, MO: Marley Cooling Tower Co., 1962).
5. Baumeister, T., *Mark's Standard Handbook for Mechanical Engineers*, eighth edition (New York, NY: McGraw-Hill, 1978), pp. 9–74.
6. Dickey, J. B., *Evaporative Cooling Towers*, Marley Cooling Tower Co., 1978.
7. Donohue, J. M., "Cooling Towers—Chemical Treatment," *Industrial Water Engineering*, May 1970.
8. Kern, D. Q., *Process Heat Transfer*, (New York, NY: McGraw-Hill, 1959), p. 563.
9. Koch, J., *Untersuchung und Berechnung und Kuehlwerken*, VDI Forschungsheft no. 404, Berlin, 1940.
10. Lipták, B. G., ed., "Cooling Towers—Their Design and Application," in *Environmental Engineer's Handbook*, vol. 1 (Radnor, PA: Chilton Book Co., 1974).
11. Lipták, B. G., *Instrument Engineer's Handbook, Process Control* (Radnor, PA: Chilton Book Co., 1985), p. 847.
12. Lipták, B. G., *Instrument Engineer's Handbook, Process Measurement*, revised edition (Radnor, PA: Chilton Book Co., 1982), p. 769, chapters 3, 4, 9.
13. Lipták, B. G., "Optimizing Controls for Chillers and Heat Pumps," *Chemical Engineering*, October 17, 1983.
14. Mathews, R. T., "Some Air Cooling Design Considerations," *Proceedings of the American Power Conference*, 1970.
15. Nottage, H. B., "Merkel's Cooling Diagram as a Performance Correlation for Air-Water Evaporative Cooling Systems," *ASHRAE Transactions*, vol. 47, 1941, p. 429.
16. Nussbaum, O. J., "Dry-type Cooling Towers for Packaged Refrigeration and Air Conditioning Systems," *Refrigeration Science and Technology*, 1972.
17. Walker, W. H., W. K. Lewis, W. H. McAdams, and E. R. Gilliand, *Principles of Chemical Engineering* (New York, NY: McGraw-Hill, 1937), p. 480.
18. Wistrom, G., "Cold Weather Operation of Crossflow Cooling Towers," *Plant Engineering*, September 8, 1975.

Chapter 8

OPTIMIZATION
OF
HEAT EXCHANGERS

THE TRANSFER OF HEAT is one of the most basic unit operations of the processing industries. Heat can be transferred between the same phases (liquid to liquid, gas to gas) or phase change can occur on either the process side (condenser, evaporator, reboiler) or the utility side (steam heater) of the heat exchanger. The control and optimization of each of these systems will be discussed in this chapter.

Degrees of Freedom

All control loops function on the basis of controlling one variable by manipulating the same or other process variable(s). The maximum number of independently acting automatic controllers that can be placed on a process is called the degree of freedom of that process.

Figure 8.1 shows a steam heater with its variables and defining parameters. The temperatures and flows are variables, and the specific and latent heats are parameters. The available degrees of freedom are determined by subtracting the number of system-defining equations from the number of variables:

Degrees of freedom = (number of variables) − (number of equations)

In this case, there are four variables and one equation, which is obtained from the first law of thermodynamics, stating the conservation of energy:

$$H_s W_s = C_p W(T_2 - T_1)$$

Therefore, this system has three degrees of freedom; thus, a maximum of three automatic controllers can be used.

In a liquid-to-liquid heat exchanger, there are four temperature and two flow variables with still only one (conservation of energy) defining equation, resulting in five degrees of freedom.

$$C_p F(T_{h1} - T_{h2}) = C_{pc} F_c(T_{2c} - T_{1c})$$

In a steam-heated reboiler or in a condenser cooled by a vaporizing refrigerant (assuming no superheating or supercooling), there are only two flow variables and one defining equation:

$$H_s W_s = H_1 W_1$$

In this case there is only a single degree of freedom and only one automatic controller.

317

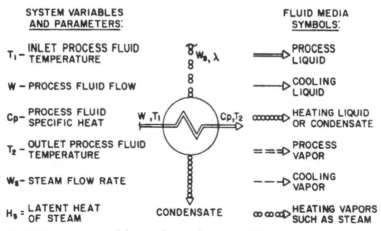

Fig. 8.1 Temperature and flow are heat exchanger variables; specific and latent heats are parameters. (Adapted from Lipták [3])

In the majority of installations, fewer controllers are used than there are degrees of freedom available, but every once in a while problems associated with overdefining a system also arise.

Liquid-to-Liquid Heat Exchangers

Figure 8.2 illustrates that the total heat transferred (Q) from the hot to the cold fluid is dependent on the overall heat transfer coefficient (U), the heat transfer area (A), and the log mean temperature difference (ΔTm). Manipulation of any of these three variables can affect the total heat transfer [8].

The dead time of a heat exchanger equals its residence time, which is its volume divided by its flow rate. As process flow increases, the process dead time is lowered and the loop gain is decreased. If the controlled variable (Th_2 in figure 8.2) is differentiated with respect to the manipulated variable Fc, the steady state process gain is found to be the following:

$$\frac{dTh_2}{dFc} = \frac{C_{pc}}{F_h C_{ph}}$$

Therefore, as the process flow (load) drops, the same amount of valve adjustment by TIC in figure 8.2 will have more effect, because the process fluid spends more time in the exchanger. As load is reduced, the exchanger becomes relatively oversized and therefore faster and more effective. As a drop in load tends to increase loop gain—that is, tends to make the loop excessively sensitive and prone to cycling—an increase in load does the opposite. As load rises, the exchanger becomes less and less effective, more and more undersized; therefore, it takes more time for the TIC to make a correction. An increase in load thus will make the loop sluggish, as the residence and dead times are reduced and the loop gain is lowered.

$$Q = UA\Delta Tm = UA \frac{(Th_1 - Tc_2) - (Th_2 - Tc_1)}{\ln((Th_1 - Tc_2)/(Th_2 - Tc_1))}$$

$$Q = Fh\, Cph\, (Th_1 - Th_2) = Fc\, Cpc\, (Tc_2 - Tc_1)$$

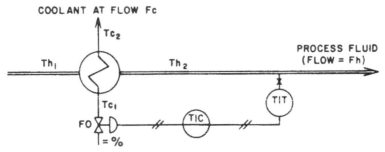

Fig. 8.2 Feedback control by throttling coolant inlet is a non-linear, variable gain process. (Adapted from Lipták [3])

As the process gain of all heat exchangers varies with load, it is not possible to tune the TIC in figure 8.2 for more than one load. In order to minimize cycling, the TIC is usually tuned for the minimum load (maximum loop gain). Therefore, it acts in a sluggish manner at higher loads. As the process gain drops with rising load, the total loop gain can be held relatively constant by the introduction of a loop component whose gain rises with load. Such element is an equal percentage valve. This will compensate the variable gain process, and conservative settings of reset and derivative will allow for the variation in the period of oscillation. This is a good solution if the temperature difference in the exchanger (Th1-Th2) is constant [5]. If it is not constant, feedforward compensation of the gain is required. This will be discussed later.

Component Selection

Figures 8.2 and 8.3 illustrate cooler and heater installations with the control valve mounted on the exchanger inlet and outlet, respectively. From a control quality point of view, it makes little difference whether the control valve is upstream or downstream to the heater. The inlet side is usually preferred, because this allows the exchanger to operate at a lower pressure than that of the return header.

It is generally recommended that positioners for these valves be provided to minimize the valve friction effects. The use of equal-percentage valve trims is also recommended, because it usually contributes to maintaining the control system gain constant under changing throughput conditions. Equal percentage trims maintain a constant relationship between valve opening and temperature change (reflecting load variations).

In the majority of installations, a three-mode controller would be used for heat exchanger service. The derivative or rate action becomes essential in long time-lag systems or when sudden changes in heat exchanger throughput are expected. Because of the relatively slow nature of these control loops, the proportional band setting must

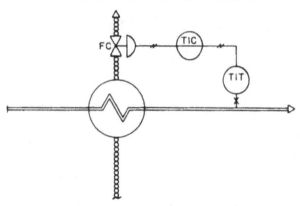

Fig. 8.3 Feedback control by throttling heating media outlet affects only the operating pressure. (From Lipták [3])

be wide to maintain stability (usually between 10 percent and 100 percent). This means that the valve will be fully stroked only as a result of a substantial deviation from desired temperature set point. The reset or integral control mode is required to correct for temperature offsets caused by process load changes. Besides the changes in process fluid flow rate, other variables can give the appearance of load changes, such as inlet temperature and header pressure changes of the heat transfer medium.

The Thermal Element

The selection and location of the thermal element are also important. This element must be placed in a representative location, without increasing measurement time lag. In reference to figure 8.2, this would mean that the bulb should be located far enough from the exchanger for adequate mixing of the process fluid but close enough so that the introduced delay will not be substantial. If the process fluid velocity is 3 ft/second (0.9 m/s), then a 1-second distance-velocity lag is introduced for each 3 ft (0.9m) of pipe between the exchanger and the bulb. This lag can be one of the factors that limits the dynamic performance of the system. This, of course, is not the only thermal lag in the system. In order to change the temperature of a sensor, heat must be introduced, and it has to enter the bulb through a fixed area.

For example, try calculating the dynamic lag of a typical filled bulb having the area of 0.02 sq ft (0.0018 m²) and the heat capacity of 0.005 BTU/°F (9.49 J/°C). If this bulb, which has a bare-bulb diameter of 3/8 inch (9.375 mm), is immersed in a fluid with a heat transfer coefficient (based on flow velocity) of 60 BTU/h/°F/ft² (1.23 MJ/h/°C/m²) and the process temperature is changed at a rate of 25°F/minute (13.9°C/min), the dynamic lag can be calculated. First the amount of heat flowing into the element under these conditions is determined:

$$q = \text{(rate of temperature change) (bulb heat capacity)}$$
$$= (25)(60)(0.005) = 7.5 \text{ BTU/h } (7.9 \text{ kJ/h})$$

The dynamic measurement error is calculated by determining the temperature differential across the fluid film surrounding the bulb that is required to produce a heat flow of 7.5 BTU/h (7.9 kJ/h):

$$q = Ah \, \Delta T$$

Therefore,

$$\Delta T = q/Ah = 7.5/(0.02)(60) = 6.25°F \, (3.47°C)$$

If the rate of process temperature change is 25°F/minute (14°C/m) and the dynamic error based on that rate is 6.25°F (3.47°C), the dynamic time lag is as follows:

$$t_o = 6.25/25 = 0.25 \text{ minutes} = 15 \text{ seconds}$$

This lag can also be calculated as

$$t_o = \frac{(\text{bulb heat capacity})}{(\text{bulb area}) \, (\text{heat transfer coefficient})} = \frac{60 \times 0.005}{0.02 \times 60} = 0.25 \text{ minutes}$$

Bulb time lags vary from a few seconds to minutes, depending on the nature and velocity of the process fluid being detected. Measurement of gases at low velocity involves the longest time lags, and measuring water (or dilute solutions) at high velocity involves the shortest lags.

The addition of a thermowell will further increase the lag time, but in most industrial installations, thermowells are necessary for reasons of safety and maintenance. When they are used, it is important to eliminate any air gaps between the bulb and the socket.

One method of reducing time lag is by miniaturizing the sensing element. Conventional thermometers are not suitable for all temperature measurement applications. For example, the accurate detection of high-temperature gases (500-2000°C) at low velocities (3-5 ft/sec) is a problem, even for thermocouples. This is because conductance through the lead wires and radiation both tend to alter the sensor temperature faster than the low velocity gas flow can resupply the lost heat. In such applications, optical fiber thermometry is a good choice. Thermocouples are usually not accurate enough for the precise measurement of temperature differences and are not fast enough to detect high-speed variations in temperature. RTDs can operate with a span for temperature difference of 10°F and can limit the measurement error to ±0.04°F. If high-speed response is desired, thermistors or infrared detectors should be considered.

In the conventional control loop, the measurement lag is only part of the total time lag of the control loop. For example, an air heater might have a total lag of 15 minutes, of which 14 minutes is the *process lag*, 50 seconds is the bulb lag, and 10 seconds is the time lag in the control valve.

Three-Way Valves

The limits within which process temperature can be controlled are a function of the nature of load changes expected and of the speed of response for the process. In

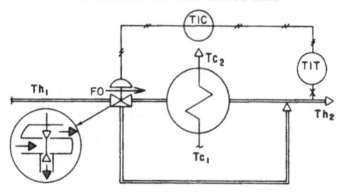

Fig. 8.4 When a diverter valve is used to control a cooler, it does not eliminate non-linearity, but does speed the response and minimize fouling. (Adapted from Lipták [3])

many installations, the process time lag in the heat exchanger is too great to allow for effective control during load changes. In such cases, it is possible to circumvent the dynamic characteristics of the exchanger by partially bypassing it and blending the warm process liquid with the cooled process fluid, as shown in figure 8.4. The resulting increased system speed of response together with some cost savings are the main motivations for considering three-way valves in such services. The bulb time lag discussed in the previous paragraph has an increased importance in these systems, because it represents a much greater percentage of total loop lag time than in the previously discussed installations.

The use of a three-way valve such as that shown in figure 8.4 does not change the nonlinear nature of the heat exchanger, yet it can still be beneficial. Although the process gain will still vary with load, the dynamic response is improved, because the bypass will shorten the time delay between a change in valve position and the response at the temperature sensor. Another benefit from the use of a three-way valve on the process side is that the coolant is not throttled, which keeps the heat transfer coefficient up while eliminating fouling.

As illustrated in figures 8.4 and 8.5, either a diverter or a mixing valve can be used as a three-way valve. Stable operation of these valves is achieved by flow tending to open the plugs in both cases. If a mixing valve is used for diverting service or if a diverting valve is used for mixing, the operation becomes unstable because of the "bathtub effect" (if you reverse flow directions, the fluid itself will try to push the valve plugs closed). Therefore, it is not good enough just to install a three-way valve; it has to be either the mixing or the diverting design to match the particular service.

Three-way valves are unbalanced designs and are normally provided with linear ports. The unbalanced nature places a limitation on allowable shut off pressure difference across the valve, and the linear ports eliminate the potential of compensating for the variable gain of the process.

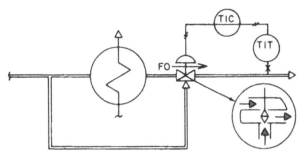

Fig. 8.5 A mixing valve can be used to control the cooler. It has
a flow to open inner valve for stability. (From Lipták [3])

Misalignment or distortion in a control valve installation can cause binding, leakage at the seats, deadband, and packing friction. Such conditions commonly arise as a result of high-temperature service on three-way valves. The valve, having been installed at ambient conditions and rigidly connected at three flanges, cannot accommodate pipe line expansion because of high process temperature, and therefore distortion results. Similarly, in mixing applications, when the temperature difference between the two ports is substantial, the resulting differential expansion can also cause distortion. For these reasons, the use of three-way valves at temperatures above 500°F (260°C) or at differential temperatures exceeding 300°F (167°C) is not recommended.

The choice of three-way valve location relative to the exchanger (figures 8.4 and 8.5) is normally based on pressure and temperature considerations, with the upstream location (figure 8.4) usually being favored for reasons of uniformity of valve temperature. When the overriding consideration is the desire to operate the exchanger at a high pressure, the downstream location might be selected.

Cooling Water Conservation

Figure 8.6 modifies the system shown in figures 8.4 and 8.5 to include the additional feature of cooling water conservation. This system tends to maximize the outlet cooling water temperature, thereby minimizing the rate of water usage. The application engineer, when using this concept, should be careful in evaluating the temperature levels involved and should make sure that the water contains chemicals to prevent tube fouling if high-temperature operation is planned. If the cooling water supply is sufficient, this system will conserve water and will also protect against excessively high outlet water temperature.

Unfortunately, this configuration will not yield stable control, because controlling temperature through manipulating the flow of the same stream results in limit cycle. The cause of this instability is that the output of TIC-01 affects the dead time of the process that it controls. This comes about as described below.

If TIC-01 detects a temperature rise, it opens its valve, causing a sudden drop in temperature as the process dead time is reduced. As TIC-01 senses this sudden drop

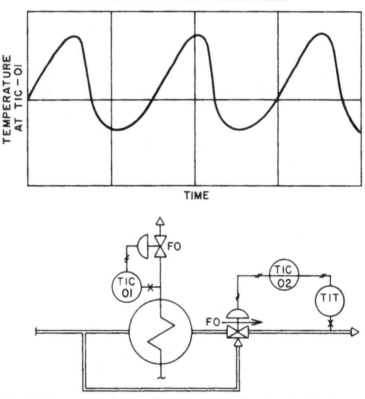

Fig. 8.6 Conservation of cooling fluid is provided at the price of stability.
(Adapted from Lipták [3])

in temperature, it closes its valve, but because this also increases the process dead time, the resulting temperature rise will be slow [6]. This limit cycle, consisting of segments of slow rise and fast fall, will also affect the performance of TIC-02 through interaction. The amplitude of the cycle can be reduced severalfold through the doubling of the proportional band, which will also double the period of oscillation [6]. The only way to eliminate this oscillation altogether is to fix the dead time. This can be done by the addition of a recirculating pump, which will return part of the heated water back to the inlet.

As can be seen, there is no easy way to make this configuration stable; therefore, the best solution is to try to avoid its use.

Balancing the Three-Way Valve

It is recommended that a manual balancing valve be installed in the exchanger bypass, as shown in figure 8.7. This valve is so adjusted that its resistance to flow equals that of the exchanger.

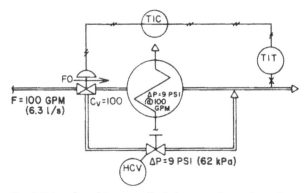

Fig. 8.7 It is desirable to install a balancing valve in the exchanger bypass. (From Lipták [3])

When the pump is sized, it should be kept in mind that the resistance to flow in such installations will be maximum when one of the paths is closed and the other is fully open, whereas minimum resistance will be experienced when the valve divides the flow equally between the two paths.

For example, assume that the flow rate of process fluid is 100 GPM (6.3 l/s), the pressure drop at full flow through either the exchanger or the balancing valve is 9 PSI (6.21 kPa), and the diverting valve has a valve coefficient of $C_v = 100$. The equivalent coefficient for the exchanger (or balancing valve) is calculated as:

$$C_e = \frac{\text{flow}}{\sqrt{\text{pressure drop}}} = \frac{100}{\sqrt{9}} = 33.3$$

Therefore, in either extreme position (closed or full bypass), the total system resistance expressed in valve coefficient units is as follows:

$$\frac{1}{(C_t)^2} = \frac{1}{(C_v)^2} + \frac{1}{(C_e)^2} = \frac{1}{(100)^2} + \frac{1}{(33.3)^2}$$

$$\therefore C_t = 31.7$$

When the valve divides the flow equally between the two paths, because of the linear characteristics of all three-way valves, its coefficient at each port will be $C_v = 50$. The equivalent coefficient ($C_e = 33.3$) of the exchanger and balancing valve being unaffected, the total system resistance in valve coefficient units is $2C_t$, where

$$\frac{1}{(C_t)^2} = \frac{1}{(C_v)^2} + \frac{1}{(C_e)^2} = \frac{1}{(50)^2} + \frac{1}{(33.3)^2}$$

$$\therefore 2C_t = 55.6$$

If the total pressure drop through the system is calculated when the valve is in its extreme and when it is in its middle position, handling the same 100 GPM (6.3 l/s) flow, the following pressure drops are found:

$$\Delta \text{Pextreme} = \left(\frac{\text{flow}}{C_t}\right)^2 = \left(\frac{100}{31.7}\right)^2 = 10 \text{ PSI (69 kPa)}$$

$$\Delta \text{Pmiddle} = \left(\frac{100}{55.6}\right)^2 = 3.25 \text{ PSI (22.4 kPa)}$$

These results indicate that the system drop in one of the extreme positions is more than three times that of the middle position.

Two Two-Way Valves

Sometimes it is desirable to improve the system response speed by the use of exchanger bypass control in situations in which, for reasons of temperature or other considerations, three-way valves cannot be used. In such situations, the installation of two two-way valves is the logical solution.

As illustrated in figure 8.8, the two valves should have opposite failure positions. Therefore, when one is open the other is closed, and at a 9 PSI (0.6 bar) signal both are halfway open. In order for these valves to give the same control as a three-way valve would, it is necessary to provide them with linear plugs.

The price of a three-way valve is about 65 percent that of two two-way valves. The installation cost is similarly lower, and when positioners are required (as is true in the majority of cases), only one needs to be purchased, instead of two. On the other hand, the capacity of a three-way valve is the same as the capacity of a single-ported two-way valve, which is only 70 percent of a double-ported one. This could mean that instead of a 10-inch (250 mm) three-way unit, two 8-inch (200 mm) double-ported two-way valves could be considered, thus reducing the cost advantage of a three-way installation. This is true if the comparison is based on double-ported two-way valves, but the inherent leakage of these units (approximately .5 percent of full capacity) makes them unsuitable for some installations. Therefore, where tight shutoff is required, only the three-way or the single-ported two-way valves can be considered, and their capacity is about the same.

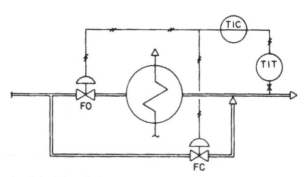

Fig. 8.8 Exchanger bypass control can be achieved using two two-way valves. (From Lipták [3])

Table 8.1
MERITS OF TWO-WAY VERSUS THREE-WAY VALVES
IN EXCHANGER BYPASS INSTALLATIONS

	One Three-Way Valve	Two Single-Seated Two-Way Valves	Two Double-Ported Two-Way Valves
Most economical	Yes		
Provides tight shutoff	Yes	Yes	
Applicable to service above 500°F (260°C)		Yes	Yes
Applicable to differential temperature service above 300°F (167°C)		Yes	Yes
Applicable to operation at high pressure and pressure differentials		Yes	Yes
Highest capacity for same valve size			Yes

Three-way valves are not recommended for high-temperature or high-pressure differential services. In addition to the reasons noted earlier, the hollow plug design of three-way valves also contributes to this limitation because it makes these valves more sensitive to thermal expansion and more difficult to harden than the solid plugs.

To summarize, bypass control is applied to circumvent the dynamic characteristics of heat exchangers, thus improving their controllability. Bypass control can be achieved by the use of either one three-way valve or two two-way valves. Table 8.1 summarizes the merits and drawbacks of using one or the other methods.

Steam Heaters

The general discussion on loop components, accessories, sensor location, and time lag considerations that was presented in connection with liquid-to-liquid heat exchangers is also applicable here.

The steam-heated exchanger shown in figure 8.9 is also non-linear. The steady state gain is the derivative of outlet temperature with respect to steam flow [8], having the dimensions of °F/(lbs/hr):

$$Kp = \frac{dT_2}{dFs} = \frac{\Delta Hs}{FCp}$$

Therefore, the process gain varies inversely with flow. In a step response test, the outlet temperature (T_2) reaches 63.2 percent of its final value after a time equivalent of its residence time, which is tube volume divided by flow. Therefore, the time constant, dead time, residence time, period of oscillation, and process gain all vary with flow. As flow (load) drops to 50 percent, gain is doubled; therefore, the TIC loops have a tendency to become unstable at low and sluggish at high loads. In order to

$$Q = UA\Delta Tm = UA \frac{(T_S - T_2) - (T_S - T_1)}{\ln((T_S - T_2)/(T_S - T_1))}$$

$$Q = F_S \Delta H_S = F \, Cp \, (T_2 - T_1)$$

WHERE Q = HEAT-TRANSFER RATE

F_S = STEAM MASS FLOW

ΔH_S = LATENT HEAT OF VAPORIZATION

F = FEED RATE

Cp = HEAT CAPACITY OF FEED

T_0 = STEAM SUPPLY TEMPERATURE

P_1 = STEAM SUPPLY PRESSURE

P_2 = STEAM VALVE OUTLET PRESSURE

P_S = CONDENSING PRESSURE

T_1 = INLET TEMPERATURE

T_2 = OUTLET TEMPERATURE

ΔTm = LOG MEAN TEMPERATURE DIFFERENCE

T_S = CONDENSING STEAM TEMPERATURE

F_S = 500 TO 2,500 lb/hr (227 kg/hr TO 1134 kg/hr)
P_1 = 200 PSIA (1.38 MPa)

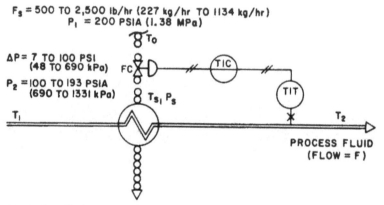

Fig. 8.9 Feedback control is frequently used on steam-heated exchangers. (Adapted from Lipták [3])

eliminate cycling, these loops are usually tuned at minimum load, resulting in sluggish response elsewhere. One way to compensate for the drop in loop gain as the flow increases is to use an equal-percentage valve whose gain increases with load. This is sufficient if the temperature rise $(T_2 - T_1)$ is constant. Otherwise, feedforward compensation is needed (this will be discussed later).

With steam heaters, the desirability of equal-percentage valve trims is even more pronounced than it was with liquid-to-liquid exchangers, because of the high rangeability required on most installations. The need for high rangeability is partially a result of the variations in condensing pressure (steam valve back-pressure) with changes in process load. This can be best visualized by an example.

In figure 8.9, both the high and the low load conditions are shown. When the steam flow demand is the greatest, the back-pressure is also the highest ($P_2 = 193$ PSIA [1331 kPa]), leaving the lowest driving force (pressure drop) for the control valve (7 PSI or 48 kPa). High flow and low pressure drop result in a large valve. The back-pressure at low loads is only 100 PSIA (690 kPa), allowing a pressure drop through the valve that is some sixteen times greater (100 PSI or 690 kPa) than at high loads. The ratio between the required valve coefficients for the high and low load conditions represents the rangeability that the valve has to furnish.

$$\text{Rangeability} = S \frac{F_{s,\,max}}{F_{s,\,min}} \sqrt{\frac{[(P_1 - P_2)(P_1 + P_2)]_{min}}{[(P_1 - P_2)(P_1 + P_2)]_{max}}}$$

$$= 1.5 \times 5 \sqrt{\frac{100 \times 300}{7 \times 393}} = 25.5$$

The letter S with the numerical value of 1.5 in the above equation represents the safety factor that is applied in selecting the control valve. A rangeability requirement of this magnitude can create some control problems. One solution is to use a large and a small valve in parallel. If this is not done, the control quality will suffer for two reasons: first, at low loads the valve will operate near to its clearance flow point, where the flow-versus-lift curve changes abruptly, contributing to unstable or possibly on-off cycling valve operation. Second, for good control the system gain should not vary with changes in load, which an equal percentage control valve can guarantee only if control valve ΔP is not a function of load. This being the case, the only way to guarantee constant system gain is to install two valves for the two load conditions, both sized to maintain the same gain.

Minimum Condensing Pressure

As has been shown in connection with figure 8.9, the condensing pressure is a function of load when the temperature is controlled by throttling the steam inlet. As long as the heat transfer area is constant, a reduction in load must result in the lowering of the log mean temperature difference across the exchanger. If T_2 is held constant by the TIC, this can occur only if the steam-side temperature (T_s) is lowered. The condensing temperature required can be calculated as follows [4]:

$$T_s = \left(\frac{T_1 + T_2}{2}\right)\left(1 - \frac{L}{100}\right) + \left(\frac{L}{100} T_0\right)$$

where L is the load in percentage of maximum. The condensing pressure corresponding to the values of T_s can be plotted as a function of load. Once this pressure drops below the trap back-pressure plus trap differential, it is no longer sufficient to discharge the condensate and it will start accumulating in the exchanger. As condensate accumulation progresses, more and more of the heat transfer area will be covered up, resulting in a corresponding increase in condensing pressure. When this pressure rises sufficiently to discharge the trap, the condensate is suddenly blown out and the effective heat transfer surface of the exchanger increases instantaneously. This can result in cycling as the exchanger surface is covered and uncovered. In addition, noise and hammering can be caused as the steam bubbles collapse on contact with the accumulated cooler condensate. The methods to remedy this situation will be discussed in the following paragraphs.

Condensate Throttling

Mounting the control valve in the condensate line, as shown in figure 8.10, is sometimes proposed as a solution to minimum condensate pressure problems. There also is a cost advantage to purchasing a small condensate valve instead of a larger one for steam service.

On the surface this appears to be a convenient solution, because the throttling of the valve causes variations only in the condensate level inside the partially flooded heater and has no effect on the steam pressure, which stays constant. Therefore, there are no problems in condensate removal. Unfortunately, the control characteristics of this loop are not so favorable.

The response times depend on both the load and on exchanger geometry. Tests have shown that a load decrease takes about three times as long as a load increase [4]. The response time for a 5-percent to 10-percent change in load is from a few seconds to a few minutes, whereas for a 50-percent change it can be from a few minutes to nearly an hour [4]. If the dead time is 5–6 seconds, the period of oscillation of the heat exchanger will be about a half minute.

When the load is decreasing, the valve is likely to close completely before the condensate builds up to a high enough level to match the new lower load with a reduced heat transfer area. In this direction, the process is slow, because steam has to condense before the level can be affected. When the load increases, the process is fast, because just a small change in control valve opening is sufficient to drain off enough condensate

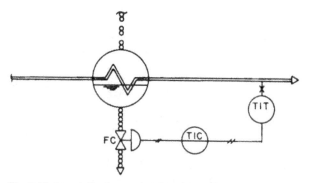

Fig. 8.10 A control valve mounted in the condensate line can
control the rate of condensate removal. (From Lipták [3])

to expose an increased heat transfer surface. When these "non-symmetrical" dynamics are present, control is bound to be poor. If the controller is tuned for the fast response speeds of the increasing load direction, sluggish performance and overshoot can result when load is decreasing, whereas if it is tuned for the slow part of the cycle, cycling can occur. Therefore, the replacement of the steam trap with an equal-percentage control valve is not a good solution.

Pumping Traps

It is possible to use lifting traps to prevent condensate accumulation in heaters operating at low condensing pressures. This device is illustrated in figure 8.11 and depends on an external pressure source for its energy.

The unit is shown in its filling position, in which the liquid head in the heater has opened the inlet check valve of the trap. Filling progresses until the condensate overflows into the bucket, which then sinks, closing the equalizer and opening the pressure source valve. As pressure builds in the trap, the inlet check valve is closed; the outlet valve opens when the pressure exceeds that of the condensate header. The discharge cycle follows; in this cycle, the bucket is emptied. When the bucket is near empty, the buoyant force raises the bucket, which then closes off the steam valve and opens the equalizer. Once the pressure in the trap is lowered, the condensate outlet check valve is closed by the header back pressure and the inlet check is opened by the liquid head in the heater, which then is the beginning of another fill cycle.

The aforementioned pumping trap guarantees condensate removal regardless of the minimum condensing pressure in the heater. If such a trap is placed on the exchanger illustrated in figure 8.9, it will make temperature control possible even when the heater is under vacuum. Of course, this does not relieve the rangeability problems discussed earlier, and the use of two valves in parallel might still be necessary.

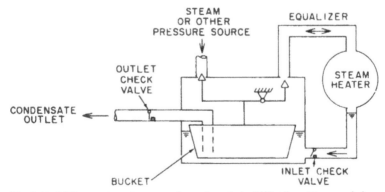

Fig. 8.11 A lifting or pumping trap, shown here in its "filling" position, can help prevent condensate accumulation in heaters operating at low condensing pressures. (From Lipták [3])

Level Controllers

The control system in figure 8.9 provided quick response but was unable to discharge its condensate at low loads. The system in figure 8.10 eliminated the condensate problem, but at the price of control response.

Because the low condensing pressure situation is a result of the combination of low load and high heat transfer surface area, it is possible to prevent it from developing by reducing the heat transfer area. One method of achieving this is shown in figure 8.12, in which the steam trap has been replaced by a level control loop. With this instrumentation provided, it is possible to adjust the size of the heater by changing the level set point to match the process load. This technique gives good temperature control if the level setting is made correctly and if there are no sudden load variations in the system.

As the required level is a function of the load, it can be adjusted automatically. This relieves the operator of the responsibility of manually changing the level controller set point whenever the load changes. More importantly, this also eliminates the potential for human error, which at low-level settings can result in condensate removal problems and at high levels can prevent temperature control.

Automatic adjustment of the level is illustrated in figure 8.13. Load is detected by the valve position controller (VPC) having a set point of, say, 50 percent. When the load rises, the steam valve opens beyond 50 percent and the VPC increases the active heat transfer area in the heat exchanger by lowering the set point of the condensate level controller. The VPC is an integral-only controller and therefore will not respond to measurement noise or valve cycling. The integral time is set to give slow floating

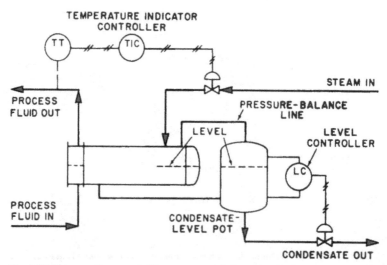

Fig. 8.12 The use of two control loops can maintain sensitive temperature control without condensate removal problems. (Adapted from Mathur [4])

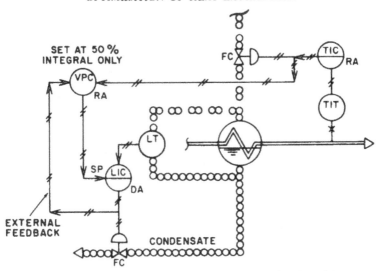

Fig. 8.13 The level set point can be floated to match the load. (Adapted from Lipták [3])

action that is fast enough to respond to anticipated load changes. The external feedback protects from reset wind-up in the VPC when LIC is switched to local set point.

In less critical applications, in which the cost of an extra level control loop cannot be justified, a continuous drainer trap such as the one shown in figure 8.14 can serve the same purpose as the level control just described. Its cost is substantially lower, but it is limited to the range within which the level setting can be varied and its control point is offset by load variations. For this reason, continuous drainer traps are unlikely to be considered for installations on vertical heaters or reboilers, where the range of level adjustment can be substantial.

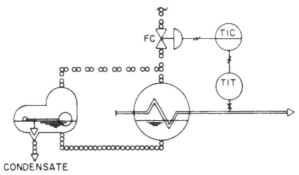

Fig. 8.14 A continuous drainer trap can serve the same purpose as level control. (From Lipták [3])

Bypass Control

Table 8.1 summarized some of the features of three-way valves when installed to circumvent the transient characteristics of coolers. Figure 8.15 shows the same concept applied to a steam heater. The advantages and limitations of this system are the same as discussed in connection with liquid-to-liquid exchangers, with one additional advantage. This additional advantage has to do with the fact that the bypass created an additional degree of freedom; therefore, steam can now be throttled as a function of some other property. The logical decision is to adjust the steam feed so that it maintains the condensing pressure constant. This then eliminates problems associated with condensate removal. It is also important to realize that in case of full bypass operation, the stagnant exchanger contents will be exposed to steam heat. Unless protection is provided, it is possible to boil this liquid.

Interaction between Parallel Heaters

If the process fluid is heated to a temperature approaching its boiling point, serious interaction can occur between parallel heaters (figure 8.16). The mechanism of developing this oscillation is as follows: a sudden drop in flow causes overheating and vaporization in one of the heaters. The vapor formation increases the back-pressure and further reduces the flow, eventually forcing all flow through the other "cold" exchanger, while the "hot" exchanger discharges slugs of liquid and vapor. After a period of noise and vibration, when the "hot" exchanger has discharged all of its liquid, the back-pressure drops and flow is resumed, drawing feed from the "cold" one. This causes the "cold" exchanger to overheat, and the cycle is repeated with the roles of the exchangers reversed [6].

This type of interaction can be eliminated easily by providing distribution controls. The control system used in making sure that the load is equally distributed between

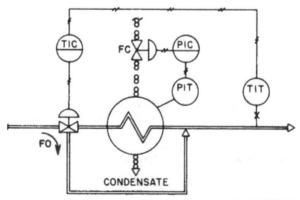

Fig. 8.15 Bypass control of steam heater provides the added degree of freedom needed for independent control of condensing pressure. (From Lipták [3])

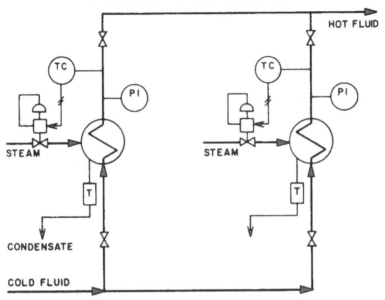

Fig. 8.16 Parallel heaters operating close to the boiling point of the process fluid
can experience serious interaction problems. (Adapted from Shinskey [6])

exchangers is the same as the system described in chapter 7 for cooling tower balancing
(see figure 7.13).

Condenser Controls

Depending on whether the control of condensate temperature or of condensing
pressure is of interest, the systems shown in figures 8.17 and 8.18 can be considered.

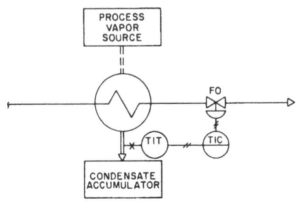

Fig. 8.17 Condensate temperature control throttles the cooling
water flow through the condenser. (From Lipták [3])

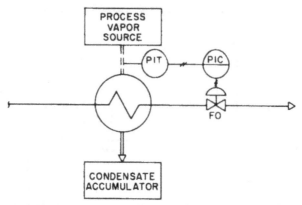

Fig. 8.18 The cooling water flow through the condenser can be throttled to control pressure. (From Lipták [3])

Both of these throttle the cooling water flow through the condenser, causing a potential for high temperature rise that is acceptable only when the water is chemically treated against fouling. For good, sensitive control, the water velocity through the condenser should be such that its residence time does not exceed one minute. Another rule of thumb is to keep the water velocity above 4.5 fps (1.35 m/s). This can be accomplished by the use of a small recirculating pump, such as P1 in figure 4.11. In this system, FSHL-3 detects the net forward velocity of the water. As long as it is under 4.5 fps, FSHL-3 will keep the recirculating pump running. When it is not desirable to throttle the cooling water, the system illustrated in figure 8.19 can be considered. Here the exposed condenser surface is varied to control the rate of condensation. Where non-condensables are present, a constant purge may serve to remove the inerts.

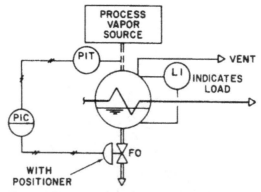

Fig. 8.19 Condenser control can be achieved by changing the wetted surface area. (Adapted from Lip-ták [3])

An important point to understand is that heat transfer efficiency is the highest when both the coolant flow and the heat transfer area are at their maximum levels. Therefore, the goal of optimization in figure 8.18 is to open the coolant valve fully; in figure 8.19, the goal of optimization is to eliminate flooding. These goals are achieved by slowly lowering the PIC set point until the cooling capacity is fully utilized and none of it is wasted through throttling (see fig 8.25).

The preferred method of condenser control is to vary the heat transfer area through partial flooding [8]. In that case, the load is indicated by the condensate level (figure 8.19). The condensate is subcooled to differing degrees as a function of residence time. Therefore, condensate temperature cannot be used for control. The controller on a partially flooded condenser is usually tuned as a level controller, but because of the noise filtering effect of the heat capacity of the system, derivative action can frequently be added [8]. The addition of valve positioners is always helpful in improving the response and stability of the loop.

Distillation Condensers

A variety of condenser designs is available for use in connection with distillation towers. The air-cooled designs are sensitive to ambient variations, particularly to rainfall. These units are throttled by manipulating their inlet louvers or by adjusting the blade pitch of the fans. If multispeed or variable-speed fans are available, the condenser rangeability can be increased while energy is saved.

When the condenser is water cooled and its inlet pressure is controlled by partial flooding (figure 8.19), the main problem is response speed. A change in valve position does not immediately affect the heat transfer area. The response speed is slower when the level needs to be increased and faster when it is to be decreased. This "non-symmetricity" has already been explained in connection with figure 8.10. In controlling distillation towers, the elimination of this non-symmetricity is one of the important goals of control.

The problem of pressure control is complicated by the presence of large percentages of inert gases. The uncondensables must be removed or they will accumulate and blanket off the condensing surface, thereby causing loss of column pressure control.

The simplest method of handling this problem is to bleed off a fixed amount of gases and vapors to a lower-pressure unit, such as to an absorption tower, if such is present in the system. If an absorber is not present, it is possible to install a vent condenser to recover the condensable vapors from this purge stream.

It is recommended that the fixed continuous purge be used wherever economically possible; however, when this is not permitted, it is possible to modulate the purge stream. This might be desirable when the amount of inerts is subject to wide variations over time (Figure 8.20).

As the uncondensables build up in the condenser, the pressure controller will tend to open the control valve to maintain the proper rate of condensation. This is done by a change in air signal to the diaphragm control valve. The air-loading pressure can also be used to signal when the pressure control valve is nearly fully open and purging is

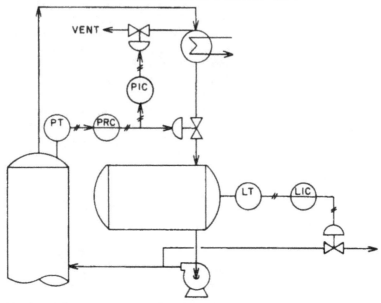

Fig. 8.20 Column pressure control can be used to also vent inerts. (From Lipták [3])

needed. This could be done by means of a calibrated valve positioner or a second pressure controller (figure 8.20).

Condenser below Receiver

Lowering the condenser below the accumulator not only reduces installation cost and makes maintenance more convenient but also eliminates the non-symmetricity from the process. It is the usual practice to elevate the bottom of the accumulator 10 to 15 feet (3 to 4.5 m) above the suction of the pump in order to provide the required suction head.

In this type of installation, the control valve is placed in a bypass from the vapor line to the accumulator (see figure 8.21). When this valve is open, it equalizes the pressure between the vapor line and the receiver. This causes the condensing surface to become flooded with condensate because of the 10 to 15 feet of head that exist in the condensate line from the condenser to the receiver. The flooding of the condensing surface causes the pressure to build up because of the decrease in the active heat transfer surface available. Under normal operating conditions, the subcooling that the condensate receives in the condenser is sufficient to reduce the vapor pressure in the receiver. The difference in pressure permits the condensate to flow up the 10 to 15 feet of pipe between the condenser and the accumulator. When the condensing pressure is to be reduced, the valve closes, resulting in an increase in the exposed condenser surface area. In order to expose more area, the condensate is transferred into the accumulator; this transfer can occur only if the accumulator vapor pressure has been

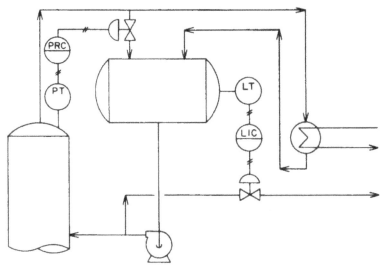

Fig. 8.21 Column pressure control can be improved using a lowered condenser. (From Lipták [3])

sufficiently lowered by condensation. Therefore, the system speed in this direction is a function of condensate supercooling. Increased supercooling increases system response speed. Based on these considerations, the unit might or might not be symmetrical in its dynamics. If high speed is desired, the controls shown in figure 8.22 can be considered. Here the pressure is controlled in the accumulator by throttling the condenser bypass flow. The column pressure is maintained by throttling the flow of vapor through the condenser. Controlling the rate of flow through the condenser gives faster pressure regulation for the column.

When the condensing temperature of the process fluid is low, water is no longer an acceptable cooling medium. One standard technique of controlling a refrigerated condenser is illustrated in figure 8.23. Here the heat transfer area is set by the level control loop and the operating temperature is maintained by the pressure controller. When process load changes, it affects the rate of refrigerant vaporization, which is compensated for by level-controlled make-up. Usually the pressure and level settings are made manually, although there is no reason why these set points could not be automatically adjusted as a function of load. This can be done by placing a valve position controller (VPC) on the PIC output signal; this VPC will detect an increase in load and will respond to it by lowering the LIC set point to increase the heat transfer area. This has been explained in more detail in connection with figure 8.13.

Vacuum Systems

For some liquid mixtures, the temperature required to vaporize the feed would need to be so high that decomposition would result. To avoid this, it is necessary to operate the column at pressures below atmospheric pressure.

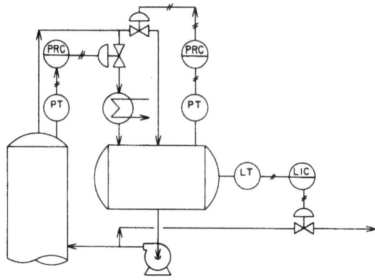

Fig. 8.22 High-speed column pressure control can result from the use of "hot gas bypass." (From Lipták [3])

The common means for creating a vacuum in distillation is to use steam jet ejectors. These can be used singly or in stages to create a wide range of vacuum conditions. Their wide acceptance is based upon their having no moving parts and requiring very little maintenance.

Most ejectors are designed for a fixed capacity and work best at one steam condition. Increasing the steam pressure above the design point will not usually increase the capacity of the ejector; as a matter of fact, it will sometimes decrease the capacity because of the choking effect of the excess steam in the diffuser throat.

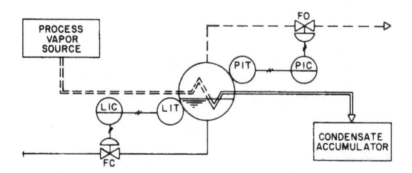

Fig. 8.23 Condensers can be controlled using refrigerant coolants. (From Lipták [3])

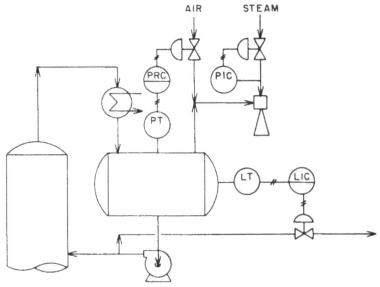

Fig. 8.24 Vacuum distillation requires operation at negative pressures. (From Lipták [3])

Steam pressure below a critical value for a jet will cause the ejector operation to be unstable. Therefore, it is recommended that a pressure controller be installed on the steam to keep it at the optimum pressure required by the ejector.

The recommended control system for vacuum distillation is shown in figure 8.24. Air or gas is bled into the vacuum line just ahead of the ejector. This makes the maximum capacity of the ejector available to handle any surges or upsets.

Because ejectors are designed for a fixed capacity, the variable load is met by air bleed into the system. At low loads, this represents a substantial waste of steam. Therefore, for processes in which load variations are expected, the operating costs can be lowered by installing a larger ejector and a smaller ejector. This makes it possible to switch to the small unit automatically when the load drops off, thereby reducing steam demand.

A control valve regulates the amount of bleed air used to maintain the pressure on the reflux accumulator. Using the pressure of the accumulator for control involves less time lag than if the column pressure were used as the control variable.

Optimization through Pressure Floating

The cost of separation is minimized when the column operates at the lowest possible pressure. Minimum pressure operation can be achieved by manual or automatic adjustment of the set point of the pressure controller to keep the condenser fully loaded in the long term. However, to prevent upsets caused by rapid set point changes, Shinskey [7] proposes the VPC scheme shown in figure 8.25. The VPC adjusts the set

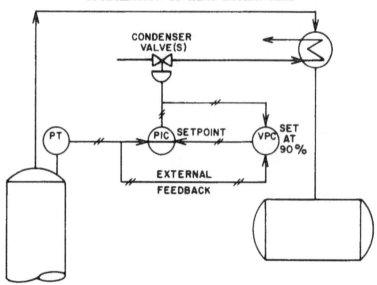

Fig. 8.25 The VPC scheme provides floating-pressure control. (From Lipták [3])

point of the pressure controller and maximizes cooling by holding the condenser control valve in a nearly fully open position.

The pressure controller should incorporate proportional plus integral action to provide rapid response to upsets, whereas the VPC should be an integral-only controller, so a rapid change in valve position will not produce a proportional change in the pressure set point. The integral time setting of the VPC should be approximately tenfold that of the overhead composition controller. In addition, it is common practice to limit the range within which the VPC can adjust the PIC set point and to provide the external feedback line to eliminate reset wind-up when the VPC output reaches one of these limits.

Reboilers and Vaporizers

As noted at the beginning of this chapter, when a steam-heated reboiler is used, only one degree of freedom is available; therefore, only one controller can be installed without overdefining the system. This one controller is usually applied to adjust the rate of steam addition. With regard to the minimum condensing pressure considerations, the same applies as has been discussed earlier in connection with liquid heaters. Figures 8.26 and 8.27 show the two basic alternatives for controlling the reboiler: either to generate vapors at a controlled superheat temperature or to generate saturated vapors at a constant rate set by the rate of heat input.

Naturally, there are other, more sophisticated reboiler control strategies, using composition, temperature difference, or derived variables as the means of control, but in their effects they are all similar to the systems in figures 8.26 and 8.27. When several

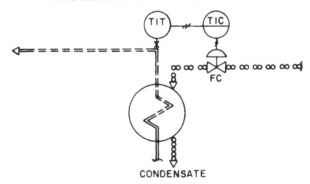

Fig. 8.26 Reboilers can be controlled by generating vapors at controlled temperatures. (From Lipták [3])

reboilers operate in parallel, it is important to balance the load distribution between them according to the type of control system, as was described in figure 7.13.

Fired Reboilers

When heat duties are great and bottom temperatures are high, fired reboilers are used. The fired reboiler heats and vaporizes the tower bottoms as this liquid circulates by natural convection through the heater tubes. The coils are generously sized to ensure adequate circulation of the bottoms liquid. Overheating the process fluid is a contingency that must be guarded against, because most tower bottoms will coke or polymerize if under excessive temperatures for some length of time.

A common control scheme is shown in figure 8.28; this figure depicts the tower bottom along with the fired reboiler. It is not usually practical to measure the flow of tower bottoms to the reboiler—first, because the liquid is near equilibrium (near the flash point), and second, because it is usually of a fouling nature, tending to plug most

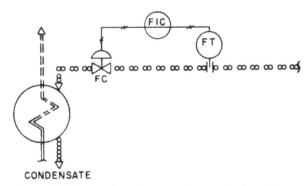

Fig. 8.27 BTU control sets the steam flow to reboilers. (From Lipták [3])

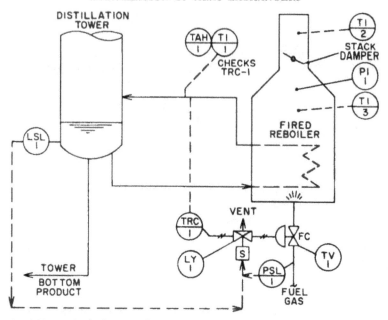

Fig. 8.28 Fired reboiler controls must protect against overheating the process fluid. (From Lipták [3])

flow elements. Proper circulation of the fluid is provided for in the careful hydraulic design of the interconnecting piping.

The other important variable is the reboiler return temperature, which is controlled by TRC-1's throttling of the fuel gas control valve (TV-1). The high-temperature alarm (TAH-1) is provided to warn the operator that the process fluid has suddenly reached an excessive temperature, indicating that manual adjustments are needed to cut back on the firing. The furnace draft is set by the operator adjusting the stack damper, while observing the draft gauge (PI-1). The stack temperature (TI-2) and the firebox temperature (TI-3) are detected as checks for excessive temperatures that may develop during periods of heavy firing.

The major dangers in this furnace type are the interruption of process fluid flow and the stoppage of fuel. The loss of process fluid can occur if the liquid level in the tower bottom is lost. If this happens, flow will stop in the reboiler tubes and a dangerous overheating of tubes may result. To protect against this, the low-level switch (LSL-1) is wired to trip the solenoid valve (LY-1), which vents the diaphragm of the control valve (TV-1). This causes the fuel valve to close.

A momentary loss of fuel can be dangerous, because the flames can be extinguished and on resumption of fuel flow a dangerous air-fuel mixture will develop in the fire box. To prevent this occurrence, the low-pressure switch (PSL-1) trips the solenoid (LY-1) on low fuel pressure, thereby closing the fuel valve (TV-1). The solenoid valve

(LY-1) should be the manual reset type that remains vented and thus causes the valve (TV-1) to remain closed even on return of fuel pressure.

The air failure action of the control valve (TV-1) is closed (FC) on loss of motivating instrument air. This action on fuel valves is the safe mode, because firing is discontinued during the emergency.

Fired Heaters and Vaporizers

The unit feed heater of a crude oil refinery is representative of the class of furnaces that includes fired heaters and vaporizers. Crude oil, prior to distillation in the "crude tower" into the various petroleum fractions (gasoline, naphtha, gas oil, heavy fuel oil, and so forth) must be heated and partially vaporized. The heating and vaporization is done in the crude heater furnace, which consists of a fire box with preheating coils and vaporizing coils. The heating is usually done in coils in the convection section of the furnace, which is the portion that does not see the flame but that is exposed to the hot flue gases on their way to the stack. The vaporizing takes place at the end of each pass in the radiant section of the furnace (where the coils see the flame and the luminous walls of the fire box). The partially vaporized effluent then enters the crude tower, where it flashes and is distilled into the desired "cuts." Other process heaters that fall into this category are refinery vacuum tower preheaters, reformer heaters, hydrocracker heaters, and dewaxing unit furnaces.

The prime variables in this process are listed below:

- Flow control of feed to the unit
- Proper splitting of flow in the parallel paths through the furnace (to prevent overheating of any one stream and its resultant coking)
- Correct amount of heat supplied to the crude tower

Figure 8.29 shows the typical process controls for this type of furnace. The crude feed rate to the unit is set by the flow controller (FRC-2). The flow is split through the parallel paths of the furnace by remote manual adjustment of the control valves (HV-1 through HV-4) via the manual stations (HIC-1 through HIC-4). The proper settings are determined by equalization of the flow indications on FI-1 through FI-4. The temperature indicators TI-1 through TI-12 are periodically observed to determine if there are any rising or falling trends in any one of the passes. If such a trend develops, the flow through that pass is altered slightly to drive the temperature back toward the norm. If a higher level of automation is desired, the flow balancing system shown in figure 7.13 can be used.

The desired heat input into the feed stream is more difficult to control, because the effluent of the furnace is partially vaporized and the feed stock varies in composition depending on its source. If the feed were only heated, with no vaporization taking place, the control would require only that an effluent temperature be maintained. If complete vaporization and superheating occurred, this too could be handled well by straight temperature control.

In the case of partial vaporization with a variable feed composition, however,

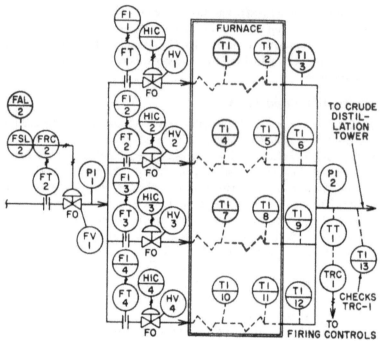

Fig. 8.29 Crude heater-vaporizer controls serve to achieve uniform heating of several parallel streams. (From Lipták [3])

effluent temperature control alone is not a reliable approach. The composition and, hence, the boiling point curve of the feed is not constant with time, and the required control temperature itself varies. Additional information is thus required, and it is obtained from the distillation downstream of the furnace. By observing the product distribution from the fractionation, the need for a change in heat input can be determined. This approach is slow and not precise, but doing a more accurate job would require a great deal of sophisticated instrumentation, involving special analyzers to measure feed composition and a computer to optimize the mathematical model and thereby determine the required heat input to the feed. Current practice is to achieve approximate control with a temperature controller (TRC-1), whose set point the unit operator periodically changes to account for feed variations. The operator depends on experience and on the results in the fractionator to determine the proper temperature setting. Naturally, in more advanced refineries, this is under computer control.

Cascade Control

By definition, cascade systems consist of two controllers in series. In heat exchangers, the master detects the process temperature and the slave is installed on a variable that may cause fluctuations in the process temperature. The master adjusts the slave set point, and the slave throttles the valve to maintain that set point. A cascade loop controls a single temperature, and the slave controller is there only to assist in achieving this. In other words, the cascade loop does not have two independent set points. In order for a cascade loop to be successful, the slave must be much faster than the master. A rule of thumb is that the time constant of the primary should be ten times that of the secondary, or that the period of oscillation of the primary should be three times that of the secondary.

Cascade loops are invariably installed to prevent outside disturbances from entering the process. An example of such a disturbance would be the header pressure variations of a steam heater. The conventional single controller system (see figure 8.9) cannot respond to a change in steam pressure until its effect is felt by the process temperature sensor. In other words, an error in the detected temperature has to develop before corrective action can be taken. The cascade loop, in contrast, responds immediately, correcting for the effect of pressure change before it can influence the process temperature (figure 8.30).

The improvement in control quality due to cascading is a function of relative speeds and time lags. A slow primary (master) variable and a secondary (slave) variable that responds quickly to disturbances represents a desirable combination for this type of control. If the slave can quickly respond to fast disturbances, these will not be allowed to enter the process and therefore will not upset the control of the primary (master) variable.

One of the best cascade slaves is the simple and inexpensive pressure regulator. Its air-loaded variety (figure 8.31) is extremely fast and can correct for steam supply pressure or load variations almost instantaneously.

In figures 8.30 and 8.32, the controlled variable is temperature and the manipulated variable is the pressure or flow of steam. The primary variable (temperature) is slow, and the secondary (manipulated) variable is capable of responding quickly to disturbances. Therefore, if disturbances occur (a sudden change in plant steam demand, for example), upsetting the manipulated variable (steam pressure), these disturbances will be sensed immediately and corrective action will be taken by the secondary controller so that the primary variable (process temperature) will not be affected.

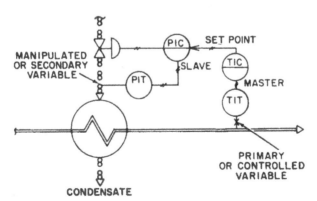

Fig. 8.30 A temperature-pressure cascade loop on a steam heater increases the speed of response. (From Lipták [3])

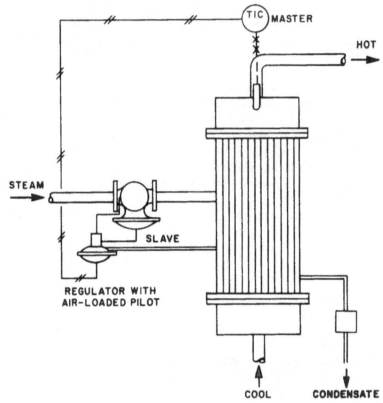

Fig. 8.31 One of the best cascade slaves is the pressure regulator. (Adapted from Driskell [1])

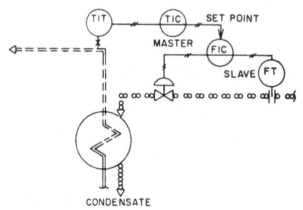

Fig. 8.32 Temperature-flow cascade loops are commonly used on steam reboilers. (From Lipták [3])

Feedforward Optimization

Feedback control involves the detection of the controlled variable (temperature) and the counteracting of changes in its value relative to a set point, by the adjustment of a manipulated variable (the flow of a heat transfer fluid). This mode of control necessitates that the disturbance variable affect the controlled variable itself before correction can take place. Hence, the term *feedback* can imply a correction "back" in terms of time—a correction that should have taken place earlier, when the disturbance occurred.

In this manner of terminology, *feedforward* is a mode of control that responds to a disturbance such that it instantaneously compensates for an error that the disturbance would otherwise have caused in the controlled variable later.

Figure 8.33 illustrates a steam heater under feedforward control. This control

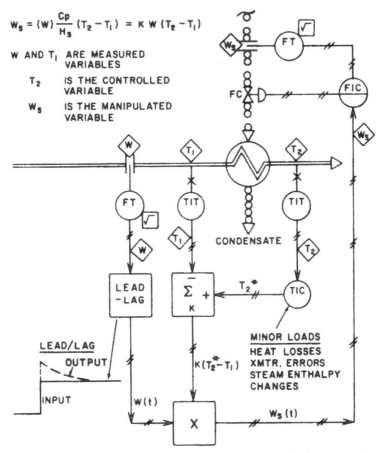

$$W_s = (W)\frac{Cp}{H_s}(T_2 - T_1) = K W (T_2 - T_1)$$

W AND T_1 ARE MEASURED VARIABLES

T_2 IS THE CONTROLLED VARIABLE

W_s IS THE MANIPULATED VARIABLE

Fig. 8.33 In feedforward optimization of steam heaters, major load variations (T_1 and W) are corrected by the feedforward portion of the loop, leaving only the minor load variables for feedback correction. (Adapted from Lipták [3])

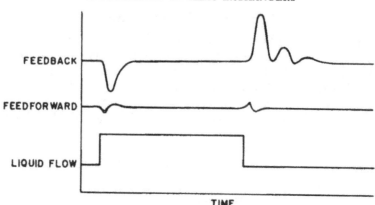

Fig. 8.34 Feedforward control is capable of reducing both the area and the duration of the load response transients. (From Shinskey [8])

system consists of two main segments. The feedforward portion of the loop detects the major load variables (the flow and temperature of the entering process fluid) and calculates the required steam flow (Ws) as a function of these variables. When the process flow increases, it should be matched with an equal increase in steam flow instantaneously. Because instantaneous response is not possible, the next best thing is to add more steam than needed as soon as possible. This is the dynamic correction function served by the lead-lag element in the feedforward loop. The resulting improvement in control performance is illustrated in figure 8.34.

The feedback portion of the loop (TIC) has to do much less work in this configuration, as it only has to correct for minor load variables, such as heat losses to the atmosphere, steam enthalpy variations, and sensor errors. The feedback and feedforward portions of the loop complement each other perfectly. Feedforward is responsive, fast, and sophisticated, but inaccurate; feedback is capable of regulating in response to unknown or poorly understood load variations, and although it is slow, it is accurate.

Adaptive Gain

As was already noted in connection with figure 8.9, the heat exchanger is a variable gain process. The steady state gain of a steam heater process is as follows:

$$dT_2/dWs = \Delta Hs/W\ Cp$$

Therefore, the process gain varies inversely with flow (W). If the temperature rise $(T_2 - T_1)$ is also a variable, even the use of an equal-percentage valve cannot correct for this nonlinearity in the process. In that case, the only way to keep the process gain constant is to use the feedforward system shown in figure 8.33. There a reduction in process flow causes a reduction in the gain of the multiplier, which exactly cancels the increase in process gain. Thus, the feedforward loop provides gain adaptation as a side benefit, because as the process gain varies inversely with flow, it causes the controller gain to vary directly with flow. The result is constant total loop gain and therefore stable loop behavior.

The feedforward concept described in figure 8.33 is not limited to steam heaters but can be used on all types of heat exchangers. The only needed modifications are the ones that are required to correctly reflect the heat balance equation of the particular heat transfer unit.

The advantages of feedforward control are similar to those of cascade control, because the load upset or supply disturbance is corrected for before its effect is felt by the controlled variable. Feedforward control contributes to stable dampened response to load changes and to fast recovery from upsets.

Multipurpose Systems

The control of isolated heat transfer units has been discussed in the earlier sections of this chapter. In the majority of critical installations, the purpose of such systems is not limited to the addition or removal of heat, but involves making use of both heating and cooling in order to maintain the process temperature constant. Such a task necessitates the application of multipurpose systems, incorporating many of the features that have been discussed individually earlier.

Figure 8.35, for example, depicts a design that uses hot oil as its heat source and water as the means of cooling, arranged in a recirculating system. The points made earlier in connection with three-way valves, cascade systems, and so forth also apply here, but a few additional considerations are worth noting.

Probably the most important single feature of this design is that it operates on a *split-range signal*. This means that when the process temperature is above the desired

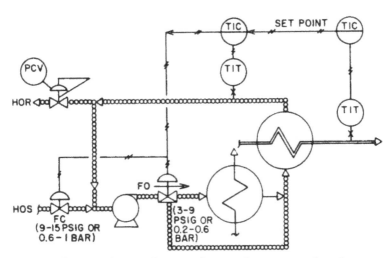

Fig. 8.35 This recirculating multipurpose heat transfer system uses hot oil as its heat source and water as its coolant. (From Lipták [3])

set point, the output signal to the valves will be reduced. When the value of this signal is between 9 and 15 PSIG (0.6 and 1 bar), the three-way valve is fully open to the exchanger bypass and the two-way hot oil supply valve is partially open. If the reduction in the two-way valve opening is not sufficient to bring the process temperature down to set point, the signal will further decrease, thereby fully closing the two-way valve at 9 PSIG (0.6 bar) and beginning to open the flow path through the cooler from the three-way valve. At a 3 PSIG (0.2 bar) signal, the total cooling capacity of the system is applied to the recirculating oil stream, which in that case flows through the cooler without bypass.

The limitations of such a split-range operation are discussed below.

First, at a signal level near 9 PSIG (0.6 bar), the system can be unstable and cycling, because this is the point at which the three-way valve is just beginning to open to the cooler, and the system might receive alternating slugs of cooling and heating because of the limited rangeability of the valves. (Zero flow is not enough; minimum flow is too much for the particular load condition.)

Second, when the signal is between 9 and 15 PSIG (0.6 and 1 bar), the cooler shell side becomes a reservoir of cold oil. This upsets the controls twice: once when the three-way valve just opens to the cooler and once when the cold oil has been completely displaced and the oil outlet temperature from the cooler suddenly changes from that of the cooling water to some much higher value.

Finally, most of these systems are non-symmetrical in that the process dynamics (lags and responses) are different for the cooling and heating phases.

To remedy these problems, several steps can be considered. As shown in figure 8.35, a cascade loop should be used so that upsets and disturbances in the circulating oil loop will be prevented from entering the process and upsetting its temperature. In addition, a slight overlapping of the two valve positioners is desirable. This will offset the beginning of cooling and the termination of heating phases so that they will not both occur at 9 PSIG (0.6 bar). The resulting sacrifice of heat energy is well justified by the improved control obtained. Finally, in order to protect against the development of an extremely cold oil reservoir in the cooler, a minimum continuous flow through this unit can be maintained.

In connection with the recirculating design shown in figure 8.35, it is also important to realize that this is a flooded system; when hot oil enters it, a corresponding volume of oil must be allowed to leave it. The pressure control valve (PCV) serves this function. The same purpose can be achieved by elevational head on the return header, the important consideration being that whatever means are used, the path of least resistance for the oil must be back to the pump suction to keep it always flooded and thereby to prevent cavitation.

Most multipurpose systems represent a compromise of various degrees. Figure 8.36, for example, illustrates a design in which low cost and rapid response to load changes are the main considerations. These characteristics are provided by using the minimum hardware and by circumventing the transient characteristics of the exchangers. The price paid in this compromise involves the full use of utilities at all times, the

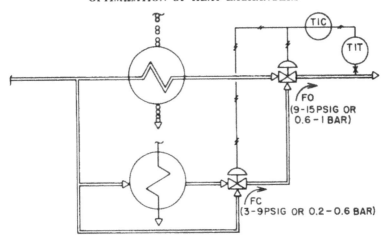

Fig. 8.36 This multipurpose temperature control system blends process streams at different temperatures. (From Lipták [3])

development of hot and cold reservoirs, and the necessity for supply disturbances to affect the process temperature before corrective action can be initiated.

Other multipurpose heat transfer systems are used in the control of jacketed chemical reactors. These can be single purpose (figures 3.5 to 3.8), double purpose (figures 3.9 and 3.10), or configured to operate with three different heat transfer fluids (figure 3.14).

Conclusions

The temperature upset resulting from the sudden doubling of the load can be reduced by an order of magnitude through the use of feedforward control (figure 8.34). Such a tenfold reduction in temperature error can make a great contribution to both safety and quality of production. Therefore, optimized heat exchanger controls are usually justified not on the basis of increased production or reduced energy costs, but on the basis of more stable, accurate, and responsive control.

Some of the considerations that always pay to investigate are listed below:

- The effect of supply disturbances on systems performance
- The response speed of the system
- Rangeability considerations
- The quality of cooling water available
- Potential problems due to non-symmetrical dynamics and to low minimum condensing pressures

In addition, it is always wise to consider the use of equal percentage control valves furnished with positioners and to evaluate the advisability of using cascade or feedforward controls.

References

1. Driskell, L. R., *Control Valve Selection and Sizing*, ISA, 1983.
2. Lipták, B. G., "Control of Heat Exchangers," *British Chemical Engineering and Process Technology* (July/August, 1972), vol. 17, no. 7/8, p. 637.
3. Lipták, B. G., *Instrument Engineer's Handbook, Process Control* (Radnor, PA: Chilton Book Co., 1985), sections 8.8, 8.12, 8.13.
4. Mathur, J., "Performance of Steam Heat-Exchangers," *Chemical Engineering*, September 3, 1973.
5. Shinskey, F. G., *Controlling Multivariable Processes*, ISA, 1981, p. 74.
6. Shinskey, F. G., *Controlling Unstable Processes*, publication 413–4, The Foxboro Co.
7. Shinskey, F. G., *Distillation Control for Productivity and Energy Conservation* (New York, NY: McGraw-Hill, 1978), p. 307.
8. Shinskey, F. G., *Process Control Systems*, second ed. (New York, NY: McGraw-Hill, 1979), pp. 186, 227.

Chapter 9

OPTIMIZATION
OF
PUMPING STATIONS

THE OPERATION of all processing plants involves the transportation of liquids. Optimized pumping can contribute to the overall optimization of the plant. This chapter describes the various pump designs and their performance characteristics, using both constant- and variable-speed drives; the subject of pump controls is also discussed. The chapter concludes with a discussion of the optimization strategies that are available to maximize the safety and energy efficiency of these systems.

Pump Types

There are two main pump designs: positive displacement pumps and centrifugal pumps. Vertical centrifugal units can pump water from depths up to 2,000 feet (600 m), and horizontal units can transport process fluids from clear water to heavy sludge at rates up to 100,000 GPM (6.3 m³/s). The centrifugal designs are of either the radial-flow or the axial-flow type. Liquid enters the radial-flow designs in the center of the impeller and is thrown out by the centrifugal force into a spiral bowl. A number of impeller designs are illustrated in figure 9.1. The axial-flow propeller pumps are designed to push rather than throw the fluid upward. Mixed-flow designs are a combination of the two.

In positive displacement pumps, a piston or plunger inside a cylinder is the driving element as it moves in reciprocating motion (figure 9.2). The stroke length and thus the volume delivered per stroke is adjustable within a 10 to 1 range. Rangeability can be increased to 100 to 1 by the addition of a variable-speed drive. The plunger designs are capable of generating higher discharge pressures than the diaphragm types, because of the strength limitation of the diaphragm. The strain on the diaphragm is reduced if it is not attached directly to the plunger but is driven indirectly through the use of a hydraulic fluid. Because solids will still settle in the pump cavities, these designs are all limited to relatively clean services. For slurry service, the hose type design is recommended (see figure 9.3). This design eliminates all the cavities, although the seating of the valves can still be a problem.

Other, less frequently used, pump designs include rotary screw pumps, which are ideally suited for sludge and slurry services, and air pumps, which use compressed air or steam to displace the accumulated liquids from tanks (figure 2.30). A special variation on pump designs are air lifts. In these, compressed air is blown into the bottom of a submerged updraft tube. As the air bubbles rise, they expand, reducing the density and therefore the hydrostatic head inside the tube. As the head is lower on the inside

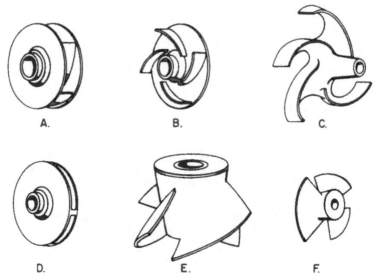

Fig. 9.1 Types of centrifugal pump impellers include: (A) closed impeller, (B) semi-open impeller, (C) open impeller, (D) diffuser, (E) mixed flow impeller, and (F) axial flow impeller.

of the tube, the manometer effect induces the surrounding fluid to enter the updraft tube. This is an efficient and maintenance-free means of lifting large volumes of slurries over short distances of elevation.

Metering Pump Installations

When the purpose of a positive displacement pump is to meter the flow rate, certain precautions are needed. These include the removal of all entrained or dissolved gases, which otherwise can destroy metering accuracy. Figure 9.4 shows how entrained gases can be returned to the supply tank.

Another concern in metering pump installations is to make sure that the discharge pressure is always higher than the suction pressure, so that fluid will not flow unrestricted through the pump. This can be guaranteed either through piping arrangement or by the addition of back pressure regulators to the pump discharge, as shown in figure 9.5. If it is desirable to dampen the pulsations, a volume chamber can also be added to the pump discharge. In order to be effective, the dampener volume must be equal to at least five times the volume displaced by each stroke.

Fig. 9.2 Various designs of positive-displacement pumps. Top: Plungers or piston-type metering pump. Center: Diaphragm pump operated with hydraulic fluid. Bottom: Diaphragm-type metering pump.

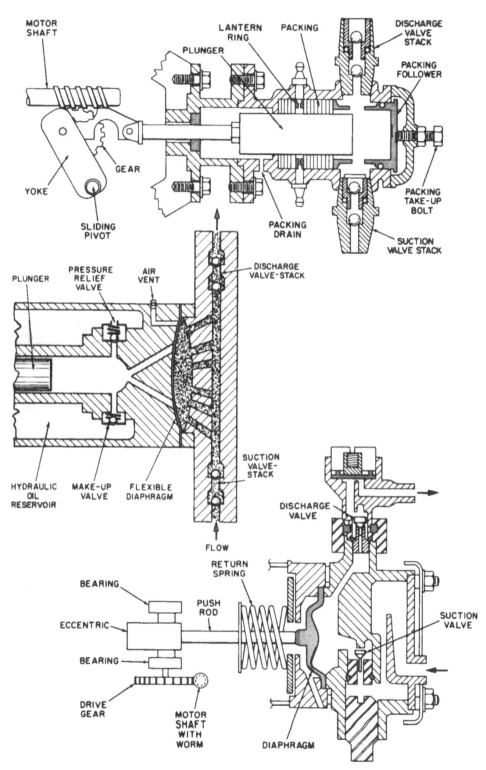

MOTOR
SHAFT

PLUNGER

LANTERN
RING

PACKING

DISCHARGE
VALVE
STACK

PACKING
FOLLOWER

GEAR

YOKE

SLIDING
PIVOT

PACKING
DRAIN

PACKING
TAKE-UP
BOLT

SUCTION
VALVE STACK

PLUNGER

PRESSURE
RELIEF
VALVE

AIR
VENT

DISCHARGE
VALVE-STACK

HYDRAULIC
OIL
RESERVOIR

MAKE-UP
VALVE

FLEXIBLE
DIAPHRAGM

SUCTION
VALVE-
STACK

FLOW

DISCHARGE
VALVE

RETURN
SPRING

BEARING

ECCENTRIC

PUSH
ROD

SUCTION
VALVE

BEARING

DRIVE
GEAR

MOTOR
SHAFT
WITH
WORM

DIAPHRAGM

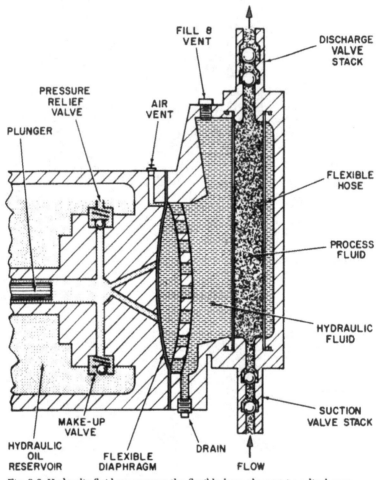

Fig. 9.3 Hydraulic fluid compresses the flexible hose element in a diaphragm pump. (From Lipták [6])

It is also important to keep the *net positive suction head* (NPSH) above 10 PSIA (69kPa). NPSH can be calculated as follows:

$$NPSH = P - P_v \pm P_h - \sqrt{\left(\frac{lvGN}{525}\right)^2 + \left(\frac{lvC}{980Gd^2}\right)^2}$$

where

P = feed tank pressure (PSIA)

P_v = liquid vapor pressure at pump inlet temperature (PSIA)

P_h = head of liquid above or below the pump center line (PSID)

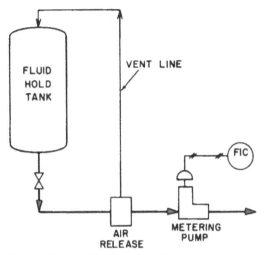

Fig. 9.4 Entrained gases can be eliminated in metering pump installations, through the use of air releases.
(From Lipták [6])

l = actual length of suction pipe (ft)
V = liquid velocity (ft/sec)
G = liquid specific gravity
N = number of pump strokes per minute
C = viscosity (centipoise)
d = inside diameter of pipe (in)

The metering pump can also be used as a final control element. In figure 9.6, for
example, the pump is used to maintain a flow ratio between fluids A and B. This system
is also provided with the capability for automatic recalibration. The calibration cycle is
started by switching the three-way valve to the "calibrate" position. When the level
has reached LSH, the three-way valve returns to the "normal" path and nitrogen enters
the tank to initiate discharge. When level drops to LSLL, discharge is terminated by
venting off the nitrogen. The counter QQI is running while the level is rising between
LSL and LSH. The total count, when compared with known calibration volume, gives
the total error. The hand switch HS initiates the calibration cycle by diverting the
three-way valve to the "calibrate" path.

Rotary Pumps

Rotary pumps include the screw, lobe, gear, and vane designs. Figure 9.7 illustrates
their nearly vertical head-capacity curve, which gives large pressure variations in response to small flow changes. These pumps are used on applications involving high-
viscosity (up to 500,000 centipoises or 500 Pa.s) or slurry services.

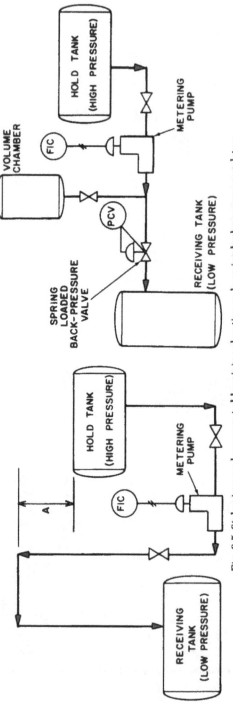

Fig. 9.5 Siphoning can be prevented by piping elevation or by using back-pressure regulators.

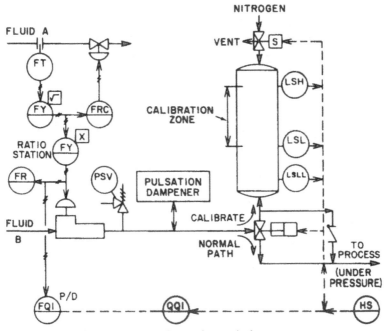

Fig. 9.6 Ratio and calibration controls are often applied to reciprocating pumps. (Adapted from Lipták [6])

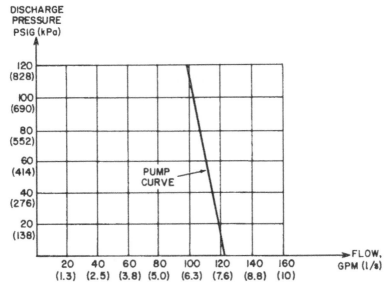

Fig. 9.7 The head-capacity curve for rotary pumps is nearly vertical. (From Lipták [6])

363

On-off control of rotary pumps is not recommended, because the settling of solids can create problems on slurry services and because frequent starting of the pump can cause overheating. In rotary pumps, the starting torque may be very large as a result of fluid viscosity, and so is the inertia load of the column of fluid in the piping, which accelerates under positive displacement each time the pump starts. For these reasons, the usual installation is similar to the one in figure 9.8, in which continuous pumping is achieved through the use of a pressure-controlled bypass back to the feed tank. The pressure controller opens this kick-back bypass valve whenever the centrifuge interlocks shut off the feed valve. The on-off control in this case is applied to the fluid rather than to the pump motor.

Manual on-off control is often applied to rotary pumps in bulk storage batch transfer services with local level indication.

A safety relief valve is always provided on a rotary pump to protect the system and pump casing from excessive pressure should the discharge line be blocked while the pump is running. The relief valve may discharge to pump suction or to the feed tank. In cases when slurries and viscous materials may not be able to pass through the relief valve, a rupture disk is placed on the discharge line (figure 9.9).

The capacity of a rotary pump is proportional to its speed, if the small losses due to slippage are disregarded. In this case, no bypass is needed. The rangeability is limited

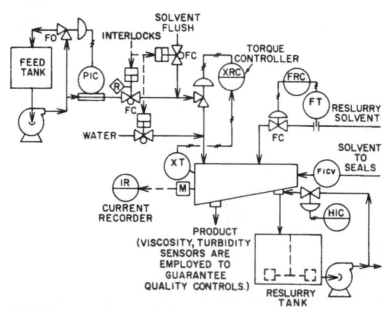

Fig. 9.8 A continuously operated rotary pump can be used to feed an intermittent centrifuge. (From Lipták [6])

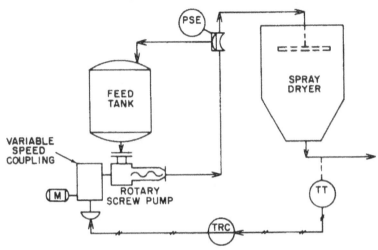

Fig. 9.9 Speed control of rotary pumps. (From Lipták [6])

by the speed-control device, which for pulleys and magnetic drives is approximately 4 to 1. The response of the loop is slower than with a control valve, but the maintenance needs are also reduced, as the potential for valve plugging is also eliminated. This is a good solution for slurry or gummy services, such as the installation shown in figure 9.9, in which a screw pump feeds latex slurry to a spray dryer.

Centrifugal Pump Characteristics

Figure 9.10 illustrates the typical pump curves of a single impeller unit and also the more common formulas used in connection with pumping. Table 9.1 defines the terms used and gives their conversion factors.

Pump efficiency is the ratio of the useful output power of the pump to its input power. It is calculated as follows [16]:

$$E_p = \frac{\text{pump output}}{P_i} = \frac{SPQH_t}{P_i} \quad \text{(SI units)}$$

$$E_p = \frac{\text{pump output}}{\text{bhp}} = \frac{SPQH_t}{\text{bhp} \times 550} \text{ (U.S. customary units)}$$

where

E_p = pump efficiency, dimensionless
P_i = power input, kW (kN · m/s)
SP = specific weight of water, lb/ft³ (kN/m³)
Q = capacity, ft³/s (m³/s)
H_t = total dynamic head, ft (m)
bhp = brake horsepower
550 = conversion factor for horsepower to ft · lb_f/s

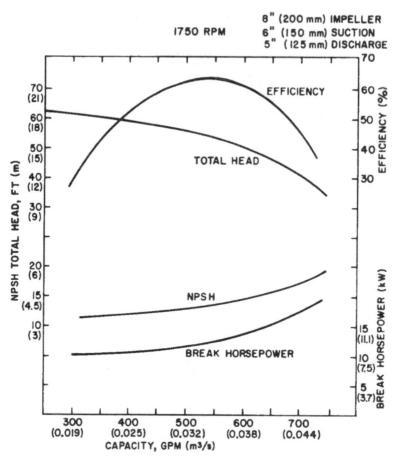

Formula for	Conventional Units	SI Units
Head	$H = psi \times 2.31/SG^* \ (ft)$	$H = kPa = 9.8023/SG^* \ (m)$
Output power	$P_o = Q_v \times H \times SG^*/3960 \ (hp)$	$P_o = Q_v \times H \times SG^*/367 \ (kW)$
Shaft power	$P_s = \dfrac{Q_v \times H \times SG^*}{39.6 \times E_p} \ (hp)$	$P_s = \dfrac{Q_v \times H \times SG^*}{3.67 \times E_p} \ (kW)$
Input power	$P_i = P_s \times 74.6/E_m \ (kW)$	$P_i = P_s \times 100/E_m \ (kW)$
Equipment efficiency, %	(Constant speed pumps)	$E_e = E_p \times E_m \times 10^{-2}$
	(Variable speed pumps)	$E_e = E_p = E_m \times E_v \times 10^{-4}$
Utilization efficiency, %	Q_D = design flow Q_A = actual flow H_D = design head H_A = actual head	$E_u = \dfrac{Q_D \times H_D}{Q_A \times H_A} \times 100$
System Efficiency Index		$SEI = E_e \times E_u \times 10^{-4}$

*SG = specific gravity.

Fig. 9.10 The operation of centrifugal pumps is described by common formulas (from ASHRAE [1]) and characteristic curves (from Lipták [6]).

Table 9.1

PUMP TERMS, ABBREVIATIONS, AND
CONVERSION FACTORS

Term	Abbreviation	Multiply	By	To Obtain
Length	l	ft	0.3048	m
Area	A	ft^2	0.0929	m^2
Velocity	v	ft/s	0.3048	m/s
Volume	V	ft^3	0.0283	m^3
Flow rate	Q$_v$	gpm	0.2272	m^3/h
		gpm	0.0631	L/s
Pressure	P	psi	6890	Pa
		psi	6.89	kPa
		psi	0.069	bar
Head (total)	H	ft	0.3048	m
NPSH	H	ft	0.3048	m
Output power (pump)	P$_o$	water hp (WHP)	0.7457	kW
Shaft power	P$_s$	BHP	0.7457	kW
Input power (driver)	P$_i$	kW	1.0	kW
Efficiency (%)				
Pump	E$_p$	—	—	—
Equipment	E$_e$	—	—	—
Electric motor	E$_m$	—	—	—
Utilization	E$_u$	—	—	—
Variable-speed drive	F$_v$	—	—	—
System efficiency index (decimal)	SEI	—	—	—
Speed	n	rpm	0.1047	rad/s
		rpm	0.1047	r/s
Density		lb/ft^3	16.0	kg/m^3
Temperature		F	—	°C

Source: From ASHRAE [1]

The typical range of pump efficiencies is from 60 to 85 percent.

The affinity laws describe the relationships among changes in speed, impeller diameter, and specific gravity. With a given impeller diameter and specific gravity, pump flow is linearly proportional to pump speed, pump discharge head relates to the square of pump speed, and pump power consumption is proportional to the cube of pump speed. This is why variable-speed pumps are so highly energy efficient.

When liquids are being pumped, it is important to keep the pressure at any point in the suction line above the vapor pressure of the fluid. The available head measured at the pump suction is called the *net positive suction head* (NPSH). A pump at sea level that is pumping 60°F water, that has a vapor pressure of $h_{vp} = 0.6$ ft, and that is operating under a barometric pressure of 33.9 ft has an available NPSH of 33.9 − 0.6 =

33.3 ft. As shown in figure 9.11, the available NPSH increases with barometric pressure and with static head, and it decreases as vapor pressure, friction or entrance losses rise. Figure 9.11 also illustrates the difference between *available* and *required* NPSH. Available NPSH is the characteristic of the process and represents the difference between the existing absolute suction head and the vapor pressure at the process temperature. The required NPSH, on the other hand, is a function of the pump design. It represents the minimum margin between suction head and vapor pressure at a particular capacity that is required for pump operation. If this minimum NPSH is not available, the pump will fail to generate the required suction lift and flow will stop.

Characteristic Curves

Figure 9.12 illustrates some typical characteristic curves of centrifugal pumps. The shape of the head-capacity curve is an important consideration in pump selection. Curve 1 is referred as a *drooping* curve, curve 2 is called a *flat* curve, and curve 3 would be considered *normal*. Curves 1 and 2 are satisfactory as long as the flow is above 100 GPM (6.3 l/s). Below this flow rate, curve 1 allows for two flows to correspond to the same head, and curve 2 can allow the flow to drop to zero as it attempts to raise the discharge head. Both are therefore unstable in this region. Curve 3 is stable for all flows and is best suited for throttling service in cases in which a wide range of flows is desired.

The system served by the pump is also represented by a head-capacity curve (figure 9.13). The head at any one capacity is the sum of the static and the friction heads. The static head does not vary with flow rate, as it is only a function of the elevation or back-pressure against which the pump is operating. The friction losses are related to the square of flow and represent the resistance to flow caused by pipe friction. A system curve tends to be flat when the piping is oversized and steep when the pipe headers are undersized. The friction losses also increase with age; therefore, the system curve for old piping tends to be steeper than for new piping.

Figure 9.14 illustrates the system curves of three different types of processes. Curve 1 corresponds to the closed-loop circulation of a fluid in a horizontal plane. Here there is no static head component at all, and the parabola that describes the system starts at zero. Curve 2 is the system curve for a condenser water circulation network. Here a limited amount of static head is present, because the pump must return the water to the top of the cooling tower. This curve also illustrates that the friction losses tend to increase when the piping is no longer new. Curve 3 gives an example of a process dominated by static head, such as the feedwater pump of a boiler. This curve is flat and is insensitive to changes in system flow.

As will be discussed later in more detail, when the system curve is flat, there is little advantage to variable or multiple-speed pumping, and the usual response to system flow variations is the stopping and starting of parallel pumps. Inversely, if the system curve is steep, substantial energy savings can be obtained from the use of booster, multiple, or variable-speed pumps.

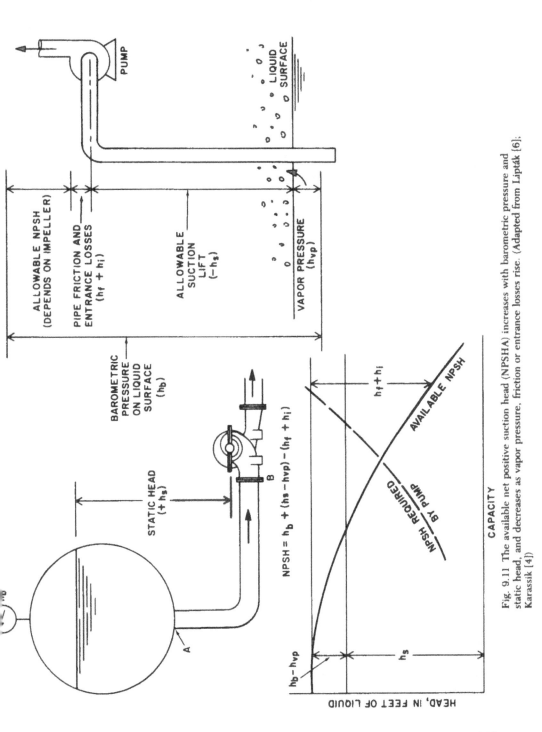

Fig. 9.11 The available net positive suction head (NPSHA) increases with barometric pressure and static head, and decreases as vapor pressure, friction or entrance losses rise. (Adapted from Lipták [6]; Karassik [4])

369

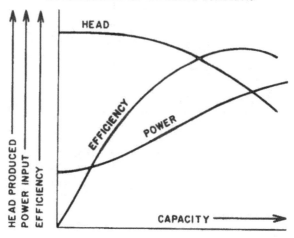

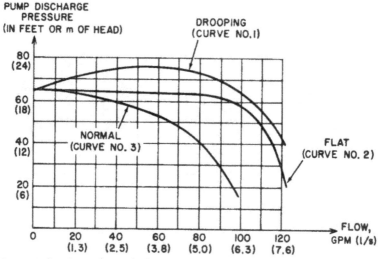

Fig. 9.12 The shape of centrifugal pump curves varies with their design (From Lipták [6])

Hydraulic systems can be open (noncirculating) or loop-type, as illustrated in figure 9.15. Water supply and distribution systems in cities and buildings are typical open systems, whereas hot-and-chilled-water heating and cooling systems are typical loop systems.

Hydraulic systems must also be evaluated as to whether flow is restricted or un-restricted. Restricted-flow systems are those that include valves to regulate the flow through the system. Plumbing systems or hot-and-chilled-water systems are restricted-

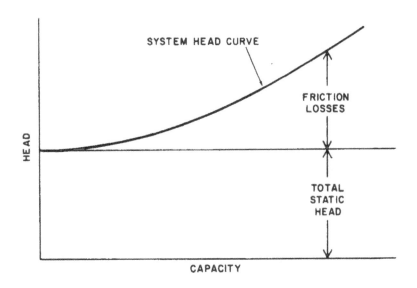

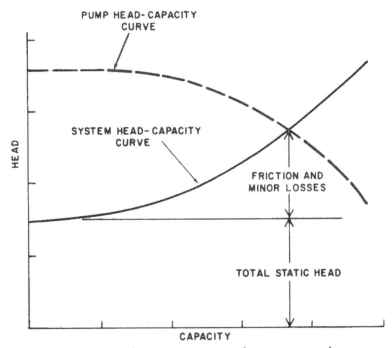

Fig. 9.13 The system head-capacity curve crosses the pump curve at the operating point.

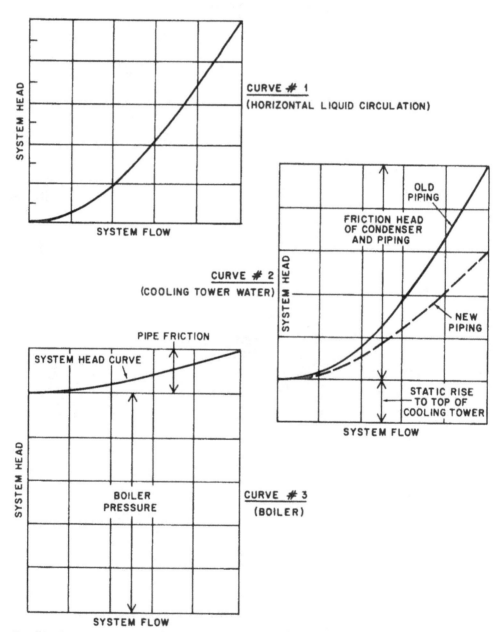

Fig. 9.14 System curves vary as the static-head components change in various processes (Adapted from ASHRAE [1])

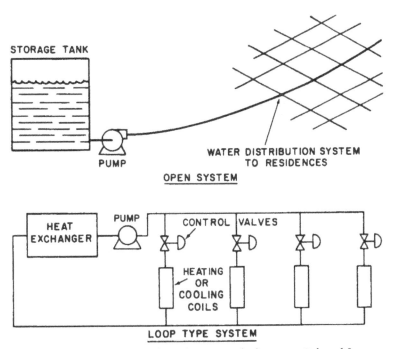

Fig. 9.15 Pumping systems can be of the open or the loop type. (Adapted from Lipták [7])

flow systems, because manual or automatic valves control the flow. Unrestricted-flow systems include sewage and storm water lift stations as well as municipal water flow into elevated storage tanks.

In actual systems, a single system head curve may not exist; often what exists is a system head band, as shown in figure 9.16. This is because the distribution of active loads will shift the curve, as this figure shows.

On-Off Control Interlocks

Figure 9.17 illustrates the interlocks used in pumping down a 1,000 gallon (3.78 m³) tank so as to keep the tank level between the settings of LSH and LSL. The two-probe conductivity level switch is operated by a relay. When conductive liquid reaches the LSH probe, current will flow through the relay, and it will pull in both of its contacts (H & I) as the relay is energized. At this point the pump starts, and although the level then drops below the upper probe, the pump keeps running because the holding contact (H) maintains the circuit. When the level drops below the lower probe, both the load (I) and the holding (H) contacts are opened, stopping the pump. When the level rises again, no action occurs when the lower electrode is contacted, because the holding

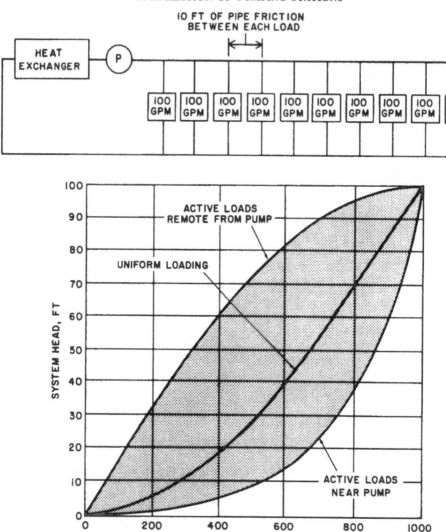

Fig. 9.16 The system head drops when active loads are near the pump and it rises when the most remote loads are active. (From Rishel [11])

contact to that probe is still open. However, when the level reaches the upper probe, electrical contact is established, and the relay closes to repeat the pumping cycle.

The bottom portion of figure 9.17 describes a simple pump starter. It is controlled by the three-position hand-off-automatic switch. When placed in automatic (contacts 3 and 4 connected), the status of the interlock contact I determines whether the circuit is energized and whether the pump is on. The purpose of the auxiliary motor contact

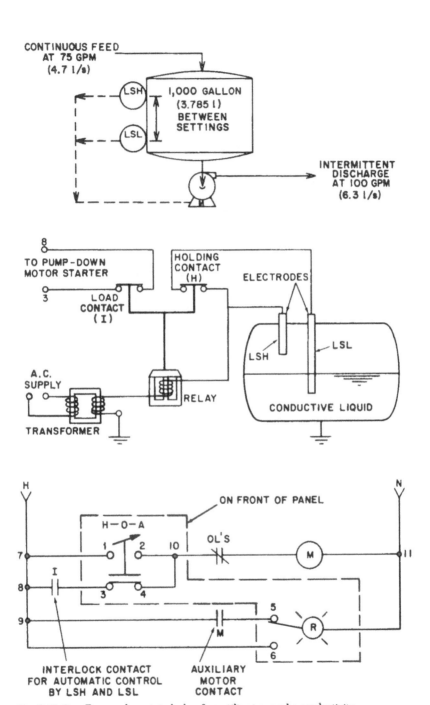

Fig. 9.17 On-off pump-down interlocks often utilize two-probe conductivity switches. (Adapted from Lipták [6])

375

M is to energize the running light R whenever the pump motor is on. The parallel hot lead to contact 6 allows the operator to check quickly to see if the light has burned out.

The amount of interlocking is usually greater than that shown in figure 9.17. Some of the additional interlock options are described in the paragraphs below.

Most pump controls include some safety overrides in addition to the overload (OL) contacts shown in figure 9.17. These overrides might stop the pump if excessive pressure or vibration is detected. They usually also provide a contact for remote alarming.

Most pump controls include a reset button that must be pressed after a safety shutdown condition is cleared before the pump can be restarted.

When a group of starters is supplied from a common feeder, it may be necessary to make sure that the starters are not overloaded by the inrush currents of simultaneously started units. If this feature is desired, a 25-second time delay is usually provided to prevent other pumps from starting until that time has passed.

When two-speed pumps are used, added interlocks are frequently provided. One interlock might guarantee that even if the operator starts the pump in high, it will operate for 0 to 30 seconds in low before advancing automatically to high. This makes the transition from off to high speed more gradual. Another interlock might guarantee that when the pump is switched from high to low, it will be off the high speed for 0 to 30 seconds before the low-speed drive is engaged. This will give time for the pump to slow before the low gears are engaged.

If the same pumps can be controlled from several locations, interlocks are needed to resolve the potential problems with conflicting requests. One method is to provide an interlock to determine the location that is in control. This can be a simple local/remote switch or a more complicated system. When many locations are involved, conflicts are resolved by interlocks that either establish priorities between control locations or select the lowest, highest, safest, or other specified status from the control requests. In such installations with multiple control centers, it is essential that feedback be provided so that the operators are always aware not only of the actual status of all pumps but also of any conflicting requests coming from the other operators or computers.

Large pumps should also be protected from overheating caused by frequent starting and stopping. This protection can be provided by a 0-to-30-minute time delay guaranteeing that the pump will not be cycled at a frequency faster than the delay time setting.

In figure 9.17, the number of starts per hour is limited not by a timer but by the sizing of the equipment. At a feed rate of 75 GPM (4.7 l/s), it takes 13 minutes to fill the 1,000 gallon (3,785 l) volume between LSL abd LSH when the discharge pump is off. The pump capacity being 100 GPM (6.3 l/s), it takes 40 minutes to discharge this same volume while the feed is on. Therefore, the pump will start approximately once an hour and will run 75 percent of the time.

Consequently, the 1,000 gallon (3,785 l) tank is satisfactory. Starting the pump only once an hour will not overheat the motor. In this type of application, the capacity of the tank is the independent variable, because the pump capacity is usually selected to be about 25 to 30 percent greater than the flow into the tank.

Throttling Control

If there is no throttling valve in the system, a constant-speed pump will always operate at the intersection of the pump and the system curve (point 1 in figure 9.18). If the actual system curve turns out to have a lesser slope than what was assumed for the system curve in the design phase, the pump will end up operating at point 2 instead of at point 1. In such a free-flowing system, the actual head will be lower and the actual flow will be higher than was designed.

If the flow is controlled by throttling a control valve on the pump discharge, the pump will operate at point 1, and the differential head between point 1 and 3 will be burned up in the form of pressure drop. If the system flow is reduced to 50 percent,

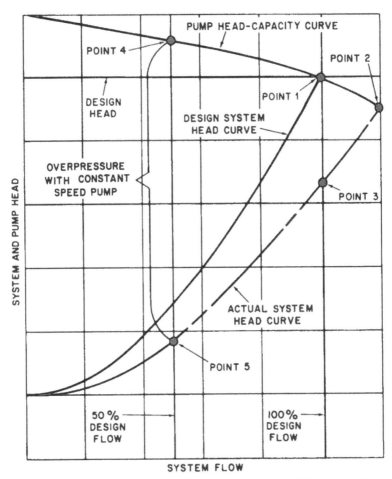

Fig. 9.18 Pump operating points are found at the intersection of the pump and the system curve (From ASHRAE [1])

the energy wasted in the form of valve pressure drop will increase to the differential between points 4 and 5.

In figure 9.19, the useful pumping head is identified as Hs, and the actual pumping head of the throttled system is given as Hp. The Hp-Hs difference identifies the energy wasted through throttling. This is only part of the total loss of efficiency. As can also be seen in figure 9.19, moving the operating point from 2 to 1 also reduces the pump efficiency from 81 percent to 70 percent.

Control Valves Waste Energy

To calculate the energy cost of throttling control, the following rule of thumb can be used: it costs $0.5/year for a flow of 1.0 GPM to overcome a differential of 1.0 PSID.

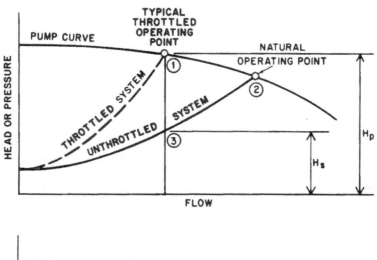

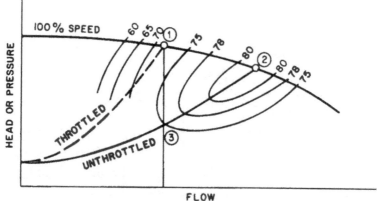

Fig. 9.19 The throttled system not only wastes pumping energy through valve pressure drop but also operates at a less efficient point on the pump curve. (Adapted from Reliance Electric [3])

The example illustrated in figure 9.20 calculates the yearly waste of pumping energy at different loads. If the FRC is set for 625 GPM (set point 2), the pump operates at point 2 on its curve and the kickback PCV is closed, because the pump discharge head at point 2 is less than the PCV set point of 90 PSIG. Therefore, the cost of throttling between points 2 and 3 can be calculated as follows:

$$\text{(FRC Set point)}(\Delta P_{2-3})(0.5 \text{ \$/YR})$$
$$= (625)(40)(0.5) = \$12{,}500/\text{yr}.$$

If, in this same example, the FRC set point is reduced to 250 GPM, the pump is throttled back on its curve to point 4. Because the pump head at point 4 is higher than the set point of the PCV, it will open and bypass enough flow to reduce the pump head to 90 PSIG. Therefore, in this case, the pump will operate at point 1. As the flow at point 1 is below that at point 2, it is fair to assume that the energy waste will be greater under these conditions. To calculate the amount of waste, the energies burned up in both the PCV and the FCV must be added as shown below:

$$\text{(FRC Set point)}(\Delta P_{1-5})(0.5 \text{ \$/YR}) + \text{(PCV Flow)}(\Delta P_{0-1})(0.5 \text{ \$/YR})$$
$$= (250)(73)(0.5) + (300)(90)(0.5) = 9{,}125 + 13{,}500 = \$22{,}625/\text{Yr}.$$

For good controllability, the control valve is usually sized to pass the design flow with a pressure drop equal to the system friction losses excluding the control valve but not less than 10 PSID (69 kPa) minimum. Pump flow is controlled by varying the pressure drop across the valve. However, the pump must not be run at zero flow or overheating will occur and the fluid will vaporize, causing the pump to cavitate. To avoid this, a bypass line can be provided with a back-pressure regulator. This regulator is set at a pressure that will guarantee minimum flow as the pump is throttled toward zero flow.

The minimum flow required to prevent overheating is calculated by assuming that the motor horsepower is converted to heat. One rule of thumb calls for a flow of 1.0 GPM for each pump horsepower to prevent overheating. If a minimum flow bypass cannot be provided, then a safety interlock should be furnished. In this interlock, a minimum flow switch is used to stop the pump on low flow.

Pump Speed Throttling

Flow control via speed control is less common than throttling with valves, because most AC electric motors are constant-speed devices. If a turbine is considered, speed control is more convenient. An instrument air signal to the governor can control speed to within ±.5 percent of the set point. In order to vary pump speeds with electric motors, it is generally necessary to use a variable-speed device in the power transmission train. This might consist of variable pulleys, gears, magnetic clutch, or hydraulic coupling. Variation of the pump speed generates a family of head-capacity curves, as shown in figure 9.21, in which the volume flow is proportional to speed if the impeller diameter is constant. The intersection of the system curve with the head curve determines the flow rate at points 1, 2, or 3.

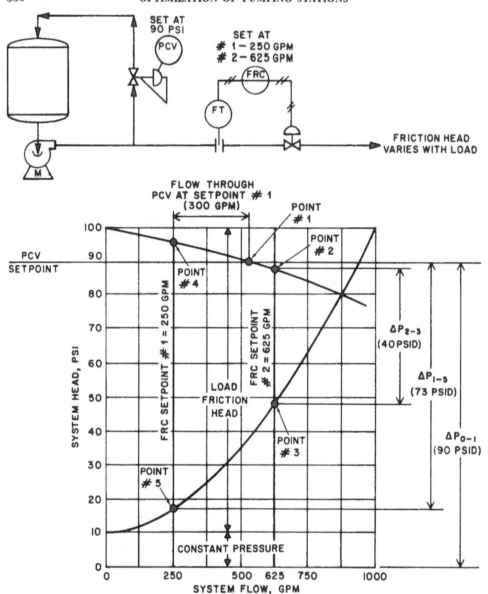

Fig. 9.20 This example calculates the energy waste due to pump discharge throttling.

Because the area of peak pump efficiency falls on a parabolic path (figure 9.22), speed throttling will usually not reduce the pump efficiency as much as valve throttling (figure 9.19). This increases the total energy saving obtained from pump speed control. As shown in figure 9.22, when the flow is reduced from F_1 to F_2, instead of wasting the excess pump head of $(P_1 - P_2)$ in pressure drop through a valve, that pump head

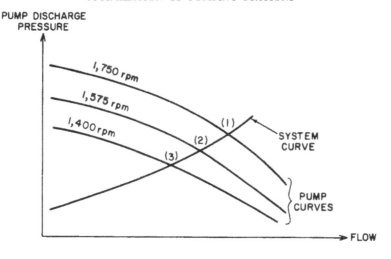

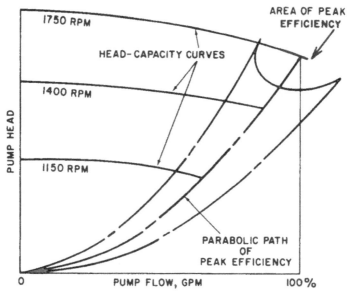

Fig. 9.21 Variable-speed pump operation can be described by a family of head-capacity curves. (From Lipták [6])

is not introduced in the first place. Thereby, speed throttling saves energy that valve throttling would have wasted.

When To Consider Variable-Speed Pumps

The shape of the system curve determines the saving potentials of variable-speed pumps. All system head curves are parabolas ($H \sim Q^2$), but they differ in the steepness

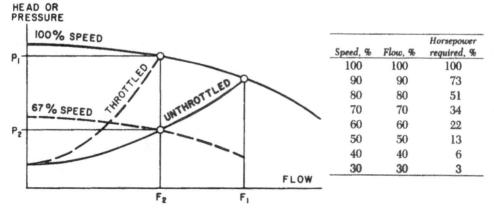

Fig. 9.22 Instead of wasting the unnecessarily introduced pump energy, speed is reduced so that such energy is not introduced in the first place. (Adapted from Lipták [8])

of these curves and in the ratio of static head to friction drop. As shown in figure 9.23, the value of variable-speed pumping increases as the system head curve becomes steeper.

Studies indicate that in *mostly friction* systems (such as zone 4 in figure 9.23), the savings represented by variable-speed pumping will increase with reduced pump loading [5]. If, on the yearly average, the pumping system operates at not more than 80 percent of design capacity, the installation of variable-speed pumps will result in a payback period of approximately three years.

The zones in figure 9.23 are defined by the H_t/H_s ratio. The higher this number, the higher the zone number and the more justifiable is the use of variable-speed pumps. Figure 9.24 illustrates how the H_t/H_s ratio is calculated. The shaded areas identify the energy-saving potentials of variable-speed pumps. The values of H_t and H_s are identified by determining the average yearly flow rate (F_a) and determining its intersections with the pump and system curves. The larger the shaded area in figure 9.24, the higher the H_t/H_s ratio will be and, therefore, the shorter the payback period for the use of variable speed pumps is likely to be.

Calculating the Savings

As was shown in figure 9.20, the savings resulting from variable-speed pumping can be calculated at any operating point on the pump curve. Based on this information, the relationship between the demand for flow and the power input required can be plotted. Figure 9.25 shows these curves for both control valve throttling and variable-speed pumping systems. The difference between the two is the savings potential.

Once the saving curve shown in figure 9.25 has been established, the next step is to determine the operating cycle. The operating cycle identifies the percentages of time when the load is 10 percent, 20 percent, etc., up to 100 percent (figure 9.26). When

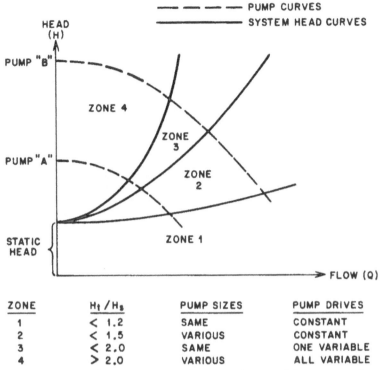

Fig. 9.23 Pumps and drives should be selected as a function of the steepness of the system curve. (Adapted from Lipták [6])

both the savings and the operating curves are established, all that needs to be done is to incrementally calculate the savings as follows:

% Flow	% of Time	% Savings	Total Savings(%)
Below 25	2.5	57	1.4
25–35	10	60	6.0
35–45	20	56	11.2
45–55	23	52	11.9
55–65	23	50	11.5
65–75	14	40	5.6
75–85	5.5	30	1.6
85–95	1.5	15	0.2
95–100	0.5	0	0
	100%		49.4%

Knowing the total horsepower of the pumps and the cost of electricity makes it possible to convert the resulting percentages into yearly costs. For example, if the cost of electricity is 6 cents per KWH, then the yearly saving of each 100-horsepower increment has the value of $39,000.

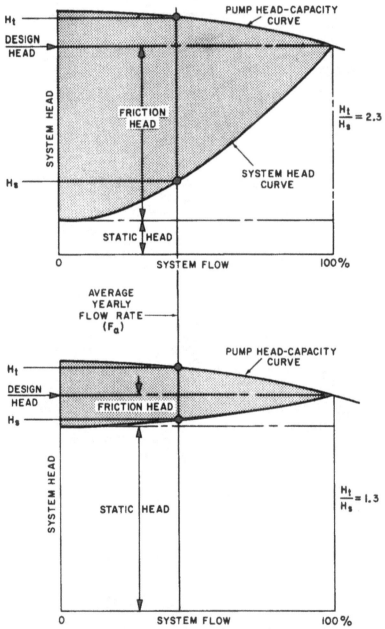

Fig. 9.24 The higher the Ht/Hs ratio, the more justifiable is the use of variable-speed pumps. (Adapted from Lipták [7])

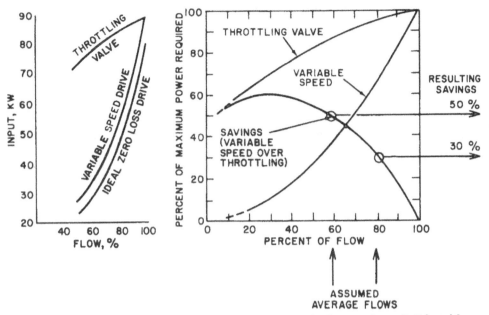

Fig. 9.25 The savings due to variable-speed pumping increase as the flow rate drops off. (Adapted from Lipták [6]; Lin [9])

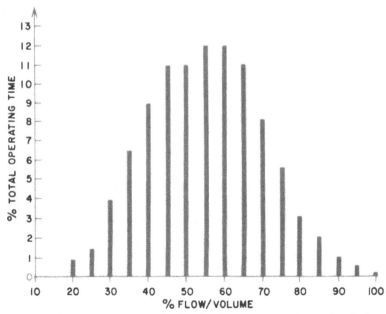

Fig. 9.26 The pump operating cycle identifies the percentages of time when load is 10 percent, 20 percent, etc.

Variable-Speed Drive Efficiencies

The overall system efficiency index (SEI) of a variable-speed pump installation is determined as follows:

$$SEI = (Ep \times Em \times Ev \times Eu)10^{-6}$$

where

Ep = the pump efficiency
Em = the motor efficiency
Ev = the variable-speed drive efficiency
Eu = the efficiency of utilization

Figure 9.27 gives some typical efficiency values for Ep, Em, and Ev. The efficiency of the variable-speed drives is represented as a range, because each design has a different efficiency. When compared to valve throttling, the energy savings are less with electromechanical or slip control drives than with solid state electrical drives (figure 9.28). At 50 percent of rated speed, the variable-speed drive efficiency can be as low as 40 percent or as high as 70 percent, depending on the design selected (figure 9.29). Naturally, the less efficient variable-speed drives are also the less expensive ones. For a 100 HP motor, the cost of a variable-speed drive can range from less than $10,000 to more than $25,000 as a function of efficiency.

The efficiency of utilization (Eu) is an indicator of the quality of the specific water distribution system design. It reflects both on safety factors and on bad design practices. For example, in the piping distribution system illustrated in figure 9.30, Qr might represent the required water flow and Hr the pressure head at the pump required to transport Qr. Because three-way valves are used in this figure, the actual flow (Qa) is much higher than the required flow, and the head required to transport this actual flow (Ha) is also greater than the required head. Therefore, the efficiency of utilization is defined as follows:

$$Eu = \frac{Qr\,Hr}{Qa\,Ha}\,100$$

The numerical value of Eu can be accurately obtained only by testing.

Once all four efficiencies are determined, they can be represented by a single system efficiency index (SEI) curve (figure 9.31). By combining different pumps, motors, drives, and system designs, it is possible to arrive at a number of SEI curves. The relative advantages of different devices and designs can be evaluated quantitatively by comparing these curves.

Types of Variable-Speed Drives

All variable-speed drives should be reliable and serviceable and should provide repeatable control and wide turndown at all loads. Variable-speed pumps must operate at least within the speed range of 50 percent to 100 percent, but the drive must also be able to function at speeds down to zero without damage or overheating.

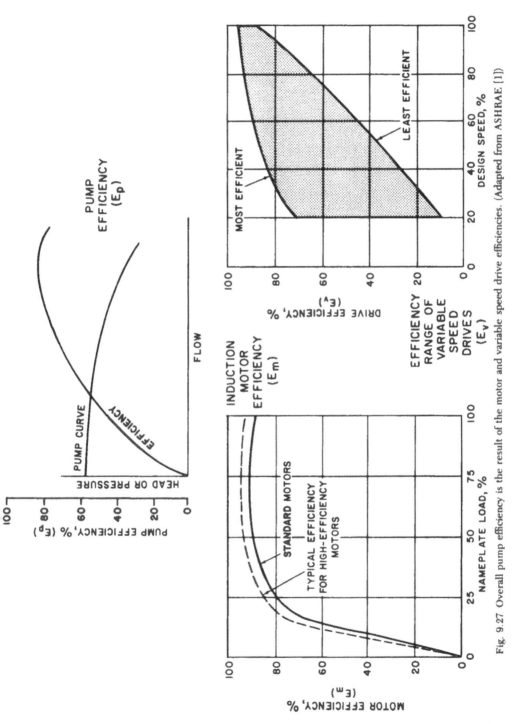

Fig. 9.27 Overall pump efficiency is the result of the motor and variable speed drive efficiencies. (Adapted from ASHRAE [1])

387

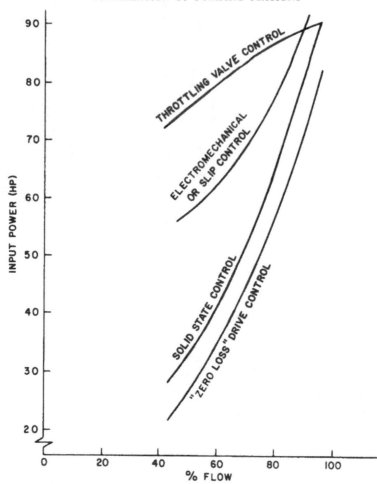

Fig. 9.28 The energy savings vary with the type of variable-speed drive used.
(Adapted from Reliance Electric [3])

Variable-speed pump drives are either electrical or electromechanical. The electrical ones can be of a number of types: two-speed, direct current, variable voltage, variable frequency (current source or pulse width modulated) and wound rotor regenerative. The common feature of all of these designs is that they depend on altering the characteristics of the electricity to the motor in order to change its speed. The electromechanical drives, on the other hand, utilize a speed-changing device that is interposed between the motor and the pump. These designs include eddy-current couplings, hydraulic or hydroviscous drives, and V-belt drives. The electromechanical drives are generally less efficient than their electrical counterparts (figure 9.29). The variable

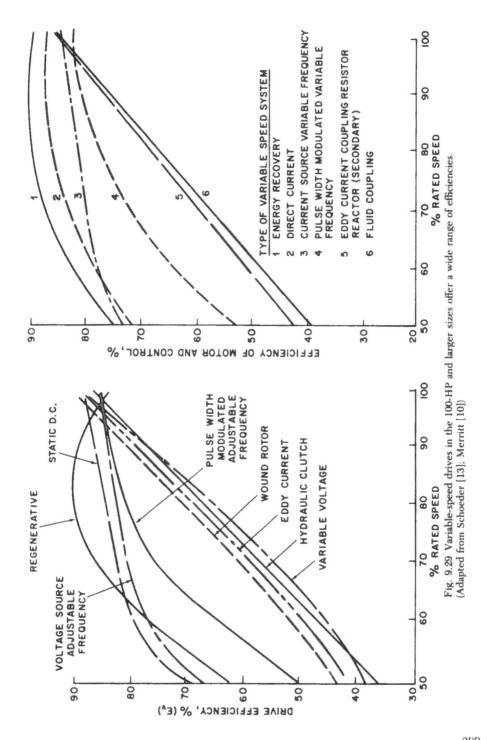

Fig. 9.29 Variable-speed drives in the 100-HP and larger sizes offer a wide range of efficiencies. (Adapted from Schoeder [13]; Merritt [10])

389

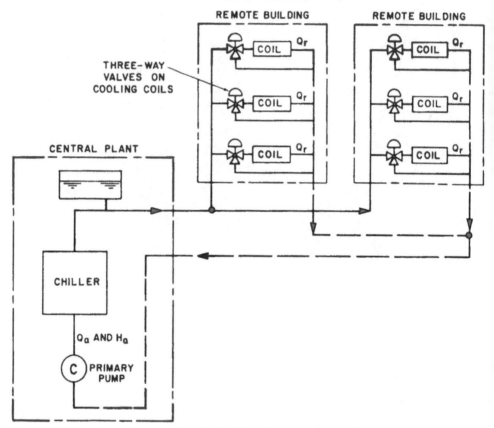

Fig. 9.30 The flow that is bypassing the coils should not have been circulated in the first place. (Adapted from Rishel [12])

voltage electric drive using high slip motors is an exception to this rule, as it is less efficient than the mechanical drives. Table 9.2 provides an overview of the various drive types.

The *variable frequency drives* are among the most reliable and efficient units, having turndown ratios of up to 3:1. They convert alternating current at 60 Hz to direct current and back to alternating at 0 to 120 Hz. As the frequency sent to the standard squirrel-cage type induction motor changes—say from 60 Hz to 40 Hz—the pump speed is correspondingly reduced from 100 percent to 60 percent. These drives are relatively expensive and sophisticated devices. They are normally available in sizes up to 250HP (185KW), with specially designed units being available in sizes up to 2000 HP (1491KW).

The *wound rotor regenerative drives* give one of the best efficiencies, although their rangeability is limited to 2:1. In the conventional wound rotor design, the motor

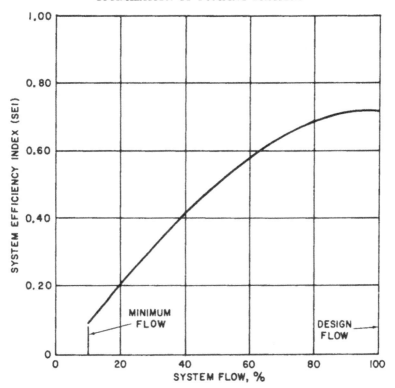

Fig. 9.31 Overall system efficiency index (SEI) curves describe the total pumping efficiency as a function of load. (From ASHRAE [1])

speed is altered by varying the resistance in the rotor. This is rather inefficient. In the regenerative design, the resistors are eliminated, and through the use of rectifiers and inverters the excess power is returned to the supply. Sizes vary from 25 to 10,000 HP (20 to 7,455 KW). The wound rotor type motor is more expensive than the induction type and the brushes used in the regenerative design require replacement every three to four years.

Direct current drives were among the first variable-speed drives using wound rotors. In combination with silicon-controlled rectifiers (SCRs), they provide stepless speed control at high efficiency and modest first cost. As the armature rotates at right angles to the magnetic field established by the field coils, rotation can be reversed easily by reversing polarity. An advantage of these drives is the ability to maintain constant speed while the load varies. Disadvantages include the limited availability of DC pump motors, the difficulty in preventing explosions resulting from sparking, and the approximately yearly replacement requirement of the carbon brushes through which the power is applied. Units are available in sizes of 1 to 500 HP (0.75 to 373 KW).

In *variable voltage drives*, a specially designed squirrel-cage, high-slip induction

Table 9.2

VARIABLE-SPEED DRIVE COMPARISON

Drive Type	Efficiency (at 70% Speed)	Turndown	Sizes (HP)	Component Requiring Replacement (Frequency in Yrs)
Wound rotor regenerative	High (85%)	2:1*	25 to 500+	Brush (3–4)
Direct current	High (80%)	Unlimited	1 to 500+	Brush (1–2)
Variable frequency	High (78%)	3:1	20 to 500+	—
Wound rotor	Med. (60%)	2:1*	25 to 500+	Brush (3–4)
Eddy current clutch	Med. (58%)	5:1	20 to 500+	—
Fluid coupling	Med. (57%)	3:1	20 to 500+	—
Variable voltage	Low (52%)	Limited	10 to 100+	—
Mechanical	Low (50%)	6:1	1 to 100	Belt or chain (1–3)

*Unstable below 50%

motor is used to reduce speed as the primary voltage is reduced. This is one of the least desirable drive designs because of poor efficiency, limited turn-down, and high rates of heat generation. This design is not recommended for centrifugal pumps.

Two-speed motors can be utilized on simple pumping systems in which accurate control of pump pressure is unnecessary. Such motors should not be used in systems with high static head and low friction, such as curve 3 in figure 9.14. On the other hand, for mostly friction systems, such as curve 2 in figure 9.14, they can offer a reasonably efficient and inexpensive alternative to variable-speed pumping. Standard two-speed motors are available with speeds of 1750/1150 rpm (29 and 19 r/s), 1750/850 rpm (29 and 14 r/s), 1150/850 rpm (19 and 14 r/s), and 3500/1750 rpm (58 and 29 r/s) (figure 9.32).

In an *Eddy current coupling*, the motor rotates a drum or rotating ring. Inside, but free from the drum, is an electromagnetic pole-type rotor assembly that is connected to the pump. The speed-control system regulates the amount of flux that exists between the drum and rotor assemblies. The amount of slip or speed difference between motor and pump increases and decreases with the flux density in the coupling. Being a slip-type, variable-speed device, the eddy current coupling does not have as high an efficiency as the variable frequency, direct current, and wound rotor regenerative types of variable-speed drives. Otherwise, however, it is a good design. It is rugged and used widely, particularly in large turbine pump applications.

Fluid couplings consist of two coupling halves, one driven by a standard electrical motor and the other connected to the centrifugal pump. The pump shaft speed increases as the amount of oil supplied to the coupling is increased, thus producing a simple control system for varying pump speed. The fluid coupling has been one of the most popular variable-speed devices for centrifugal pumps because of its relatively low initial cost, high reliability, and ease of maintenance. This design is particularly useful for installations in which efficiency is of no serious concern.

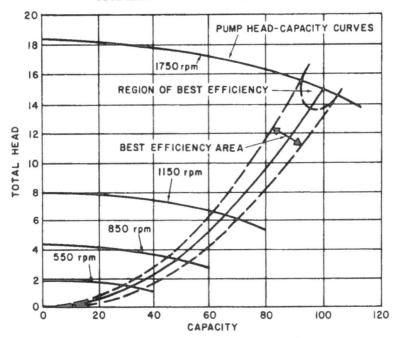

Fig. 9.32 Several speed combinations are available in two-speed pumps. (From ASHRAE [1])

Mechanical drives have been used only on pumps smaller than 100 HP (75 KW). They use rubber belts or metal chains to adjust the speed within a range of 6:1. Because of their relatively low reliability and efficiency, they are not very popular.

Multiple Pump Stations

When two or more pumps operate in parallel, the combined head-capacity curve is obtained by adding up the capacities at the same discharge head, as illustrated in figure 9.33. The total capacity of the pump station is found at the intersection of the combined head-capacity curve with the system head curve. This point also gives the head at which each of the pumps is operating. If the selection is to be very accurate, the head-capacity curves should be modified by substituting the station losses (the friction losses at the suction and discharge of the individual pumps) so that the resulting "modified head" curve will represent the pump plus its valving and fittings.

When constant-speed pumps are used in parallel, the second pump can be started and stopped automatically on the basis of flow. In this case, an adjustable dead band is provided in the flow switch (FSH in figure 9.34); this dead band starts the second pump when the flow rises to 120 GPM (7.6 l/s) but will not stop it until it drops to 100 GPM (6.3 l/s). This prevents excessive cycling. The addition of an alternator can equalize

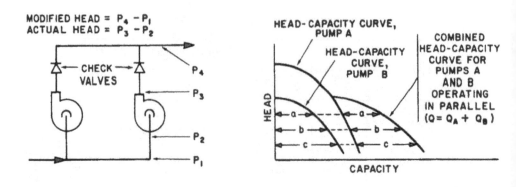

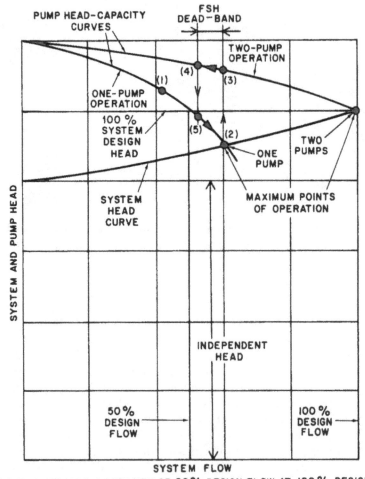

Fig. 9.33 Multiple pumps in parallel can eliminate overpressure at low flows. (Adapted from ASHRAE [1]; Tchobanoglous [16])

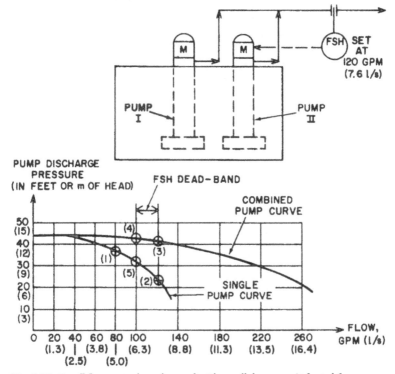

Fig. 9.34 On-off flow control can be used with parallel pumps. (Adapted from Lipták [6])

the running times of the two pumps. In figure 9.34, the normal operation of the system is represented by point 1. If the load rises, the single pump meets it until point 2 is reached. At this load, the second pump is started and the system operates at point 3. When the load drops off, it will pass through point 3 without any effect and the second pump is not stopped until the load drops to point 4. At this load, the second pump is turned off and the system operation is returned to point 5 using only one pump.

Booster Pumps

When two or more pumps operate in series the total head-capacity curve is obtained by summing up the heads at each capacity. When a booster pump is added to a main fed by several *parallel* pumps, the total head-capacity curve is obtained by adding the booster curve to the modified head of the parallel pumps at each capacity point. Series pumping is most effective when the system head curve is steep, such as in figure 9.35. With such mostly-friction loads, series pumping can substantially reduce the over-pressure at low loads. Therefore, booster pumps or two-speed pumps (figure 9.32) can both be considered for the same kind of system curve. Multiple pumps in series are

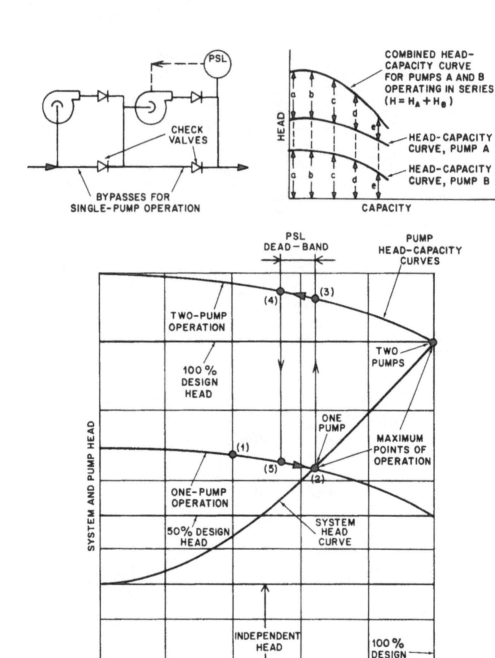

TWO PUMPS, EACH WITH A CAPACITY OF 100% DESIGN FLOW AT 50% DESIGN HEAD

Fig. 9.35 Multiple pumps in series are effective when the system head curve is steep. The two pumps illustrated by the lower graph are each capable of generating 100-percent design flow at 50 percent design head. (Adapted from ASHRAE [1]; Tchobanoglous [16])

preferred from an operating cost point of view, but a single two-speed pump represents a lower capital investment.

When constant-speed pumps are used, the booster pump can be started and stopped automatically on the basis of pressure. In this case, an adjustable dead band is provided in the pressure switch. The normal system operation can be represented by point 1 in figure 9.35. As the load increases, the pump discharge head drops to the set point of PSL at point 2 and the booster pump is automatically started. Now the system operates to the right of point 3 until the load drops off again. The booster stays on as the load drops below point 3 until the PSL turns it off at point 4. At this point, the system is returned automatically to the single pump operation at point 5. The dead band in the PSL prevents the on-off cycling of the booster pump at any particular load. The width of the dead band is a compromise: as the band is narrowed, the probability of cycling increases, and widening the band results in extending the periods when the booster is operated unnecessarily. If the pumps are identical, their running times can be equalized by alternating between them, so that the pump with the higher running time will be the one that is stopped.

Water Distribution System Design

The overall efficiency of the water distribution system shown in figure 9.30 is shown in figure 9.36 by curve 1. A substantial increase in efficiency (reduction in operating cost) can be obtained by replacing the three-way valves with two-way ones and by replacing the single large pump with several smaller ones. Figure 9.37 shows such a system. Here a small and a large primary pump are provided in the central plant and a small and a large booster pump are furnished in each of the user buildings. When the load is low, the small pumps are operating; when it is high, the large pumps take their place. The minimum flow requirements of the chiller are guaranteed by a bypass valve, and the chilled water make-up into the recirculating loop of each building is under temperature control (TC). The resulting improvement in overall efficiency is shown by curve 2 in Figure 9.36.

The highest overall efficiency can be obtained through the use of variable-volume load-following. Figure 9.38 illustrates such a system, utilizing variable-speed pumps in two sizes. In this system, all waste is eliminated except the small minimum flow bypass around the chiller, which is guaranteed by a small constant-speed pump. The resulting increase in overall efficiency is illustrated by curve 3 in Figure 9.36.

Water Hammer

If a valve opening is suddenly reduced in a moving water column, this causes a pressure wave to travel in the opposite direction to the water flow. When this pressure wave reaches a solid surface (elbow, tee, etc.), it is reflected and travels back to the valve. If, in the meantime, the valve has closed, a series of shocks, sounding like hammer blows, results. An example can illustrate this phemomenon.

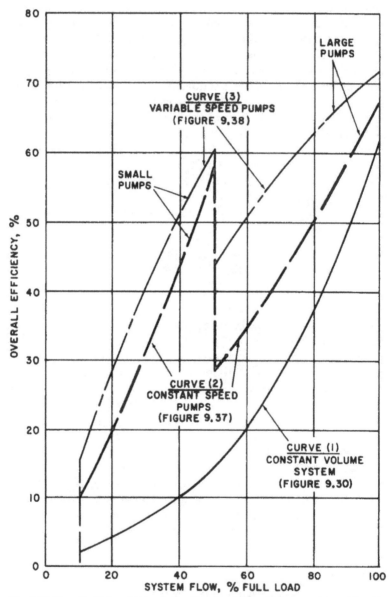

Fig. 9.36 Overall efficiency is maximum if two variable speed pumps are used and is minimum with a constant volume installation. (Adapted from Rishel [12])

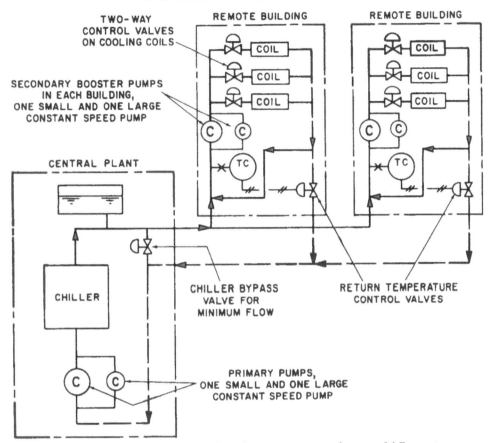

TWO-WAY
CONTROL VALVES
ON COOLING COILS

REMOTE BUILDING

REMOTE BUILDING

SECONDARY BOOSTER PUMPS
IN EACH BUILDING,
ONE SMALL AND ONE LARGE
CONSTANT SPEED PUMP

CENTRAL PLANT

CHILLER BYPASS
VALVE FOR
MINIMUM FLOW

CHILLER

RETURN TEMPERATURE
CONTROL VALVES

PRIMARY PUMPS,
ONE SMALL AND ONE LARGE
CONSTANT SPEED PUMP

Fig. 9.37 Supply-demand matching can be achieved using constant-speed pumps of different sizes.
(Adapted from Rishel [12])

Assume that 60°F (20°C) water is flowing at a velocity of 10 ft/s in a 3 inch Schedule 40 pipe, and a valve located 200 ft downstream is suddenly closed. The pressure rise and the minimum acceptable time for valve closure can be calculated. If the valve closes faster than this limit, water hammer will result.

In rigid pipe, the pressure rise (ΔP) is the product of water density (ρ in units of slugs/ft^3), the velocity of sound (c in units of ft/s), and the change in water velocity (ΔV in units of ft/s) [2]. Therefore, the pressure rise can be calculated as follows:

$$\Delta P = -\rho c\, \Delta V = (1.937)(4,860)(-10)$$
$$= 94,138 \text{ lbf/ft}^2 = 653.8 \text{ PSI}$$

In order to prevent water hammer, the valve closure time (t) must exceed the ratio of two pipe lengths (2L) divided by the speed of sound:

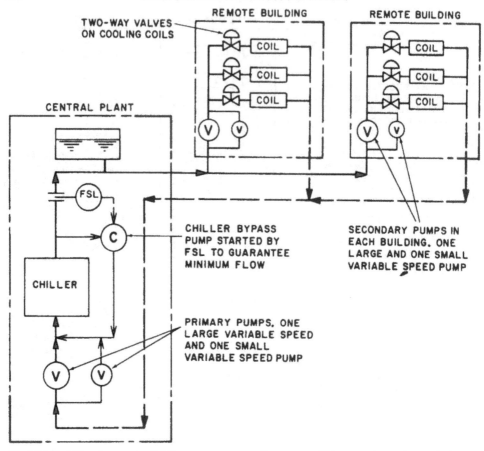

Fig. 9.38 Variable-volume water distribution systems provide maximum efficiency. (Adapted from Rishel [12])

$$t = 2L/c = (2)(200)/4,860 = 0.0823 \text{ seconds}$$

Therefore, in this example, the valve closure time must exceed 0.1 second.

The possible methods of preventing water hammer include [15]:

- Designing the system with low velocities
- Using valves with slow closure rates
- Providing slow-closing bypasses around fast-closing valves, such as check valves

When water hammer is already present and the cause of it cannot be corrected, its symptoms can be treated as follows:

- By adding air chambers, accumulators, or surge tanks
- By using surge suppressors, such as positively controlled relief valves

• When water flows are split or combined, by using vacuum breakers to admit air and thereby cushion the shock resulting from the sudden opening or closing of the second split stream

Optimized Pump Controls

A pumping system is optimized when it meets the process demand for liquid transportation at minimum pumping cost and in a safe and stable manner. Once the equipment is installed, the potential for optimization is limited by the capabilities of the selected equipment and piping configuration. As an example, figure 9.39 illustrates the optimization of a pump station consisting of a variable-speed and a constant-speed pump, which is the correct equipment selection if the system curve falls in zone 3 in figure 9.23.

Figure 9.39 illustrates the instrumentation requirements of an optimized, variable-speed pumping system. PDIC-01 maintains a minimum of 10 PSID (69 kPa) pressure difference between supply and return from each user. Therefore, no user will be denied more fluid if its control valve opens further. On the other hand, if demand drops, PDIC-

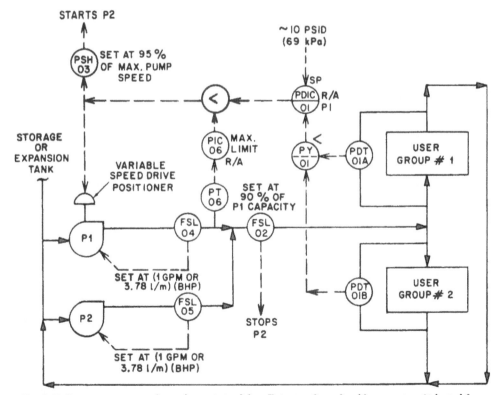

Fig. 9.39 Pump station controls can be optimized for efficient, safe, and stable operation. (Adapted from Lipták [6])

01 will slow down the variable-speed pump to keep the differential at the users from rising beyond 10 PSID (69 kPa).

When the variable-speed pump approaches its maximum speed, PSH-03 will automatically start pump 2. When the load drops down to the set point of FSL-02, the second pump is stopped. One important feature to remember in connection with multiple-pump station controls is that extra increments of pumps are started by pressure but stopped by flow controls.

FSL-04 and 05 are safety devices that will protect the pumps from overheating or cavitation, which might occur if the pump capacity is very low. As long as the pump flow rate is greater than 1 GPM (3.78 l/m) for each pump break horsepower, the operation is safe. If the flow rate drops below this limit, the pumps are stopped by the low flow switches.

Variable-speed pumping will never completely replace the use of control valves. However, its use will increase, because variable-speed pumps do represent a more energy-efficient method of fluid transportation and distribution than did the constant-

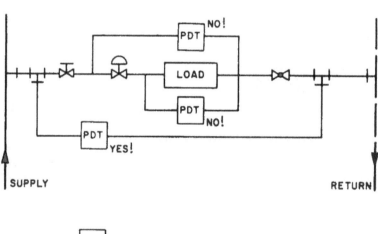

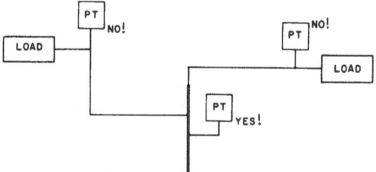

Fig. 9.40 The pressure and differential pressure transmitters must be correctly located in the pipe distribution network.

speed pumps used in combination with control valves. Therefore, if the fluid is being distributed in a mostly-friction system and the yearly average loading is under 80 percent of pump capacity, the installation of variable-speed pumps will reduce operating costs.

As is shown in figure 9.39, the pump speed is set to keep the lowest user pressure drop above some minimum limit; thus, it is important that the pressure drop across the users (PDT) be accurately detected. Figure 9.40 illustrates both the correct and the incorrect location of these transmitters. The differential pressure transmitter should measure the head loss not only of the load but also of the control valve, hand valves, and piping.

Similarly, if the header pressure is to be controlled (or limited, as in figure 9.39), the pressure transmitters should be located on main raisers or headers. They should not be located near major on-off loads, as shown in figure 9.40.

Valve Position Optimization

One of the best methods of finding the optimum pump discharge pressure is illustrated in figure 9.41. The optimum pressure is that which will keep the most-open user valve at a 90-percent opening. As the pressure rises, all user valves close; as it drops, they open. Therefore, opening the most-open valve to 90 percent causes all others to be opened also. This keeps the pump discharge pressure and the use of pumping energy at a minimum.

As the valve position controller (VPC-02) opens the valves, it not only minimizes the valve pressure drops but also reduces valve cycling and maintenance. This is because cycling is more likely to occur when the valve is nearly closed, and maintenance is high when the pressure drop is high. The other important advantage of this control system is that no user can ever run out of coolant, as no valve is ever allowed to be 100 percent open. This increases plant safety.

In order to make sure that the pressure controller (PC-01) set point is changed slowly and in a stable manner, the valve position controller (VPC-02) is provided with integral action only, and its integral time is set to be about ten times that of PC-01. In order to prevent reset wind-up when the PC-01 is switched from cascade to automatic, the valve position controller is also provided with an external feedback.

The pump station in figure 9.41 consists of two variable-speed pumps. Their speed is set by PC-01. When only one pump is in operation and the PC-01 output reaches 100 percent, PSH-03 is actuated and the second pump is started, as shown in the interlock table in figure 9.41. When both pumps are in operation and the flow drops to 90 percent of the capacity of one pump, the second pump is stopped if TD-04 has timed out. The purpose of the time delay (TD-04) is to make sure that the pump is not started and stopped too often.

The cycling of the second pump is also illustrated on the pump curves on the top of figure 9.41. As the demand for water rises, the speed of the one operating pump increases until 100 percent is reached at point A. Here PSH-3 starts the second pump. However, if the speed-control signal was unchanged when the second pump was started, an upset would occur, as the operating point would instantaneously jump from point

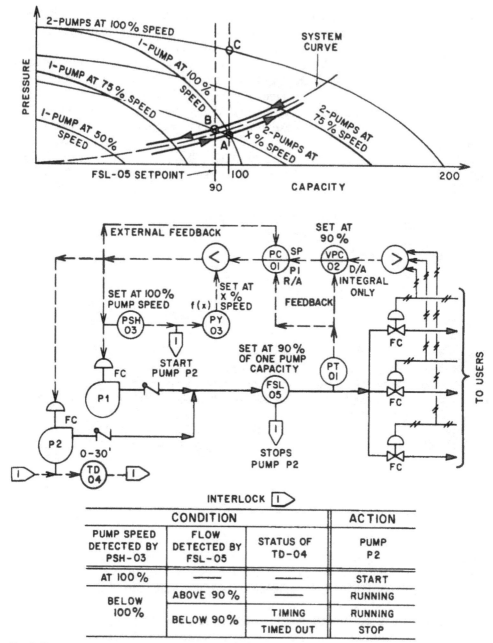

Fig. 9.41 A pump station optimization system can be based on valve position control of two equal-size variable-speed pumps.

A to C. In order to eliminate this temporary surge in pressure, which otherwise could shut down the station, PY-03 is introduced. This is a signal generator that, upon actuation by interlock 1, drops its output to x. This is the required speed for the two-pump operation at point A. The low signal selector immediately selects this signal x for control, thereby avoiding the upset. After actuation, the output signal of PY-03 slowly rises to full scale. As soon as it rises above the output of PIC-01, it is disregarded and control is returned to PIC-01.

Once both pumps are operating smoothly, the next control task is to stop the second pump when the load drops to the point where it can be met by a single pump. This is controlled by the low flow switch FSL-05, which is set at 90 percent of the capacity of one pump (point B on the system curve).

The pump-cycling controls described here can be used for any number of pumps. For each additional pump, another PY and FSL must be added, but otherwise, the system is the same.

Alternative Optimization Strategies

In pump distributions systems in which the number of users served is large, Shinskey [14] suggests a simpler system than the one described in figure 9.41.

The logic behind the control loop shown in figure 9.42 is that the terminal pressure will be kept constant if pump discharge pressure is varied in proportion to orifice drop. The pressure drop through an orifice varies in the same way as does the pressure drop in a mostly friction system. Therefore, if the orifice drop (h_2) is measured, this measurement can be related to the pipeline pressure losses by multiplying the drop by a set resistance ratio (K). Therefore, a set terminal pressure at the user can be maintained under variable load conditions by adjusting the pump speed, as shown in figure 9.42.

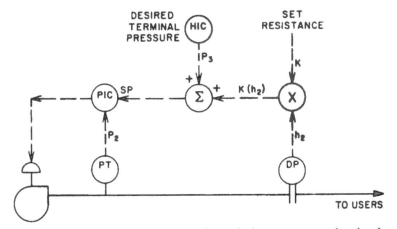

Fig. 9.42 This system saves energy by reducing discharge pressure with reduced flow, while maintaining a constant terminal pressure. (Adapted from Shinskey [14])

It is also advisable to monitor the electric power consumption of the pumps continuously, so that pump efficiency data will be empirical and up to date. This allows any load to be met with the most efficient pump or pump combination. In addition, the continuous monitoring of efficiency can also be used to signal the need for pump maintenance.

Opposed Centrifugal Pumps

The opposed centrifugal pump is not specifically designed as a control element but represents an adaptation of a centrifugal pump to flow control. This method of control is particularly suitable for coarse, rapidly settling slurries at low flow rates. In such instances, the conflicting requirements of control at low flow and large free area to pass the solids may make it impossible to find a suitable control valve. A system that requires a small quantity of slurry to be fed to a receiving vessel under controlled conditions is depicted in figure 9.43. Pump P_1 continuously circulates the slurry from the feed tank at high velocity. A branch line from the discharge of P_1 is run to the opposed centrifugal pump P_2. Pump P_2 is connected in opposition to the direction of slurry flow, and pressure drop to throttle flow is obtained by means of the mechanical energy supplied to the pump. A variable-speed driver on pump P_2 permits the pump pressure drop to be changed. At full speed, the pressure difference across P_2 is sufficient to stop the branch line slurry flow completely. A magnetic flow meter or some other suitable device can be used to measure the slurry flow. An SCR control can be used to vary pump speed in response to the flow controller output signal.

Conclusions

As was shown in figures 9.25 and 9.36, the use of optimized variable-speed pumps can easily cut the yearly operating costs in half when the average load is low [8]. In

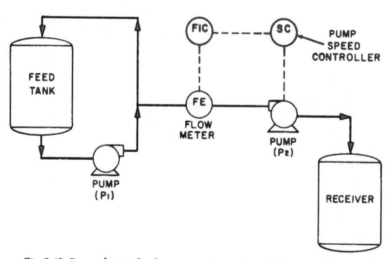

Fig. 9.43 Opposed centrifugal pumps can be used as a final control element.

addition, optimized supply-demand matching can protect from loss of control while reducing maintenance and cycling. The full automation of pumping stations—including automatic start-up and shut-down—not only will reduce operating costs but will also increase operating safety as human errors are eliminated.

References

1. American Society of Heating, Refrigerating, and Air-Conditioning Engineers, "Centrifugal Pumps," in *ASHRAE Handbook, 1983 Equipment Volume* (Atlanta, GA: ASHRAE, 1983), chapter 31.

2. Baumeister, T., *Mark's Standard Handbook for Mechanical Engineers*, eighth edition (New York, NY: McGraw-Hill, 1978), p. 3.69.

3. Conzett, J. C., "Adjustable Speed Drives," Bulletin D-7100, Reliance Electric, July 1981.

4. Karassik, I., *Centrifugal Pumps*, F. W. Dodge Corp., 1960.

5. Langfeldt, M. K., "Economic Consideration of Variable Speed Drives," ASME Paper 80-PET-81, 1980.

6. Lipták, B. G., *Instrument Engineer's Handbook, Process Control* (Radnor, PA: Chilton Book Co., 1985), sections 5.8, 8.17.

7. Lipták, B. G., *Instrument Engineer's Handbook, Process Measurement* (Radnor, PA: Chilton Book Co., 1982), section 9.12.

8. Lipták, B. G., "Save Energy by Optimizing Your Boilers, Chillers, and Pumps," *Intech*, March 1981.

9. Liu, T., "Controlling Pipeline Pumps for Energy Efficiency," *Intech*, June 1979.

10. Merritt, R., "What's Happening with Pumps," *Instruments and Control Systems*, September 1980.

11. Rishel, J. B., "Water System Head Analysis," *Plant Engineering*, October 13, 1977.

12. Rishel, J. B., "Wire to Water Efficiency of Pumping Systems," Central Chilled Water Conference, 1975, Purdue University.

13. Schroeder, E. G., "Choose Variable Speed Drives for Pump and Fan Efficiency," *Intech*, September 1980.

14. Shinskey, F. G., *Energy Conservation Control* (New York: Academic Press, 1978).

15. Systecon (Division of CEC), Bulletin No. 10-320-1.

16. Tchobanoglous, G. "Wastewater Engineering." Metcalf and Eddy, Inc. (New York, NY: McGraw-Hill, 1981).

Index

Page numbers in *italic* indicate information in illustrations;
page numbers followed by "t" indicate information in tables.